ZOOLOGY

ZOOLOGY

Vinh Luu
Oksana Savenko

Kruger Brentt
Publishers
2025

Kruger Brentt Publishers UK. LTD.
Company Number 9728962

Regd. Office: 68 St Margarets Road, Edgware, Middlesex HA8 9UU

ISBN: 978-1-78715-393-6

For information on all our publications visit our website at http://krugerbrentt.com/

PREFACE

"**Zoology**" serves as an essential primer on the fundamental principles, concepts, and methodologies of zoology, offering readers a comprehensive overview of the diverse world of animals and their interactions with their environments. This handbook is designed to be accessible to students, educators, researchers, and enthusiasts alike, providing a solid foundation in the field of zoology.

Zoology is the scientific study of animals, encompassing their anatomy, physiology, behavior, ecology, evolution, and conservation. It seeks to understand the diversity of animal life, from microscopic organisms to complex vertebrates, and the roles they play in ecosystems around the world. Through the lens of zoology, we gain insights into the complexities of life on Earth and the interconnectedness of all living organisms.

As editors of this volume, we have endeavored to compile a diverse array of perspectives, methodologies, and case studies from leading experts and practitioners in the field of zoology. Through a blend of theoretical foundations, practical insights, and real-world examples, this book covers a wide range of topics, including animal taxonomy and classification, comparative anatomy and physiology, animal behavior, ecology, evolution, and conservation biology.

Our overarching goal in presenting this book is to provide readers with a comprehensive understanding of the basic principles and concepts of zoology. By offering insights into the rich diversity of animal life and the methods used to study it, we aim to inspire curiosity, appreciation, and stewardship of the natural world.

We extend our sincere gratitude to all the contributors who have generously shared their expertise, experiences, and insights in the field of zoology. It is our hope that this handbook will serve as a valuable reference and guide for students, educators, researchers, and enthusiasts as they explore the fascinating realm of animal biology and its relevance to our understanding of life on Earth.

Vinh Luu | Oksana Savenko

CONTENTS

CHAPTER

MAKING A PLAN FOR THE FUTURE OF SOUTHEAST ASIA'S BIODIVERSITY

1

1. INTRODUCTION: AN OVERVIEW

Biodiversity levels worldwide are presently undergoing extraordinary changes due to numerous anthropogenic sources (Sala et al., 2000). In the coming century, the adverse trends in biodiversity are expected to persist, necessitating the selection of areas for conservation prioritization to optimally save regions of highest diversity (Brook et al., 2006). Although research has employed varying criteria for prioritizing regions, certain studies have amalgamated multiple criteria (Myers et al., 2000), resulting in the identification of 25 global biodiversity and endemic species hotspots. These areas constitute merely 1.4% of the Earth's land surface yet harbor 44% of all known plant species and 35% of currently recognized vertebrates. This chapter will primarily focus on three biodiversity hotspots that collectively constitute Southeast Asia (SEA). This section outlines the region's biodiversity and subsequently addresses the risks to it. To effectively conserve existing species, it is vital to understand their distributions and identify them; therefore, suitable strategies will be explained and shown. An investigation of techniques to assess the impacts of these threats will follow, aiming to identify the most effective strategy for conserving the maximum number of species in the region. This chapter mostly examines Southeast Asia, although numerous areas worldwide are presently encountering analogous circumstances. The methodologies and ideas presented will be widely applicable to additional regions and species beyond those addressed herein.

1.1 Southeast Asia's biodiversity

Some of the greatest biodiversity and endemicity on the planet are found in Southeast Asia (SEA), which contains a number of the biodiversity hotspots identified by Myers et al. (2000) (Gaston, 1995a). The Indo-Burmese region, Wallacea, Sundaland, and the Philippines are among the biotas that comprise the region. In terms of the number of endemic plants and vertebrates, three Southeast Asian regions are ranked in the global top ten (Sundaland-2nd, Indo-Burma-8th, Philippines-9th). Additionally, when the ratio of endemic species relative to area is measured, these three regions are in the top five (Phillipines-2nd, Sundaland-3rd, Indo-Burma-5th) (Myers et al., 2000). Additionally, the SEA is home to a significant amount of endemic evolutionary diversity at the species, family, and clade levels. Sundaland is ranked second on a global scale, Wallacea is third, and Indo-Burma is fifth in terms of unique evolutionary history. Each region has a unique evolutionary history spanning 65 to 40 million years (Sechrest, et al., 2002). Consequently, the SEA is a priority conservation area due to its irreplaceable biodiversity. In fact, biologists have identified the forests of the SEA as one of the most critical conservation priorities (Laurance, 2007). The landscape of SEA is also diverse and varied and comprises a large number of ecoregions (Olson et al., 2001). Stibig (2007) categorised sixteen native forest types, in addition to woodland, savannah, two types of thorn scrub and forest, alpine grassland and cold desert among the native vegetation types. Such diversity in vegetation cover also creates very varied ecosystems with very different animal and plant communities. Karsts (limestone outcrops) make up around 400,000 km^2 of SEA, and though they only make up one percent of the land area, around two percent of Malaysian species are endemic to karst landscapes (Clements et al., 2006). Globally karsts also harbour a great proportion of endemic species, and therefore contribute significantly to landscape diversity and heterogeneity throughout SEA.

One reason for the high levels of diversity and endemicity in SEA is the dynamic and complex geo-physical history of the region, which has been described as a biogeographic theatre (Woodruff, 2003). Some of the landmasses that form SEA only joined as little as 15 Mya (Million years ago), and the addition of new landmasses caused faults and regional instability in many regions (Hall, 2002), which in turn contributed to the formation of unique biotas. Even at only five Mya SEA had not taken its present shape and landmasses within it were still subject to small but significant movements (Hall, 2002). Since this time glacial cycles have periodically transformed SEA, both in terms of shape and vegetation cover (Woodruff, 2003). During successive glaciations mainland and insular areas of SEA have been joined, and glaciers existed as recently as 10 Kya (thousand years ago) in Borneo and Sumatra (Morley & Flenley, 1987). This dynamic geophysical history has led to a highly complex pattern of species distributions and the area contains no less than three zoo/floro– geographic boundaries: Wallace's line, the Kangar-Pattani line and the Isthmus of Kra (Whitmore, 1981; Baltzer, 2008; Cox & Moore, 2010; A.C. Hughes et al., 2011). Therefore, the region has a rich and highly varied biota, and thus represents a priority region for conservation.

2. BIODIVERSITY THREATS

Southeast Asia is widely acknowledged to be experiencing a crisis with biodiversity loss (Laurance, 2007).South East Asia exhibits the largest global deforestation rate, exceeding double that recorded in other regions (Laurance, 2007).Thailand retains approximately 17.6% of its potential forest, while Peninsula Malaysia maintains over 50%, despite its considerable biodiversity (Witmer, 2005).The rates of change in vegetation cover in Southeast Asia from 1981 to 2000 were the highest worldwide (Lepers et al., 2005), and these rates are further increasing (Hansen & DeFries, 2004). Loss of habitat and deforestation are not the only threats to the biodiversity of SEA. The Convention on International Trade in Endangered Species (CITES) listed that at least 35 million animals in addition to 18 million pieces of coral (and 2 million kg live coral) were exported from SEA between 1998 and 2007 (Nijman, 2010). Many species are also hunted for recreation (Epstein et al., 2009) in addition to bushmeat (Brodie et al., 2009). Furthermore the Chinese medicine trade is stated to be the "single major threat" for some species (EIA, 2004; Ellis, 2005). These problems are not limited to "unprotected areas" as even National Parks fail to offer protection from either illegal logging (Sodhi et al., 2010) or high levels of hunting (Brodie et al., 2009).

The above mentioned factors affecting biodiversity loss are further complicated by the effects of climate change (Figs.1-2), which may act to amplify other threats, and which itself may be amplified by other threat factors (such as wood burning and subsequent release of greenhouse gases-Brook et al., 2006). Fires present a major threat to biodiversity in the region, and during the past decade major fires have started progressively further north in response to climate change (Taylor et al., 1999). Even without considering of many of these factors, projections of the number of extinctions have been made, which project the extinction of 43% of endemic Indo-Burmese fauna within the next century (Malcolm et al., 2006). Thus despite harbouring considerable biodiversity, few areas in SEA have sufficient levels of protection, and with many new species still to be found (as demonstrated by the rapid rate of discovery (Giam et al., 2010)) it is currently almost impossible to determine the most effective means of conservation prioritisation within SEA given the level of knowledge of much of the fauna, and high levels of corruption (Global Witness, 2007).

Some conservation biologists have advocated the use of "indicator species" to monitor more general threats to biodiversity (Carignan & Villard, 2002). Chosen species must obviously be sensitive to the potential threats in the area, and such species must be possible to monitor in a standardised and repeatable way to generate meaningful and comparable data over large spatial and temporal scales. Indicator species can also be used to indicate trends in overall biodiversity (Mace & Baillie, 2007) and therefore provide a gauge of biodiversity change at large regional scales over time.

Bats provide an ideal indicator group (G. Jones et al., 2009), and their diversity means that species can be susceptible to a wide variety of different threats. Bats form a large component of bush-meat through SEA (Mickleburgh et al., 2009), and many of these species often perform vital roles within ecosystems and their loss could have negative implications for a wide range of interacting taxa (Mohd-Azlan et al., 2001). A number of ecosystem services are provided by bat species, including pollination, seed dispersal and insect control, and therefore bats are frequently keystone species (Myers, 1987; Fujita and Tuttle, 1991; Hodgkison et al., 2004). Effective conservation of these keystone species is crucial not only for their survival, but for the ecosystems dependent upon them. Furthermore many bat species are either dependent on forests or caves for foraging and roosting, and some species have limited dispersal ability (Kingston et al., 2003), suggesting that their status may be indicative of destruction and consequent fragmentation of both karst and forest areas.

Mitigating the effects of the Southeast Asian biodiversity crisis necessitates many actions: quantifying species distribution and its alterations, analyzing the dangers confronting each species, and assessing the potential impact of these threats. Effective impact mitigation methods can only be formulated if these basic steps have been accomplished. While this chapter primarily focuses on bats as a case study due to their efficacy as indicator species, the concepts discussed are broadly relevant to biodiversity conservation across Southeast Asia and in formulating strategies to mitigate species loss in other regions facing analogous challenges.

3. IDENTIFYING SPECIES AND DISTRIBUTIONS

Although over 320 species of bat are currently described from SEA (Simmons, 2005; Kingston, 2010) research in the area has been sporadic and the rate of species discovery is now high for not only bats (Bumrungsri et al., 2006), but across many other taxa (Duckworth & Hedges, 1998; Bain et al., 2003; Giam et al., 2010). Recent research has revealed that many bats previously regarded as one species are in reality complexes, comprising a number of cryptic species (Soisook et al., 2008, 2010; Francis et al., 2010). Therefore before any conservation measures can be put in place the distribution and status of current species must first be established. SEA has some of the highest diversity of bats on the planet in addition rate of species discovery (Simmons & Wetterer, 2011). A projection of the species richness of 171 species throughout SEA (Fig.1) shows that most forested regions still retain high species richness, and therefore present priority regions for research.

However recent research has clearly demonstrated that currently known SEA bat species only represent a fraction of total species numbers (Francis et al., 2010; Giam et al., 2010; A.C.Hughes et al., in prep a). Both recent taxonomic and genetic research show that much further work is needed in order to identify all species in the region, and similar trends are liable to exist across biotic groups. Species identification is clearly a priority, because it is impossible to try to develop effective conservation strategies when there is little understanding of the true ranges of many species; and when species currently classified as showing large distributions are in actuality made up of a number of cryptic species with small ranges and much smaller populations (A.C.Hughes et al., in prep a). Both taxonomic (Soisook et al., 2008, 2010) and genetic work (Francis et al., 2010) demonstrate that there are many currently undescribed and potentially cryptic species throughout SEA.

Methods used to determine species present obviously involve detailed taxonomic surveys (as advocated by Webb et al., 2010), in addition to genetic analyses where possible. However other protocols for species identification and monitoring may also be valuable components of species discovery in some taxa, such as the use of call analysis to identify cryptic bat species (e.g. G. Jones & Van Parijs, 1993). In such cases the identification of potentially cryptic species may begin with call analysis, as was recently found to be the case in Hipposideros bicolor, (Douangboubpha et al., 2010). Acoustic monitoring also provides a means of potentially monitoring population trends as well as identifying possible cryptic species (K. E. Jones et al., 2011). Two protocols have recently been developed which describe the potential for using localised call libraries for identifying bat species in SEA (A.C.Hughes et al., 2010, in press). Once acoustic identification libraries have been developed then acoustic surveys and inventories of surrounding regions (e.g. 1° of the areas used to develop the library) can be made to identify species present (using discriminant function analysis) and the presence of species outside their known range. The presence of novel call variants could cue and promote further research to determine if sub-species or cryptic species are present, and the spatial distributions of call variants of some species suggests spatial segregation which could denote cryptic species (A.C.Hughes et al., in prep a). Monitoring surveys are also essential to determine distribution and population trends, however funds and specialists are not always available to carry out this valuable work when it requires repeated taxonomic surveys and specialist knowledge. Acoustic analysis and monitoring only requires specialists initially, during the creation of acoustic libraries, and surveys can then be carried out by non-specialists or automated software programs (K.E. Jones et al., 2011). Thus protocols such as these provide a viable means of both identifying species present and subsequently monitoring trends, and may be able to detect variation over shorter periods than in trapping-based monitoring which has been previously been advocated (Meyer et al., 2010). Acoustic surveys are currently limited in species coverage, and are biased towards bat taxa that use high-intensity echolocation calls. Acoustic surveys are therefore best used side-by-side with conventional survey techniques such as using mist- nets and harp traps in a standardised manner (MacSwiney G et al., 2008). However invasive trapping techniques are expensive and require highly trained experts, whereas acoustic surveys can be carried out with little training and recordings can then be forwarded to highly trained researchers for analysis, or analysed by software to provide standardised and comparable data for any region. If initially surveys combine both trapping and acoustic techniques to establish acoustic libraries within a given area then those libraries can subsequently be employed to monitor trends in many species across wide areas. The use of common species as indicators for abundance and distribution of rarer species has been found to be accurate in previous studies, as correlations have been found in the trends of common species with other species present (Pearman et al., 2010). Therefore even if acoustic surveys cannot cover all species, the trends in the distributions and populations of common species may still be more widely applicable.

Logistical constraints also mean that it is not always possible to survey all areas in a region, and thus methods which determine range based on limited spatial knowledge of an organism's total distribution provides a valuable tool when applied properly (i.e. predictive modelling approaches, Box 1, Fig. 1). Former distributions of species and zoogeographic constraints must also be considered and included in analyses of species distributions. Within SEA the geophysical history is to a large extent responsible for the current patterns of diversity and species' distributions, and thus analyses of present species distributions cannot be conducted without by making reference to the past (Woodruff, 2003). The connections and separations of the various parts of SEA during past time periods not only influence current distributions but further constrain possible responses to future change. A zoogeographic transition in the distributions of some animal groups centred around the Isthmus of Kra has persisted for over a million years (De Bruyn et al., 2004). Recent analyses (A.C. Hughes et al., 2011) show that although breaks in the distribution patterns of bats are apparent along the Thai peninsula, they occur not only at the Isthmus of Kra and are influenced by climatic discontinuities in conjunction with biogeographic consequences associated with the narrow breadth of the peninsula; and it is probable that these circumstances have also caused divisions known to occur in the distributions of other taxa in the region (J.B. Hughes, et al., 2003). Zoogeographic transitions have persisted over long time periods along the peninsula because the position of climatic boundaries appears remarkably constant. Climatic discontinuities continue to affect the distributions of species, and will also affect how effectively species can respond to climatic change in the future.

Identification of species present, their ranges and trends in distribution and population form an important first step in the development of effective conservation plans. Once these steps have been fulfilled then threats to current distributions and diversity can be analysed (Fig. 1) and necessary conservation actions planned.

3.1 Evaluating and measuring threats to existing diversity, and assessing impacts

Analyses have previously shown that species richness is negatively related to human population density (A.C. Hughes et al., in prep b), and therefore further increase in human population size is likely to have detrimental effects on bat biodiversity. Projections suggest that human populations will continue to increase until at least 2050 and further urbanisation is likely throughout SEA (CIESEN, 2002; Gaffin et al., 2003; United Nations Population Division, 2008; Seto et al., 2010). Larger human populations impinge on biodiversity in a number of ways: through increased demand for wild-sourced products and via higher pollution (Corlett, 2009; Peh, 2010). Urbanisation and increasing deforestation also increase the potential for invasive species to spread throughout SEA (Riley et al., 2005) and further work is necessary to determine the effects of invasive species on the native fauna.

Forest fragment size correlates positively with bat species richness (Struebig et al., 2008; A.C. Hughes et al., in prep b). As deforestation is projected to increase throughout most of SEA, including in "protected areas" (Fuller et al., 2003), this trend is likely to lead to progressive loss of species richness in many areas due to the increased fragmentation of large forest patches. Currently many protected areas fail to offer protection, and are subject to both high hunting pressure (Steinmetz, et al., 2006) and deforestation (Fuller et al., 2003). Heightened accessibility of parks and involvement of rangers may indeed lead to greater pressures within National Parks than in other forested regions. Many regions were predicted to have high species richness during this study, however many forests have been described as showing "empty forest syndrome" (Redford, 1992; Tungittiplakorn & Dearden 2002). Therefore although many areas may be suitable for certain species, they are overexploited by humans, and do not contain the native fauna previously held. Empty forest syndrome and overexploitation have serious implications for a wide range of species: rodents are the most "harvested" taxa, followed by bats, and almost all bat species in SEA are eaten (Mickleburgh et al., 2009). The loss of species due to hunting has implications for the entire ecosystem. Frugivorous and nectarivorous bats, large bodied mammals and birds all have essential functions in seed dispersal and pollination and fulfil vital ecosystem services, yet such species are often the most threatened by human hunting activities. If such species are lost, there may be negative consequences for the entire ecosystem. Yet these animals are among the most hunted organisms in the region (Wright, et al., 2007; Corlett, 2008; Brodie et al., 2009).

When projections of the distribution of bats under future climatic scenarios are made, three broad outcomes can be noted (Fig. 2) (A.C. Hughes et al., in review). First almost all species are projected to show reductions in original range under future scenarios and second, most species are projected to move north. The third probable outcome is the large projected loss of species (up to 44) from areas currently predicted to have the highest levels of species richness (figs 1-2). Though some species were projected to show expansions in original range, this is unlikely to be logistically possible due to the limited dispersal abilities of many species (Struebig et al., 2008). This loss in species richness is based on climate change alone, and therefore is a conservative projection, and though it is possible to prevent the loss of species due to deforestation in protected areas it is not possible to prevent species loss due to climatic change. Forest is becoming increasingly fragmented even within "protected areas", and mining rates in SEA are the highest in the tropics (Day & Ulrich, 2000). Mining not only destroys important roost sites (Clements et al., 2006), but also degrades areas and increases accessibility to previously remote areas (which in turn facilitates deforestation, McMahon, et al., 2000; Laurance, 2008a). Therefore not only are current suitable habitat and roosts being destroyed, but the distance between suitable areas may actually be increasing for the same reasons. Other factors such as fires are also prevalent through SEA, and fires have increasingly been found to move north in response to climatic change, therefore posing an increasing threat to the biodiversity of SEA (Taylor et al., 1999). Projections of total biodiversity loss currently estimate the extinction of up to 85% of current biodiversity in SEA within this century (Sodhi et al., 2010). However the estimates of undiscovered species show that we may potentially have only discovered around half of the species in many orders (Giam et al., 2010) and only around 40% of bat species (A.C. Hughes et al., in prep a). Groups containing cryptic species are likely to have particularly high numbers of undiscovered species, and this is highlighted in bats by recent genetic work (Francis et al., 2010). Species with smaller distributions are more likely to have specialist requirements (limiting overall distribution), and will be more susceptible to loss of range and therefore have a higher probability of extinction (Kotiaho et al., 2005). Hence many species currently regarded as widespread, and thus of "Least Concern" by the IUCN may comprise a complex of cryptic species each of which will show higher categories of threat. As almost all species analysed here (fig. 2) showed a loss in original habitat in all scenarios, and many of those species may be species complexes it is likely that impacts for many of the species will be worse than estimated during this study (fig. 2). Projections here (Fig. 2) only account for climate change, but cannot consider hunting, fires, mining and the plethora of other threats. Fungal diseases have recently devastated populations of North American bats (Blehert et al., 2009), in addition to South American frogs (Berger et al., 1999). Moreover the spread of pathogens has been associated with temperature change, for example the spread of chytrid fungus is believed to be related to global warming (Pounds et al., 2006; Boyles & Willis, 2010). Therefore the effect of climate change on species is dynamic and complex, as it has both direct and indirect implications for distributions and populations of all species. Furthermore climatic changes have already been shown to cause changes to the distribution of different biomes (Salazar et al., 2007), and hence has profound implications for species within those biomes.

SEA is currently in the midst of a biodiversity crisis which has been described as a 6th mass extinction (Myers, 1988). There are some undeniable implications of the current threats, and others such as the possibility of 'no-analogue' communities (Stralberg et al., 2009) and the effect of invasive species, which are less certain. However native species are likely to attempt to either migrate north spatially, or move to higher altitudes (Malcolm et al., 2006). Continued decreases in the patch sizes of rainforest will decrease species richness, and increasing accessibility for humans will increase the probability of hunting within areas. Increases in human population will negatively affect biodiversity, if current unsustainable practices continue. Not only is the modification of human activities necessary to decrease further species loss, but human intervention is necessary to allow species any opportunity to respond effectively to climatic changes. The methods to mitigate possible threats require detailed evaluation to try to curb species extinctions.

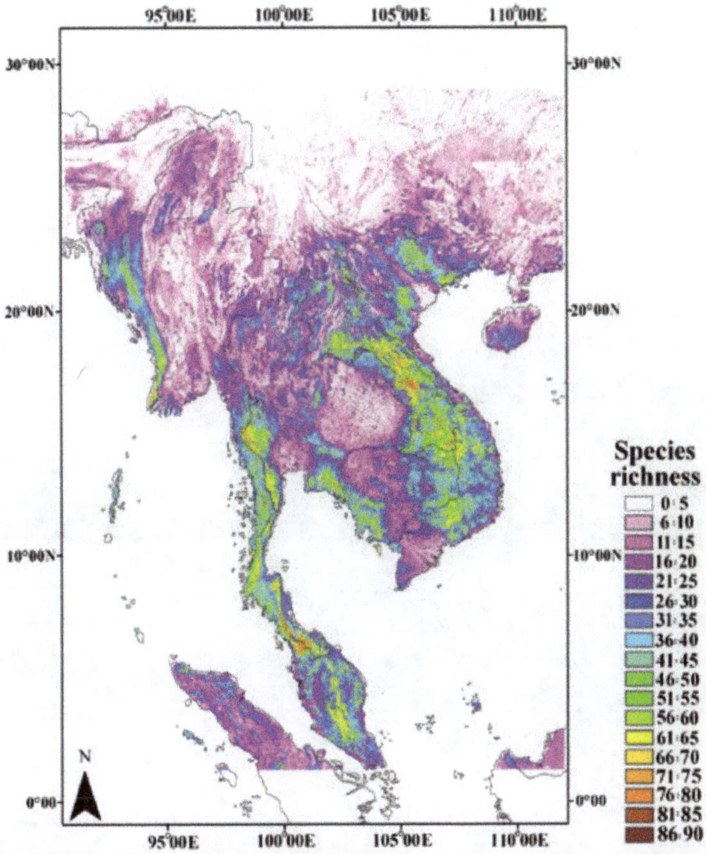

Fig. 1. The current projected species richness for 171 Southeast Asian bat species on a km² basis. Projections were generated using Maxent, methods are shown in Box 1.

Environmental variables used in projections are included in Appendix 1.

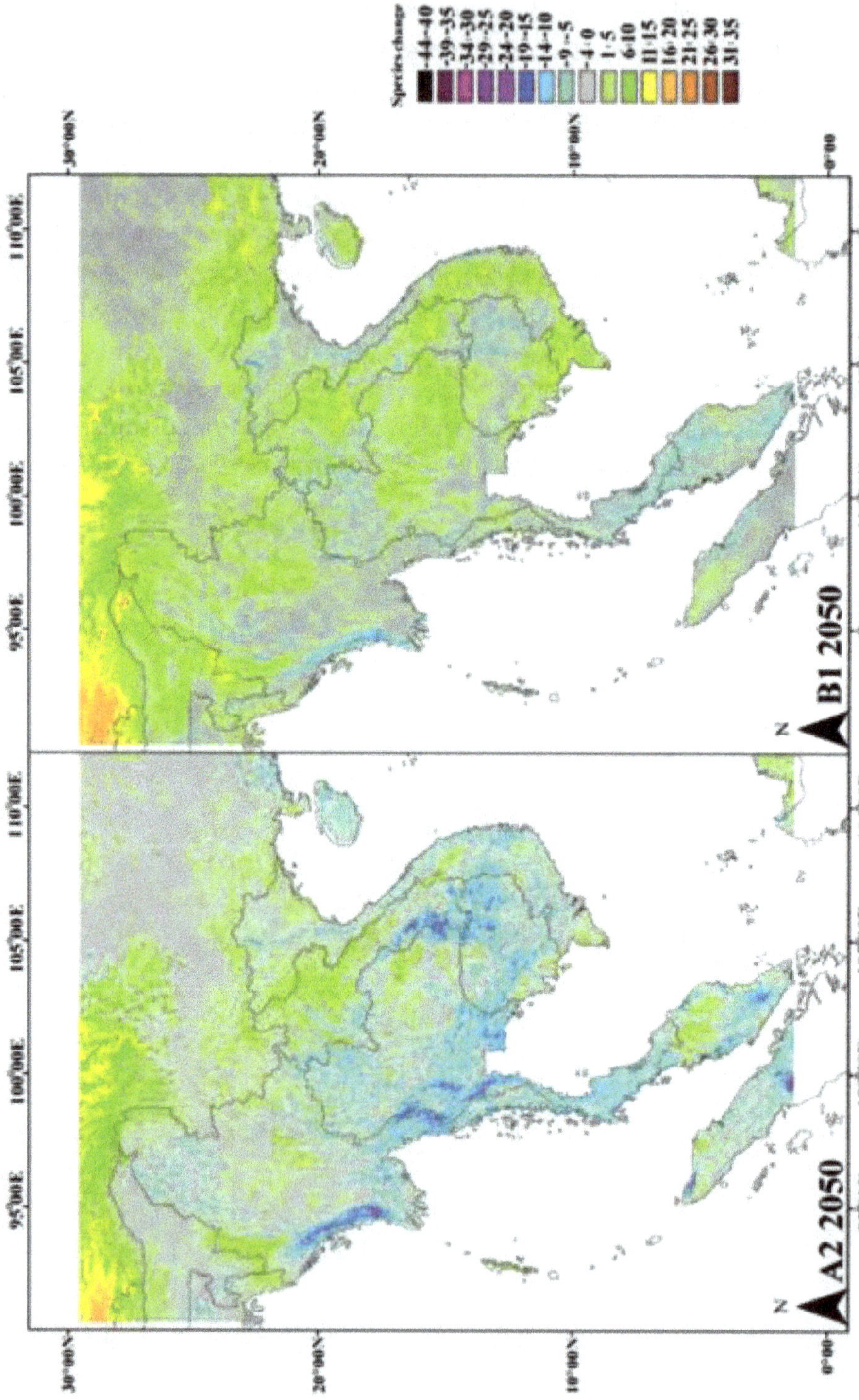

Fig. 2. A-B.

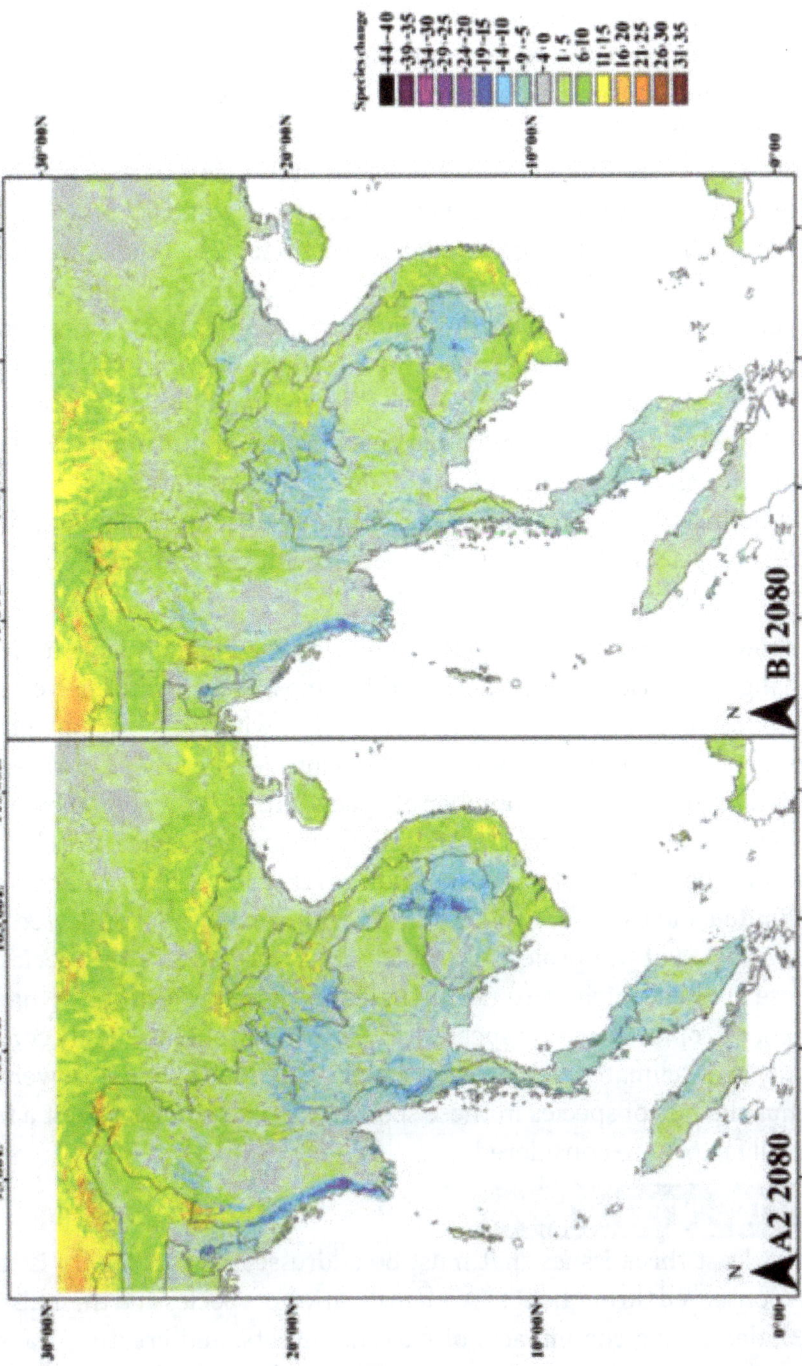

Fig. 2. C-D. These maps display the projected change in the number of bat species under the A2 and B1 climate change scenarios produced by the IPPC. A2 represents the most severe of the climate change scenarios and B1 the mildest. Many regions are projected to lose between five to nineteen species, with some regions projected to lose up to forty-four species.

Predicting species richness by pairing the known distribution of each species with environmental parameters to determine the habitat requirements of each species, (and thus distribution) can help inform and target research and conservation (Box 1). Such models can also aid conservation planning under probable future scenarios, but are targeted to specific questions and can only incorporate some dimensions of ecosystems and must therefore only be interpreted while acknowledging inputs, assumptions and limitations. Models are a powerful tool for predicting the effects of climatic, land-cover and direct anthropogenic change on species richness (if these anthropogenic drivers have been projected). What such models are less good at is incorporating the biotic dimension, the inter-dependence of some species, and temporal interactions such as the flowering of trees and breeding of organisms (L. Hughes, 2000) which can cause resource asynchrony. More complicated ecological interactions and phenomena cannot yet be incorporated in the building of models, but should be included in the interpretation of results; and thus both dimensions of possible ecosystem change can be used to inform and develop appropriate conservation strategies.

Figure 2 shows projections of the effects of climatic change under two potential climate change scenarios (the mildest, and the most severe). Extensive regions are projected to lose between five to nineteen species, with some regions projected to lose up to forty-four species. Though some regions, especially in Northern regions are projected to increase in the number of species, this result should be interpreted with caution for a number of reasons. Firstly these projections only include climatic change, and do not reflect changes in land- cover, and secondly many species are dispersal limited and therefore will not to show the expansions projected here. Also the Northern areas of the projection, which are predicted to gain species here are liable to lose species which were not included in these projections, (which include predominantly tropical dwelling species). These projections are highly conservative, for they only show climate mediated loss of species, this is only one driver of species loss and thus the loss of species in these scenarios is liable to represent a fraction of that when all factors are considered.

4. REDUCING SPECIES LOSS

There are at least three issues that must be addressed if biodiversity is to be most effectively conserved throughout SEA: identification of species and their distributions (Section 3), decreasing the impacts of current threats, and creating ways to allow species to respond to climate change (because halting further climate change is considered impossible, Bowen & Ranger, 2009; Vistor et al., 2009). Each issue requires different actions in order to respond effectively.

As stated previously, accurate species identification requires thorough systematic surveys, taxonomic and acoustic studies, and where possible genetic research.

However the scale of this work requires the use of university researchers, students and park rangers. The use of citizen science for survey and monitoring has been advocated by some researchers (Webb et al., 2010). However citizen science is plagued with potential problems in SEA: not only is hunting exceedingly popular, but in many taxonomic groups' cryptic species and the lack of adequate taxonomic knowledge precludes species surveys by non-specialists. However education, and enthusing of the population could allow some citizen science in distinct and recognisable species. School children in some parts of SEA also must complete science projects whilst at high school, and with little training such projects could contribute to this knowledge pool (Sara Bumrungsri pers. comm.). However, for successful citizen science to be conducted, people must be educated to the importance of species such as bats (as throughout SEA bats are generally viewed negatively by the public, Kingston et al., 2006). Nature recreation has been implemented in schools, and provides an important means of enthusing the next generations about biodiversity and engendering greater respect for the environment (Pergams & Zaradac, 2008). Education is of paramount importance in the realisation of any level of conservation or mitigation. Without the support and backing of local people no changes to current activities will take place. Projections of species distributions, like those within this study - and subsequent ground truthing (validation and testing) by trained surveyors - can also provide a focus for further research and conservation activity.

Many strategies have attempted to decrease anthropogenic impacts on biodiversity. Recent studies have projected the species richness patterns of bats throughout SEA (Fig 1) (Hughes et al., in review; in prep b), and the regions of high species richness obviously provide a focus for conservation efforts. However within the scientific community there is great dispute as to what criteria should be used to assign conservation priorities. There is debate as to whether regions, species richness, evolutionary uniqueness and richness, numbers of threatened species or specific species should be used in area prioritisation (Corlett, 2009). Under the current circumstances it is not feasible to conserve on a single species basis, because this approach is financially unviable, and it ignores interactions within ecosystems. Furthermore there is currently inadequate knowledge to reliably designate IUCN threat levels for many species throughout SEA, due to the lack of knowledge about the distributions and population sizes of many species, and the presence of cryptic species. However SEA is regarded as an area of both evolutionary, and species richness (Gaston, 1995b; Sechrest, et al., 2002). Most currently species-rich areas are also liable to contain high levels of intraspecific genetic diversity as populations of most species have been predicted to have expanded during the last glacial maximum (LGM), and because current ranges may be restricted compared with ranges occupied during the LGM in tropical areas and may overlie former glacial distributions (Woodruff, 2010). Current species populations within areas of former glacial refugia often contain high genetic diversity in comparison to those in non- refugial areas (Anthony et al., 2007), and genetic heterogeneity and diversity is known to make populations more robust to environmental change and therefore to increase the capacity of such populations to adapt (Aitken et al., 2008). Therefore former refugial areas, many of which fall within current National Parks deserve prioritisation on all grounds. However the current system of National Parks fails to function in many regions (Fuller et al., 2003), and with the high levels of corruption (Global Witness, 2007; EIA/Telepak, 2008) the enforcement of laws such as those governing reserves is difficult.

A variety of schemes and approaches have been developed to try to promote biodiversity conservation and decrease deforestation. The following section evaluates some of these methods in an attempt to formulate a viable method of mitigating biodiversity loss.

Paying for Environmental Services (PES) is one scheme suggested for conservation (Blackman & Woodward, 2010). PES schemes use money generated by environmental service users to pay people who own an area which is (in part) responsible for the service, in order to maintain forest/ biodiversity within the area responsible for that service. For example, forest cover in watersheds may be preserved by using the income generated by hydro-electric dams. Such approaches have potential but must be closely tailored to each site and country, to be economically viable both for those responsible for the maintenance of the area which provides the ecosystem service and those who profit from the service. These schemes have a great number of potential pitfalls which have prevented their success in some areas (Wunder, 2006, 2007). PES-type schemes are obviously unsuitable when the service users earn less than the ecosystem service users, such as the cases of guano miners and durian growers (both of whom have income streams dependent on cave bat populations) and people whom mine karsts (and therefore are responsible for the resource). In situations involving the mining of karsts, determining who should pay for environmental services is difficult, as the income of the miners (who may own the karst) may be higher than those who benefit from bat related services (Wunder, 2006). However when PES-type schemes are well-tailored and targeted to specific areas, they can effectively protect forests, and stipulations can place more emphasis on biodiversity rather than solely forest, as in the case of some "engineered PES" schemes (Wunscher et al., 2006).

Carbon offsets (carbon credits) and the REDD (Reduced Emissions from Deforestation and Forest Degradation) systems also provide a means to fund forest protection (Laurance, 2008b). Afforestation can also be part of such schemes, but in some existing schemes this has included the use of non-native trees. If biodiversity is to be protected it is important that afforestation uses only native species (Corlett, 2009). Afforestation schemes are currently the subject of much debate, however when well applied they have the potential to both decrease rates of biodiversity loss and to mitigate climate change (Canadell & Raupach, 2008). Biodiversity offsets have also been used in some regions (i.e. Uganda), however in some areas (i.e. the USA) the heightened protection of one area has led to greater biodiversity declines elsewhere and thus yielding no net benefit to conservation overall (Ten Kate et al., 2004). Therefore education is necessary alongside offset schemes in order to attempt to prevent greater pressures being deflected elsewhere as a result of conservation within one area (conservation leakage, Gan & McCarl, 2007). Problems also arise when it comes to prioritising areas for conservation based on current risks alone. Hence it is valuable to predict future scenarios based on land cover and climate in assessing conservation priorities (Fig. 2). Although future risk should be part of any assessment criterion, assessment must also analyse other factors, so even if environmental pressure is deflected to other areas as a result of conservation in a particular area: that the most important areas (in terms of biodiversity/uniqueness) are adequately protected (Laumonier et al., 2010). Risk and enforcement can also be projected together and the combined effects predicted to generate the most effective means of minimising deforestation or biodiversity loss within an area (Linkie et al., 2010).

The protection of specific areas still requires funding, as National Parks are currently ineffective in many regions of SEA (Fuller et al., 2003; Steinmetz, et al., 2006). In other countries (i.e. Costa Rica) ecotourism has provided a highly successful means of funding biodiversity protection and educating local people about the value of biodiversity (Jacobson & Robles, 1992; Aylward et al., 1996). Currently although ethnotourism is popular (Zeppel, 2006), ecotourism in mainland SEA is mainly dominated by bird watching tours (Mollmann, 2008). Ecotourism has been shown to work well in parts of Malaysia and Indonesia (Hill et al., 2007, Pearce et al., 2008), and if it were to develop throughout SEA it could provide a viable means of conservation.

Multiple models exist to spatially project species probable distributions used limited spatial data, and in recent years the use of such models has increased dramatically; in 1999-2004 only 74 published studies used species/niche distribution models, however between 2005-2010 this increased to over 850 (Beale and Lennon, in review). Clearly such models represent useful tools for projecting species distributions, and can further allow targeted conservation to either species habitat requirements or the prioritization of areas for research or conservation (Pawar et al., 2007; Sergio et. al., 2007). Recent developments in habitat suitability modelling allow the prediction of a species' potential distribution based on presence-only records (e.g. Hirzel et al., 2002; Phillips et al., 2006). Presence-only modelling is a valuable tool in contemporary conservation biology, and has been applied to a wide range of taxa, from bryophytes (Sergio et. al., 2007) to reptiles (Pawar et al., 2007). Presence-only modelling may be more reliable than presence-absence models for species in which absence records cannot be reliably gathered (i.e. failure to capture a species at a site does not necessarily mean the species is absent- Wintle et al., 2004; MacKenzie, 2005; Elith et al., 2006; Jime´nez-Valverde et al., 2008). One presence-only modelling method that is used widely (Maxent – Phillips et al., 2006) involves maximum entropy modelling and has been used successfully to predict the distributions of bat species in both present day conditions (e.g. Lamb et al., 2008; Rebelo and G. Jones, 2010) and under projected climate change scenarios (Rebelo et al., 2010). Additionally Maxent has been found to be robust to changes in sample size, and still have good predictive ability at low sample sizes, making it the ideal model for the prediction of distributions for rare species (Hernandez et al., 2006; Wisz et al., 2008).

Figures 1 and 2 both use Maxent to project the distributions of 171 bat species for a number of time periods. By pairing known distribution coordinates each species has been recorded at with appropriate environmental variables it is possible to project the probable distribution of each species for any time period for which spatial data exists, and to combine these to calculate species richness (see Hughes et al., In review, for a full account of methods used).

Using projections of future climatic change it is possible to project the probable impacts and develop targeted solutions and effective conservation methods (Prentice et al., 1992; Beerling et al., 1995; Huntley et al., 1995; Sykes et al., 1996; Berry et al., 2001, 2002; Hannah et al., 2002; Midgley et al., 2002). Though improvements in modelling approaches in the future will allow further insights to be generated, such models will take time to be developed and refined. In many areas (such as Southeast Asia) with rapid rates of deforestation, prioritisation of key areas is required to protect areas of high conservation value from deforestation and modelling can facilitate the determination of these priority areas in a region of high conservation importance (Pawar et al., 2007; Sergio et. al., 2007; Gibson et al., 2010).

Other countries (e.g. Brazil) with large export markets have also started to produce certified wood for a large proportion of their exports, however few certifications have sufficient biodiversity emphasis (McNeely, 2007). Attempts at certification programs throughout much of SEA have met with little success, as most logged wood is used within the country, and people are not prepared to pay increased prices involved with establishing and maintaining certification (Cashore, et al., 2006; Laurance, 2008b). Until local people value the natural environment, or exports increase, certification will remain an unsuitable scheme for much of SEA. It may be for similar reasons that previous integrated conservation and development projects have met with little success (in terms of impact) despite large-scale investment throughout SEA (Terborgh et al., 2002; McShane & Wells, 2004). Community- based conservation schemes have also been little used outside marine national parks, and their use may be unsuitable for many areas (Gray et al., 2007).

Certification is unlikely to work within SEA, and logging is liable to continue within natural forests (Fredericksen & Putz, 2003). The use of "reduced impact logging" could at least provide a means of providing both humans and biodiversity with a means of existence (Sessions, 2007; Putz et al., 2008). Reduced impact logging would require less human behavioural modification than stopping altogether or certification, and if local people can be educated to perceive it as an efficient way of logging, which preserves ecosystem services then it may provide a means of conservation. However as most logging which takes place in SEA is illegal, enforcement of laws is first essential (EIA/Telepak, 2008). Enforcement is also necessary to restrict hunting, and requires not only education but an enforced system of permits to control it. Logging programmes must also consider that the removal of the most mature trees may have negative consequences for those bat species that roost under bark and in other species which dwell in holes of mature tree (Gibbons & Lindenmayer, 2001; Kunz & Lumsden, 2003; Barclay & Kurta, 2007). As these forest-dwelling species are often the most limited in dispersal abilities, they are liable to suffer most from deforestation (Kingston et al., 2003).

Therefore in the protection of existing highly biodiverse areas, and to prevent an increasingly fragmented landscape further reducing biodiversity (A.C. Hughes et al., in prep b) education and law enforcement are paramount. Well considered funding systems also provide a good opportunity for decreasing biodiversity loss, and ecotourism if well developed could remedy both habitat destruction and overhunting. These are the primary means for protecting areas from anthropogenic direct threats.

4.1 Mitigating the effect of climatic changes on biodiversity

Recently developed models forecasted that bat species will both lose adequate habitat within their original range and frequently need to migrate northward to maintain comparable ecological niches in response to climate change (A.C. Hughes et al., under review). To adapt, organisms must access appropriate habitats. Translocation and aided migrations are frequently proposed as methods to achieve this (McLachlan et al., 2007). Numerous species encounter identical risks; thus, what criteria should be employed for selecting species for translocation: uniqueness, charismatic megafauna, ecological significance, or risk of extinction? A multitude of species confront analogous circumstances, yet insufficient knowledge is available on many to render translocation a feasible solution. Some species designated as candidates for translocation, based on IUCN status or other criteria, exhibit adverse reactions and demonstrate low survival rates post-translocation (Weinberger et al., 2009). Consequently, translocation is not a viable solution, owing to the multitude of species at risk, the variety in species' responses to transfer, and the associated financial costs. Human-mediated adaptive techniques should enable species to alter their ranges in response to climate change, akin to natural shifts observed during prior climatic transitions (Hickling et al., 2006; Lenoir et al., 2008). Enhancing landscape connectivity can facilitate movements by establishing corridors of natural forest between existing forest patches, especially in a north-south alignment (Heller & Zavaleta, 2008). These regions must be sufficiently expansive to avoid functioning as population sinks and must exhibit variability in both species and genetic variation to ensure viability and sustainability (Lamb et al., 2008; Lamb & Erskine, 2008; Kettle, 2010). Therefore, it is essential to meticulously match tree species to soil types and inter-site areas, and corridors should incorporate site-appropriate flora, particularly nitrogen-fixing legumes, to enhance canopy density (Siddique et al., 2008; Suzuki et al., 2009). Afforestation has commenced in numerous countries (UK, Vietnam; McNamara et al., 2006; McNamara et al., 2008), and if utilized to link disparate areas, it will enhance species' likelihood of adapting to climatic change by facilitating their northward range expansion, as outlined in predictive models (fig. 2) and corroborated by studies in other regions (Malcolm et al., 2006), thereby mitigating significant reductions in overall range.

5. CONCISE SUMMARY

South East Asia is one of the most biodiverse regions globally; yet, species in this area are endangered due to complex interactions among several stressors, encompassing both direct and indirect factors contributing to human-induced biodiversity decline. To preserve a portion of the existing fauna, significant alterations in human activities are essential, necessitating the education of individuals across Southeast Asia and the reduction of corruption at all levels. Current biodiversity can only be sustained if individuals derive benefits from its preservation; thus, initiatives that promote sustainable environmental practices offer avenues for instigating change. Under even moderate impact scenarios, nearly all bat species experienced a loss of native habitat (up to 99%), and many will be incapable of accessing new appropriate places (A.C. Hughes et al., in review). Facilitating species' adaptation to climate change and preventing extinction necessitates not only the cessation of harmful practices but also proactive human intervention to establish forested corridors connecting existing forests, thereby enabling species to access suitable habitats amid shifting conditions. It is unequivocal that, despite direct conservation efforts, climate change and environmental alterations will result in species extinction, even those not yet identified. The number of species that will become extinct in the next century remains unquantifiable, as extinctions are directly influenced by current human decisions and activities. Currently, humanity has the opportunity to diminish the repercussions of harmful activities and alleviate the effects of climate change through deliberate conservation efforts; however, continued inaction will exacerbate the extinction of numerous species due to our unsustainable practices.

6. APPENDIX

Variables included in species distribution models:

Vegetation cover: Globcover-Ionia (http://ionia1.esrin.esa.int/)

Mean annual temperature, minimum and maximum annual temperature, minimum and maximum mouthy precipitation, total annual precipitation, isothermality: www.worldclim.org

Humidity: New et al. 1999 (http://atlas.sage.wisc.edu/)

Elevation: NGDC http://www.ngdc.noaa.gov/mgg/topo/globe.html) Soil pH :ISRIC-WISE (www.isric.org/)

Distance from waterways and distance from roads: Edited from U.S. Geological Survey (USGS- www.usgs.gov/)

Karsts: Karst portal- School of Environment, University of Auckland, (http://web.env.auckland.ac.nz/our_research/karst/)

Geology :CCOP-Coordinating Committee of Geoscience Programmes in Asia and Southeast Asia (www.ccop.or.th/),Prince of Songkla University's GIS centre, Ministry of Mining in Myanmar.

Human population density: Ciesen (Grump v1: http://sedac.ciesin.columbia.edu/gpw/) A2 and B1 future climate scenarios: CIAT-GCM (Centro Internacional de Agricultura

Tropical-Global Climate Model, - CSIRO-Mk2.0 model: http://ccafs-climate.org/)

7. REFERENCES

Aitken, S.N.; Yeaman, S.; Holliday, J.A.; Wang, T. & Curtis-McLane, S. (2008). Adaptation, migration or extirpation: climate change outcomes for tree populations, *Evolutionary Applications* Vol.1, No.1, pp.95–111, ISSN 17524563

Anthony, N.M.; Johnson-Bawe, M.; Jeffery, K.; Clifford, S.L.; Abernethy, K.A.; Tutin, C.E.; Lahm, S.A.; White, L.J.T.; Utley, J.F.; Wickings, E.J. & Bruford, M.W. (2007). The role of Pleistocene refugia and rivers in shaping gorilla genetic diversity in central Africa. *Proceeding of the National Academy of Sciences USA,* Vol.104, No.51, pp.20432– 20436, ISSN 0027-8424

Aylward, B.; Allen K.; Echeverria J. & Tosi, J. (1996). Sustainable Ecotourism in Costa Rica: The Monteverde Cloud Forest. *Biodiversity and Conservation,* Vol.5, No.3, pp.315-43, ISSN 0960-3115

Bain, R.H.; Lathrop, A.; Murphy, R.W, Orlov, N.L. & Ho, C.T. (2003). Cryptic species of a cascade frog from Southeast Asia: taxonomic revisions and descriptions of six new species. *American Museum Noviates,* Vol.3417, pp.1-60, ISSN 0003-008

Baltzer, J.L.; Davies, S.J.; Bunyavejchewin, S. & Noor, N.S.M. (2008). The role of desiccation tolerance in determining tree species distributions along the Malay-Thai Peninsula. *Functional Ecology*, Vol.22, No. 2, pp.221-231, ISSN 0269-8463

Barclay, R.M.R.; Kurta, A. (2007). Ecology and behaviour of bats roosting in tree cavities and under bark. In: *Conservation and management of bats in forests*, Lacki, M.J.; Kurta, A.; Hayes, J.P. (Eds) pp 17-49. IBSN-10: 0-8018-8499-3, Baltimore, MD Johns Hopkins University Press.

Beale, C. M. & Lennon, J. J. (*In review*). Incorporating uncertainty in predictive species distribution modelling. *Philosophical transactions of the Royal society B.*

Beerling, D. J.; Huntley, B. & Bailey, J. P. (1995). Climate and the distribution of *Fallopia japonica*: use of an introduced species to test the predictive capacity of response surfaces. *Journal of Vegetation Science*, Vol.6, No.6, pp.269–282, ISSN 1100-9233

Berger, L.; Speare, R. & Hyatt, A. D. (1999). Chytrid fungi and amphibian declines: overview, implications and future directions. In: *Declines and disappearances of Australian frogs. Environment Australia*, A. Campbell (Ed.) pp. 23-33. Canberra.

Berry, P. M.; Dawson, T. P.; Harrison, P. A. & Pearson, R. G. (2002). Modelling potential impacts of climate change on the bioclimatic envelope of species in Britain and Ireland. Global Ecology and Biogeography, Vol.11, No.6, pp.453–462, ISSN 1466- 822X

Berry, P. M.; Vanhinsberg, D.; Viles, H. A.; Harrison, P. A.; Pearson, R. G.; Fuller, R.; Butt, N. & Miller, F. (2001). Impacts on terrestrial environments. In: Climate change and nature conservation in Britain and Ireland: modelling natural resourse responses to climate change (the MONARCH Project). P. A. Harrison, P. M. Berry and T. P. Dawson (Ed), pp. 43–149. ISBN 978-1-84754-026- 3, UKCIP technical report, Oxford. http://www.ukcip.org.uk/model_nat_res/model_nat_res.html

Blackman, A. & Woodward, R. (2010). *User financing in a national payments for environmental services program: Costa Rican hydropower, resources for the future*, Washington, DC RFFDP09-04-RE. http://ideas.repec.org/a/eee/ecolec/v69y2010i8p1626-1638.html

Blehert, D.S.; Hicks, A.C.; Behr, M.; Meteyer, C.U.; Berlowski-Zier, B.M.; Buckles, E.L.; Coleman, J.T.H.; Darling, S.R.; Gargas, A.; Niver, R.; Okoniewski, J.C.; Rudd, R.J. & Stone W.B. (2009). Bat white-nose syndrome: an emerging fungal pathogen? *Science*, Vol.323, No.5911, pp.227. ISSN 1095-9203

Bowen, A. & Ranger, N. (2009). Mitigating climate change through reductions in greenhouse gas emissions: the science and economics of future paths for global annual emissions, *Grantham Research Institute on Climate Change and the Environment*. London. http://personal.lse.ac.uk/RANGERN/Bowen-Ranger%20 policy%20brief.pdf

Boyles, J.G. & Willis, C.K.R. (2010). Could localized warm areas in cold caves reduce mortality of hibernating bats affected by white-nose syndrome? *Frontiers in Ecology and Environment*, Vol.8, No. 2, pp.92-98, doi:10.1890/080187

Brodie, J.F.; Helmy, O.E.; Brockelman, W.Y. & Maron, J.L. (2009). Bushmeat poaching reduces the seed dispersal and population growth rate of a mammal-dispersed tree. *Ecological Applications,* Vol.19, No.4, pp.854–863, ISSN 1051-0761

Brook, B.W.; Bradshaw. C.J.A.; Koh, L.P. & Sodhi, N.S. (2006). Momentum drives the crash: mass extinction in the tropics. *Biotropica*, Vol.38, No.3, pp.302-305, ISSN 0006- 3606

Bumrungsri, S.; Harrison, D.L.; Satasook, C.; Prajukjitr, A.; Thong-Aree, S. & Bates, P.J.J (2006). A review of bat research in Thailand with eight new species records for the country. *Acta Chiropterologica*, Vol.8, No.2, pp.325-359, ISSN 1733-5329 Canadell, J.G. & Raupach, M.R. (2008). Managing forests for climate change mitigation. *Science*, 320, No.5882, pp.1456–1457, ISSN 1095-920

Carignan, V. & Villard, M. (2002) Selecting indicator species to monitor ecological integrity: a review. *Environmental Monitoring and Assessment*, Vol.78, No.1, pp.45-61, ISSN 0167-6369

Cashore, B.; Gale, F.; Meidinger, E. & Newsom, D. (2006). Confronting Sustainability: Forest Certification in Developing and Transitioning Countries, ISBN 10: 0970788258, Yale FandES Publication Series, New Haven.

Center for International Earth Science Information Network (CIESIN) (2002). Country-level Population and Downscaled Projections based on the B2 Scenario, 1990-2100, Palisades, NY: CIESIN, Columbia University. http://www.ciesin.columbia.edu/datasets/downscaled

Clements, R.; Sodhi, N.S.; Ng, P.K.L. & Schilthuizen, M. (2006). Limestone karsts of Southeast Asia: imperiled arks of biodiversity. *Bioscience*, Vol.56, No.9, pp.733-742, ISSN 0006-3568

Corlett, R.T. (2008). Frugivory and seed dispersal by vertebrates in the Oriental (Indomalayan) region. *Biological Reviews*, Vol.73, No.4, pp.413-448, ISSN 1464-7931

Corlett, R. T. (2009). The ecology of tropical East Asia. ISBN-10: 019953246X, Oxford University Press, New York.

Cox, C. B. & Moore, P. D. (2010). *Biogeography: An ecological and evolutionary approach*, 8th Edn. ISBN-10: 0470637943, Oxford: Blackwell.

Day, M.J. & Urich, P.B. (2000). An assessment of protected karst landscapes in Southeast Asia. *Cave and Karst Science*, Vol.27, No. 2, pp.61-70, ISSN 1356191X

De Bruyn, M.E.; Nugroho, M.M. Hossain, J.C.; Wilson, J.C. & Mather, P.B. (2004). Phylogeographic evidence for the existence of an ancient biogeographic barrier: the Isthmus of Kra seaway. *Heredity*, Vol.94, pp.370-378, ISSN 1365-2540

Douangboubpha, B.; Bumrungsri, S.; Satasook, C.; Hammond, N. & Bates, P.J.J. (2010). A taxonomic review of *Hipposideros bicolor* and *H. pomona* (Chiroptera: Hipposideridae) in Thailand. *Acta Chiropterologica*, Vol.12, No.1, pp.415-438, ISSN 1733-5329

Duckworth, J.W. Hedges, S. (1998). Bird records from Cambodia in 1997, including records of sixteen species new for the country. *Forktail,* Vol.14, pp.29-36.

EIA/Telapak. (2008). Borderlines. Vietnam's Booming Furniture Industry and Timber Smuggling in the Mekong Region. Environmental Investigation Agency and Telapak, ISBN 0-9540768-6-9, London.

EIA. (2004). The tiger skin trail. Environmental Investigation Agency, London, UK. http://www.eia-international.org/files/reports85-1.pdf

Elith, J.; Graham, C. H.; Anderson, R. P.; Dudík, M.; Ferrier, S.; Guisan, A.; Hijmans, R. J.; Huettman, F.; Leathwick, J. R.; Lehmann, A.; Li, J.; Lohmann, L. G.; Loiselle, B. A.; Manion, G.; Moritz, C.; Nakamura, M.; Nakazawa, Y.; Overton, J. M.; Peterson, A. T.; Phillips, S. J.; Richardson, K. S.; Scachetti Pereira, R.; Schapire, R. E.; Soberón, J.; Williams, S.; Wisz, M. S. & Zimmermann, N. E. (2006). Novel methods improve prediction of species' distributions from occurrence data. *Ecography,* Vol.29, No.2, pp.129–151, ISSN 0906-7590

Ellis, R. (2005). Tiger bone and rhino horn: The destruction of wildlife for traditional Chinese medicine. ISBN-10: 1559635320, Island Press, Washington, DC.

Epstein, J.H.; Olival, K.J.; Pulliam, J.R.C.; Smith, C.; Westrum, J.; Hughes, T.; Dobson, A.P.; Zubaid, A.; Rahman, S.A.; Basir, M.M.; Field, H.E. & Daszak, P. (2009). *Pteropus vampyrus*, a hunted migratory species with a multinational home-range and a need for regional management. *Journal of Applied Ecology*, Vol.46, No.5, pp.991-1002, ISSN 0021-8901

Francis, C.M.; Borisenko, A.V.; Ivanova, N.V.; Judith, L.; Eger, E.R.; Burton, J.L.; Lim, K.; Guillén-Servent, A.; Sergei V. Kruskop, S.C.; Mackie, I. & Hebert, P. D.N. (2010). The Role of DNA Barcodes in Understanding and Conservation of Mammal Diversity in Southeast Asia. *PLoS ONE,* 9: e12575. doi:10.1371/journal.pone.0012575

Fredericksen, T.S. & Putz, F.E.; (2003). Silvicultural intensification for tropical forest conservation. *Biodiversity and Conservation,* Vol.12, No.7, pp.1445–1453, ISSN 0960- 3115

Fujita, M.S. & Tuttle, M.D. (1991). Flying foxes (Chiroptera: Pteropodidae): threatened animals of key ecological and economic importance. *Conservation Biology*, Vol.4, No.6, pp.455-463, ISSN 0888-8892

Fuller, D.O.; Jessup, T.C.; Salim, A. (2003). Loss of forest cover in Kalimantan, Indonesia, since the 1997–1998 El Niño. *Conservation Biology*, Vol.18, No.1, pp.249–254, ISSN 0888-8892

Gan, J. & McCarl, B.A. (2007). Measuring transnational leakage of forest conservation. *Ecological Economics*, Vol.64, No.2, pp.423–432, ISSN 0921-8009

Gaffin, S.R.; Xing, X. & Yetman, G. (2003). Downscaling and geo-spatial gridding of socio- economic projections from the IPCC Special Report on Emissions Scenarios (SRES), *Global Environmental Change Part A*, Vol.14, No.2, pp.105-123, doi:10.1016/j.physletb.2003.10.071

Gaston, K.J.; Williams, P.H.; Eggleton. P.J. & Humphries, C.J. (1995a). Large scale patterns of biodiversity: spatial variation in family richness. *Proceedings of the Royal Society. London* B, Vol.260, No.1358, pp.149-154, ISSN 09628452

Gaston, K.J. & Blackburn, T.M. (1995b). Mapping biodiversity using surrogates for species richness: macro-scales and New World birds. *Proceedings of the Royal Society of London B,* Vol.262, No.1365, pp.335-341, ISSN 0962-8452

Giam, X.; Ng, T.H.; Yap, V.B.; Tan, H.T.W. (2010). The extent of undiscovered species in Southeast Asia. *Biological Conservation*, Vol.19, No.4, pp.943-954, ISSN 1572-9710

Global Witness (2007). Cambodia's family trees: Illegal logging and the stripping of public assets by Cambodia's elite, Washington DC.

Gibbons, P. & Lindenmayer, D. (2001). Tree hollows and wildlife conservation in Australia. pp. 240, ISBN: 9780643090033 Canberra: CSIRO.

Gibson. L.; McNeill, A.; De Tores, P.; Waynec, A. & Yates, C. (2010). Will future climate change threaten a range restricted endemic species, the quokka (*Setonix brachyurus*), in south west Australia? *Biological Conservation*, Vol.143, No.11, pp.2453-2461, ISSN 0006-3207

Gray, T.N.E.; Chamnan, H.; Borey, B.; Collar, J. & Dolman, P.M. (2007). Habitat preferences of a globally threatened bustard provide support for community-based conservation in Cambodia. *Biological Conservation,* Vol.138, No3-4, pp.341–350, ISSN 0006-3207

Hall, R. (2002). Cenozoic geological and plate tectonic evolution of SE Asia and the SW Pacific: computer-based reconstructions and animations. *Journal of Asian Earth Sciences*, Vol.20, No.4, pp.353-434, ISSN 1367-9120

Hannah, L.; Midgley, G. F.; Lovejoy, T.; Bond, W. J.; Bush, M.; Lovett, J. C.; Scott, D. & Woodward, F. I. (2002). Conservation of biodiversity in a changing climate. *Conservation Biology*, Vol.16, No.1, pp.264–268, ISSN 0888-8892

Hansen, M.C. & DeFries, R. (2004). Detecting long-term global forest change using continuous fields of tree-cover maps from 8-km advanced very high resolution radiometer (AVHRR) data for the years 1982-99. *Ecosystems*, Vol.7, No.7, pp.695-716, ISSN 1432-9840

Heller, N.E. & Zavaleta, E.S. (2009). Biodiversity management in the face of climate change: a review of 22 years of recommendations. *Biological Conservation*, Vol.142, No.1, pp.14-32, ISSN 0006-3207

Hernandez, P. A.; Graham, C. H.; Master, L. L. & Albert, D. L. (2006). The effect of sample size and species characteristics on performance of different species distribution modelling methods. *Ecography*, Vol.29, No.5, pp. 773–785, ISSN 0906-7590

Hickling, R.; Roy, D.B.; Hill, J.K.; Fox, R. & Thomas, C.D. (2006). The distributions of a wide range of taxonomic groups are expanding polewards. *Global Change Biology*, Vol.12, No.3, pp.450–455, ISSN I1354-1013

Hill, J.; Woodland, W. & Gough, G. (2007). Can visitor satisfaction and knowledge about tropical rainforests be enhanced through biodiversity interpretation, and does this promote a positive attitude towards ecosystem conservation? *Journal of Ecotourism*, Vol.6, No.1, pp.75–85, ISSN 1472-4049

Hirzel, A. H.; Hausser, J.; Chessel, D. & Perrin, N. (2002). Ecological-niche factor analysis: how to compute habitat suitability maps without absence data. *Ecology*, Vol.83, No.7, pp2027-2036, ISSN: 00129658

Hodgkison, R.; Balding, S.T.; Zubaid, A. & Kunz, T.H. (2004). Temporal variation in the relative abundance of fruit bats (Megachiroptera: Pteropodidae) in relation to the availability of food in a lowland Malaysian rain forest. *Biotropica*, Vol.36, No.4, pp.522-533, ISSN 0006-3606

Hughes, A.C.; Satasook, C.; Bates, P.J.J.; Bumrungsri, S. & Jones, G. (2011). Explaining the causes of the zoogeographic divide at the Isthmus of Kra: using bats as a case study. *Journal of Biogeography*, doi/10.1111/j.1365-2699.2011.02568.x/pdf

Hughes, A.C.; Satasook, C.; Bates, P.J.J.; Soisook, P.; Sritongchuay, T.; Jones, G. & Bumrungsri, S. (2010). Echolocation call analysis and presence-only modelling as conservation monitoring tools for rhinolophoid bats in Thailand. *Acta Chiropterologica*, Vol.12, No.2, pp.311-327, ISSN 1733-5329

Hughes, A.C.; Satasook, C.; Bates, P.J.J.; Bumrungsri, S. & Jones, G. (*in press*). Using echolocation call structure to identify Thai bat species: Vespertilionidae, Emballonuridae, Nycteridae and Megadermatidae. *Acta Chiropterologica,*

Hughes, A.C.; Satasook, C.; Bates, P.J.J.; Bumrungsri, S. & Jones, G. (*In review*). The projected effects of climatic and vegetation changes on the distribution and diversity of Southeast Asian bats.

Hughes, A.C.; Satasook, C.; Bates, P.J.J.; Soisook, P. & Bumrungsri, S. (*In prep a*). Cryptic clues to a complex problem.

Hughes, A.C.; Satasook, C.; Bates, P.J.J.; Bumrungsri, S. & Jones, G. (*In prep b*). Southeast Asian bat biodiversity, and an analysis of the relationship between anthropogenic threat factors.

Hughes, J.B.; Round, P.D. & Woodruff, D.S. (2003). The Sundaland-Asian faunal transition at the Isthmus of Kra: an analysis of resident forest bird species distributions. *Journal of Biogeography*, Vol.30, No.4, pp.569-580, ISSN 0305-0270

Hughes, L. (2000). Biological consequences of global warming: is the signal already apparent? *Trends in Ecology and Evolution*, Vol.15, No.2, pp.56–61, ISSN 0169-5347

Huntley, B.; Berry, P. M.; Cramer, W. & Mcdonald, A. P. (1995). Modelling present and potential future ranges of some European higher plants using climate response surfaces. *Journal of Biogeography*, Vol.22, No.6, pp.967–1001, ISSN 0305-0270

Jacobson, S.K. & Robles, R. (1992). Ecotourism, sustainable development, and conservation education: development of a tour guide training program in Tortuguero, Costa Rica. *Environmental Management*, Vol.16, No.6, pp. 701-713, ISSN: 0364152X

Jime´nez-Valverde, A.; Lobo, J. M. & Hortal, J. (2008). Not as good as they seem: the importance of concepts in species distribution modelling. *Diversity and Distributions*, Vol.14, No.6, pp.885–890 ISSN 1366-9516

Jones, G.; Jacobs, D.S.; Kunz, T.H.; Willig, M.R. & Racey, P.A. (2009). Carpe noctem: the importance of bats as bioindicators. *Endangered Species Research*, Vol.8, No.1-2, pp.93-115, doi: 10.3354/esr0018

Jones, G.; Van Parijs, S.M. (1993). Bimodal echolocation in pipistrelle bats: are cryptic species present? *Proceedings of the Royal Society of London, B. Biological Sciences*, Vol.251, No.1331, pp.119-125, ISSN 0962-8452

Jones, K. E.; Russ, J. A.; Bilhari, Z.; Catto, C.; Csosz, I.; Gorbachev, A.; Gyorfi, P.; Hughes, A.; Ivashkiv, I.; Koryagina, N.; Kurali, A.; Langton, S.; Maltby, A.; Margina, G.; Pandourski, I.; Parsons, S.; Prokofev, I.; Szodoray-Paradi, A.; Szodoray-Paradi, F.; Taras-Bashta, A.; Tilova, E.; Walters, C.; Weatherill, A. & Zavarzin, O. (2011). Monitoring ultrasonic biodiversity: using bats as biodiversity indicators.

Biodiversity monitoring and conservation: bridging the gaps between global commitment and local action, B. P. Collen, N. Pettorelli, S. M. Durant, L. Krueger, J. Baillie (Eds). ISBN-10: 1444332929, Blackwell Press, London

Kettle, C.J. (2010). Ecological considerations for using dipterocarps for restoration of lowland rainforest in Southeast Asia. *Biodiversity and Conservation*, Vol.19, No.4, pp.1137-1151, ISSN 1572-9710

Kingston, T. (2010). Research priorities for bat conservation in Southeast Asia: a consensus approach. *Biodiversity and conservation*, Vol.19, No.2, pp.471-484, ISSN 1572-9710

Kingston, T.; Francis, C.M.; Zubaid, A. & Kunz, T.H. (2003). Species richness in an insectivorous bat assemblage from Malaysia. *Journal of Tropical Ecology*, Vol.19, No.1, pp.67-79, ISSN 0266-4674

Kingston, T.; Juliana, S.; Rakhmad, S. K, Fletcher, C. D.; Benton-Browne, A.; Struebig, M.; Wood, A.; Murray, S. W.; Kunz, T. H. & Zubaid, A. (2006). The Malaysian Bat Conservation Research Unit: research, capacity building and education in an Old World hotspot. 41-60 In: *Proceedings of the National Seminar On Protected Areas*, Othman, S. H. Yatim, S. Elaguipillay, Sh. Md. Nor, N. A. Sharul and A. M. Sah. (Eds) Department of Wildlife and National Parks.

Kotiaho, J.S.; Kaitala, V.; Kolmonen, A. & Paivinen, J. (2005). Predicting the risk of extinction from shared ecological characteristics. *Proceedings of the National Academy of Sciences USA*, Vol.102, No.6, pp.1963–1967, ISSN 0027-8424

Kunz, T.H. & Lumsden, L.F. (2003). Ecology of cavity and foliage roosting bats. In: *Bat ecology,* T. H. Kunz & M. B. Fenton (Eds.) pp. 3-90. Chicago, IL University of Chicago Press.

Lamb, D. & Erskine, P. (2008). Forest restoration on a landscape scale. In: *Living in a dynamic forest environment,* N.E. Stork and S.M. Turton (Eds.) pp.469-484, ISBN 978-1-4051- 5643-1, Blackwell, Malden.

Lamb, J. M.; Ralph, T.M.C.; Goodman, S. M.; Bogdanowicz, B, Fahr, J.; Gajewska, M.; Bates, P. J. J.; Eger, J.; Benda, P. & Taylor, P. J. (2008). Phylogeography and predicted distribution of African-Arabian and Malagasy populations of giant mastiff bats, *Otomops* spp. (Chiroptera: Molossidae). *Acta Chiropterologica,* Vol.10, No.1, pp21-40, ISSN 1508-1109

Laumonier, Y.; Uryu, Y.; Stüwe, M.; Budiman, A.; Setiabudi, B. & Hadian, O. (2010). Eco- floristic sectors and deforestation threats in Sumatra: identifying new conservation area network priorities for ecosystem-based land use planning. *Biodiversity and Conservation*, Vol.19, No.4, pp.1153-1174, ISSN 1572-9710

Laurance, W.F. (2007). Have we overstated the tropical biodiversity crisis? *Trends in Ecology and Evolution*, Vol.22, No.2, pp.65-70, ISSN 0169-5347

Laurance, W.F. (2008a). The real cost of minerals. *The New Scientist*, Vol.199, pp.16. http://hdl.handle.net/10088/12008

Laurance, W.F. (2008b). Can carbon trading save vanishing forests? *BioScience*, Vol.58, No.8, pp.286–287, ISSN 0006-3568 Lepers, E.; Lambin, E.F.; Janetos, A.C.; DeFries, R.; Achard, F.; Ramankutty, N. & Scholes, R. J. (2005). A synthesis of information on rapid land-cover change for the period 1981-2000. *BioScience*, Vol.55, No.2, pp.115-124, ISSN 0006-3568

Lenoir, J.; Gegout, J.C.; Marquet, P.A.; De Ruffray, P. & Brisse, H. (2008). A significant upward shift in plant species optimum elevation during the 20th century. *Science*, Vol.320, No.5884, pp.1768–1771, ISSN 0036-8075

Linkie, M.; Rood, E. & Smith, R.J. (2010). Modelling the effectiveness of enforcement strategies for avoiding tropical deforestation in Kerinci Seblat National Park, Sumatra. *Biodiversity and Conservation*, Vol.19, No.4, pp.973-984, ISSN 1572-9710

Mace, G.M. & Baillie, J.E.M. (2007). The 2010 biodiversity indicators: challenges for science and policy. *Conservation Biology*, Vol.21, No.6, pp.1406-1413, doi/10.1111/j.1523- 1739.2007.00830.x/full

MacKenzie, D. I. (2005). Was it there? Dealing with imperfect detection for species presence/absence data. *Australia and New Zealand Journal of Statistics*, Vol.47, No.1, pp.65–74, ISSN 13691473

Macswiney, G. M.; Clarke, C. F. & Racey, P. A. (2008). What you see is not what you get: the role of ultrasonic detectors in maximising inventory completeness in Neotropical forests. *Journal of Applied Ecology*, Vol.45, No.5, pp.1364-1371, ISSN: 00218901

Malcolm, J.R.; Liu, C.; Neilson, R.P.; Hansen. L. & Hannah, L. (2006). Global warming and extinctions of endemic species from biodiversity hotspots. *Conservation Biology*, Vol.20, No.2, pp.538–548, ISSN 0888-8892

McLachlan, J.S.; Hellmann, J. & Schwartz, M. (2007). A framework for debate of assisted migration in an era of climate change. *Conservation Biology*, Vol.21, No.2, pp.297– 302, ISSN 0888-8892

McMahon, G.; Subdibjo, E.R.; Eden, J.; Bouzaher, A.; Dore, G. & Kunanayagan, R. (2000). Mining and the environment in Indonesia: Long-term trends and repercussions of the Asian economic crisis (EASES Discussion Paper Series). Washington, DC: World Bank, East Asia Environment and Social Development Unit, www.natural- resources.org/minerals/cd/docs/twb/mining_indonesia.pdf

McNamara, S.; Duong Viet Tinh, Erskine, P. D.; Lamb, D.; Yates, D. & Brown, S. (2006). Rehabilitating degraded forest land in central Vietnam with mixed native species plantings. *Forest Ecology and Management*, Vol.233, No.2, pp.358-365, DOI: 10.1111/j.1749-4877.2007.00048.x

McNamara, N.P.; Black, H.I.J, Piearce, T.G.; Reay, D.S. & Ineson, P. (2008). The influence of afforestation and tree species on soil methane fluxes from shallow organic soils at the UK gisburn forest experiment. *Soil Use Management*, Vol.24, No.1, pp.1–7, ISSN 0266-0032

McNeely, J.A. (2007). A zoological perspective on payments for ecosystem services. *Integrative Zoology*, Vol.2, No.2, pp.68–78, DOI: 10.1111/j.1749-4877.2007.00048.x McShane, T.O. & Wells, M.P. (2004). *Getting biodiversity projects to work: towards more effective conservation and development.* New York: Columbia University Press.

Meyer, C. F. J.; Aguiar, M. W. S.; Aguirre, L. F.; Baumgarten, J.; Clarke, F. M.; Cosson, J. F.; Villegas, S. E.; Fahr, J.; Faria, D.; Furey, N.; Henry, M.; Hodgkison, R.; Jenkins, R. K. B.; Jung, K. G.; Kingston, T.; Kunz, T. H.; M. MacSwiney G, M. C.; Moya, I.; Pons, J-M.; Racey, P. A.; Rex, K.; Sampaio, E. M.; Stoner, K. E.; Voigt, C. C.; von Staden, D.; Weise, C. D. & Kalko, E. K. V. (2010). Long-term monitoring of tropical bats for anthropogenic impact assessment: Gauging the statistical power to detect population change, *Biological Conservation*, Vol.143, No.11, pp.2797-2807, ISSN 0006- 3207

Mickleburgh, S.; Waylen, K. & Racey, P.A. (2009). Bats as bushmeat – a global review, *Oryx*, Vol.43, No.2, pp.217–234, DOI: 10.1017/S0030605308000938

Midgley, G. F.; Hannah, L.; Millar, D.; Rutherford, M. C. & Powerie, L. W. (2002). Assessing the vulnerability of species richness to anthropogenic climate change in a biodiversity hotspot. *Global Ecology and Biogeography*, Vol.11, No.6, pp.445–451, ISSN 1466-822X

Mollmann, S. (2008). Birders flock east. *The Wall Street Journal,* September 12, W1. http://online.wsj.com/article/SB122115338586124101.html

Mohd-Azlan, J.; Zubaid, A. & Kunz, T.H. (2001). Distribution, relative abundance, and conservation status of the large flying fox, *Pteropus vampyrus*, in peninsular Malaysia: a preliminary assessment. *Acta Chiropterologica*, Vol.3, No.2, pp.149-162, ISSN 1508-1109

Morley, R.J. & Flenley, J.R. (1987). Late Cainozoic Vegetational and Environmental Changes in the Malay Archipelago', In: *Biogeographical Evolution in the Malay Archipelago, Biogeography*, T. C. Whitmore (Ed.), pp.50-59, ISBN-10 0198541856, Oxford Monographs.

Myers, N. (1987) The extinction spasm impending: synergisms at work. *Journal of Conservation Biology*, Vol.1, No.1, pp.14-21, ISSN 0888-8892

Myers, N. (1988). Mass extinction-profound problem, splendid opportunity. *Oryx,* Vol.22, pp.205-15, DOI: 10.1017/S003060530002233X

Myers, N.; Mittermeier, R. A.; Mittermeier, C. G.; Da Fonseca, G. A. B. & Kent, J. (2000). Biodiversity hotspots for conservation priorities. *Nature*, Vol.403, No.6772, pp. 853- 858, ISSN 0028-0836

Nijman, V. (2010). An overview of international wildlife trade from Southeast Asia. *Biodiversity and Conservation,* Vol.19, No.4, pp.1101-1114, ISSN 1572-9710

Olson, D.M.; Dinerstein, E.; Wikramanayake, E.D.; Burgess, N.D.; Powell, G.V.N.; Underwood, E.C. D'Amico, J.A.; Itoua, I.; Strand, H.E.; Morrison, J.C.; Loucks, C.J.; Allnutt, T.F.; Ricketss, T.H.; Kura, Y.L.J.F.; Wettengel, W.W.; Hedao, P. & Kassem, K.R. (2001). Terrestrial ecoregions of the World: a new map of life on Earth. *BioScience*, Vol.51, No.11, pp.933-938, ISSN 0006-3568

Pawar, S.; Koo, M. S.; Kelley, C.; Firoz Ahmed, M.; Chaudhuri, S. & Sarkar, S. (2007). Conservation assessment and prioritization of areas in Northeast India: priorities for amphibians and reptiles. *Biological Conservation,* Vol.136, No.3, pp.346-361, ISSN 0006-3207

Pearce, P.L. (2008). The nature of rainforest tourism: insights from a tourism social science research programme.In: *Living in a dynamic forest environment,* N.E. Stork and S.M. Turton (Eds.) pp.94-104. ISBN 978-1-4051-5643-1, Blackwell, Malden.

Pearman, P. B.; Guisan, A. & Zimmermann, N.E. (2010). Impacts of climate change on Swiss biodiversity: an indicator taxa approach. *Biological Conservation*, Vol.144, No.2, pp.866-875, doi:10.1016/j.biocon.2010.11.020

Peh, K.S.H. (2010). Invasive species in Southeast Asia: the knowledge so far. *Biodiversity and Conservation*, Vol.19, No.4, pp.1083-1099, ISSN 1572-9710

Pergams, O.R.W. & Zaradic, P.A. (2008). Evidence for a fundamental and pervasive shift away from nature-based recreation. *Proceedings of the National Academy of Sciences*, Vol.105, No.7, pp.2295–2300, ISSN 0027-8424

Phillips, S.J.; Anderson, R.P. & Schapire, R.E. (2006). Maximum entropy modelling of species geographic distributions. *Ecological Modelling*, Vol.190, No3-4, pp.231-259, ISSN 0304-3800

Pounds, J.A.; Bustamente, M.R.; Coloma, L.A.; Consuegra, J.A.; Fogden, M.P.L.; Foster, P. N.; La Marca, E.; Masters, K.L.; Merino-Viteri A.; Puschendorf, R.; Santiago R.R.; Sánchez-Azofeifa, G.A.; Still, C.J. & Young, B.E. (2006). Widespread amphibian extinctions from epidemic disease driven by global warming. *Nature,* Vol.439, No.7073, pp.161–67, ISSN 0028-0836

Prentice, I. C.; Cramer, W.; Harrison, S. P.; Leemans, R.; Monserud, R. A. & Solomon, A. M. (1992). A global biome model based on plant physiology and dominance, soil properties and climate. *Journal of Biogeography*, Vol.19, No.2, pp.117–134, ISSN: 03050270

Putz, F.E.; Sist P.; Fredericksen T.S. & Dykstra D. (2008). Reduced-impact logging: challenges and opportunities. *Forest Ecology and Management,* Vol.256, No.7, pp.1427–1433, ISSN 0378-1127

Rebelo, H. & Jones, G. (2010). Ground validation of presence-only modelling with rare species: a case study on barbastelles *Barbastella barbastellus* (Chiroptera: Vespertilionidae). *Journal of Applied Ecology*, Vol.47, No.2, pp.410-420, ISSN 0021-8901

Rebelo, H.; Tarroso, P, Jones, G. (2010). Predicted impact of climate change on European bats in relation to their biogeographic patterns. *Global Change Biology*, Vol.16, No.2, pp.561-576, ISSN 1354-1013

Redford, K.H. (1992). The empty forest. *BioScience*, Vol.42, No.6, pp.412–422, ISSN 0006-3568 Riley, S.P.D.; Busteed, G.T.; Kats, L.B.; Vandergon, T.L.; Lee, L.F.S.; Dagit, R.G.; Kerby, J.L.; Fisher, R.N. & Sauvajot, R.M. (2005). Effects of urbanization on the distribution and abundance of amphibians and invasive species in southern California streams. *Conservation Biology*, Vol.19, No.6, pp.1894–1907, ISSN 0888-8892

Salazar, L.F.; Nobre, C.A. & Oyama, M.D. (2007). Climate change consequences on the biome distribution in tropical South America. *Geophysical Research Letters*, Vol.34, No.9, pp.1–6, ISSN 0094-8276

Sechrest, W.; Brooks, T.M.; da Fonseca, G.A.B.; Konstant, W.R.; Mittermeier, R.A.; Purvis, A.; Rylands, A.B. & Gittleman, J.L. (2002). Hotspots and the conservation of evolutionary history. *Proceedings of the National Academy of Sciences USA*, Vol.99, No.4, pp.2067–2071, ISSN 0027-8424

Sérgio, C.; Figueira, R.; Draper, D.; Menezes, R. & Sousa, A.J. (2007). Modelling bryophyte distribution based on ecological information for extent of occurrence assessment. *Biological Conservation*, Vol.135, No.3, pp.341-351, ISSN 0006-3207

Sessions, J. (2007). Forest Road Operations in the Tropics. Tropical Forestry Series, Springer- Verlag, pp. 168, ISBN-10 3642079776, Berlin.

Seto, K.; Sanchez-RodrJguez, R. & Fragkias, M. (2010). The new geography of contemporary urbanization and the environment. *Annual review of environment and resources*, Vol.35, pp.167–94, ISSN 1543-5938

Siddique, I.; Engel, V.L.; Parrotta, J.A.; Lamb, D.; Nardoto, G.B.; Ometto, J.; Martinelli, L.A. & Schmidt, S. (2008). Dominance of legume trees alters nutrient relations in mixed species forest restoration plantings within seven years. *Biogeochemistry,* Vol.88, No.1, pp.89–101, ISSN 0168-2563

Simmons, N. (2005). Order Chiroptera, In: *Mammal species of the World: a taxonomic and geographic reference*, Wilson and D. M. Reeder(Eds) pp 312-529. ISBN-10 0801882214, John Hopkins University Press, Baltimore.

Simmons, N. B. & Wetterer, A.L (2011) Estimating diversity: how many bat species are there? The Second International South-East Asian Bat Conference Bogor, 6-9 June 2011.

Sodhi, N.S.; Posa, M.R.; Lee, T.M.; Bickford, D.; Koh, L.P. & Brook, B.W. (2010). The state and conservation of Southeast Asian biodiversity. *Biodiversity and Conservation*, Vol.19, No.2, pp.317-328, ISSN 1572-9710

Soisook, P.; Bumrungsri, S.; Satasook, C.; Thong, V.D.; Bu, S.S.H.; Harrison, D.L. & Bates,

P.J.J (2008). A taxonomic review of *Rhinolophus stheno* and *R. malayanus* (Chiroptera: Rhinolophidae) from continental Southeast Asia: an evaluation of echolocation call frequency in discriminating between cryptic species. *Acta Chiropterologica*, Vol.10, No.2, pp.221-242, ISSN 1508-1109

Soisook. P.; Niyomwan. P.; Srikrachang. M.; Srithongchuay, F. & Bates, P.J.J. (2010). Discovery of *Rhinolophus beddomei* (Chiroptera: Rhinolophidae) from Thailand with a brief comparison to other related taxa. *Tropical Natural History*, Vol.10, No.1, pp.67-79.

Stibig, H.J.; Belward, A.S.; Roy, P.S.; Rosalinea-Wasrin, U.; Agrawal, S.; Joshi, P.K.; Hildanus Beuchle, S.; Fritz, S.; Mubareka, S. & Giri, C. (2007). A land-cover map for South and Southeast Asia derived from SPOT-VEGETATION data. *Journal of Biogeography*, Vol.34, No.4, pp.625-637, ISSN 0305-0270

Steinmetz, R.; Chutpong, W. & Seuaturien, N. (2006). Collaborating to conserve large mammals in Southeast Asia. *Conservation Biology*, Vol.20, No.5, pp.1391–1401, ISSN 0888-8892

Stralberg, D.; Jongsomjit, D.; Howell, C.A.; Snyder, M.A.; Alexander, J.D.; Wiens, J.A.& Root, T.L. (2009). Re-shuffling of species with climate disruption: a no-analog future for California birds? PLoS One, Vol.4, No.9, e6825, doi:10.1371/journal.pone.0006825.

Struebig, M.; Kingston, T.; Zubaid, A.; Adnan, A. & Rossiter, S. (2008). Conservation value of forest fragments to Palaeotropical bats. *Biological Conservation*, Vol.141, No.8, pp.2112-2126, ISSN 0006-3207

Suzuki, R.; Numata, S.; Okuda, T.; Supardi, M.N.; Kachi, N. (2009). Growth strategies differentiate the spatial patterns of 11 dipterocarp species coexisting in a Malaysian tropical rain forest. *Journal of Plant Research*, Vol.122, No.1, pp.81–93, ISSN 1618- 0860

Sykes, M. T.; Prentice, I. C. & Cramer, W. (1996). A bioclimatic model for the potential distributions of north European tree species under present and future climates. *Journal of Biogeography*, Vol.23, No.2, pp.203–233, ISSN 0305-0270

Taylor, D.; Saksena, P.; Sanderson, P.G. & Kucera, K. (1999). Environmental change and rain forests on the Sunda Shelf: drought, fire and the biological cooling of biodiversity hotspots. *Biodiversity and Conservation*, Vol.8, No.8, pp.1159–1177, ISSN 0960-3115

Ten Kate, K.; Bishop, J.; Bayron, R. (2004). *Biodiversity Offsets: Views, Experience and the Business Case. International Union for Conservation of Nature and Natural Resources and Insight Investment, London.* http://www.eldis.org/static/DOC16610.htm

Terborgh, J.; Van Schaik, C.P.; Davenport, L. & Rao, M. (2002). *Making parks work: strategies for preserving tropical nature.* Island Press, Washington , D.C.

Tungittiplakorn, W. & Dearden, P. (2002). Biodiversity conservation and cash crop development in Northern Thailand. *Biodiversity and Conservation,* Vol.11, No.11, pp.2007–2025, DOI: 10.1023/A:1020812316749

United Nations Population Division (2008). World Urbanization Prospects: The 2007 Revision Population Database (http://esa.un.org/unup/)

Vistor, D. G.; Morgan, M. G.; Apt, J.; Steinbruner, J. & Ricke, K. (2009) The geoengineering option: a last resort against Global warming? *Foreign Affairs*, March/April http://iis-db.stanford.edu/pubs/22456/The_Geoengineering_Option.pdf

Webb, C.O.; Slik, J.W.F. & Triono, T. (2010). Biodiversity inventory and informatics in Southeast Asia. *Biodiversity and Conservation,* Vol.19, No.4, pp.955-972, ISSN 1572- 9710

Whitmore, T.C. (1981). Palaeoclimate and vegetation history.In: *Wallace's Line and Plate Tectonics*, T.C. Whitmore (Ed.), pp.36-42. ISBN-10: 0198545452, Clarendon Press, Oxford.

Wintle, B. A.; McCarthy, M. A.; Parris, K. M. & Burgman, M. A. (2004). Precision and bias of methods for estimating point survey detection probabilities. *Ecological Applications*, Vol.14, No.3, pp.703–712, http://www.jstor.org/stable/4493574

Wisz, M. S.; Hijmans, R. J.; Li, J.; Peterson, A. T.; Graham, C. H. & Guisan, A. (2008). Effects of sample size on the performance of species distribution models. *Diversity and Distributions*, Vol.14, No.5, pp.763–773, ISSN 1366-9516

Witmer, F. (2005). Simulating future global deforestation using geo-graphically explicit models, IIASA *Interim Report* IR-05-010

Woodruff, D.S. (2003). Neogene marine transgressions, paleogeography and biogeographic transitions on the Thai-Malay Peninsula. *Journal of Biogeography*, Vol.30, No.4, pp.551-567, ISSN 0305-0270

Woodruff, D.S (2010). Biogeography and conservation in Southeast Asia: how 2.7 million years of repeated environmental fluctuations affect today's patterns and the future of the remaining refugial-phase biodiversity. *Biodiversity and Conservation*, Vol.19, No.4, pp.919-941, ISSN 1572-9710

Wright, S.J.; Stoner, K.E.; Beckman, N.; Corlett, R.T.; Dirzo, R.; Muller-Landau, H.C.; Nunez- Iturri, G.; Peres C.A. & Wang, B.C. (2007). The plight of large animals in tropical forests and the consequences for plant regeneration. *Biotropica,* Vol.39, No.3, pp.289–291, DOI: 10.1111/j.1744-7429.2007.00293.

Wunder, S. (2006). Are direct payments for environmental services spelling doom for sustainable forest management in the tropics? *Ecology and Society*, Vol.11, No.2, pp.25, http://www.ecologyandsociety.org/vol11/iss2/art23/

Wunder, S. (2007). The efficiency of payments for environmental services in tropical conservation. *Conservation Biology*, Vol.21, No.1, pp.48–58, ISSN 0888-8892

Wunscher, T.; Engel S. & Wunder, S. (2006). Payments for environmental services in Costa Rica: increasing efficiency through spatial differentiation. *Quarterly Journal of International Agriculture*, Vol.45, No.4, pp.317-335, http://131.220.109.9/module/register/media/a9b0_Wuenscher%20Engel%20Wunder%202006.pdf

Zeppel, H. (2006). Indigenous ecotourism: Sustainable development and management, pp.272, ISBN-10: 1845931246, CABI, Wallingford, UK.

CHAPTER 2

PROTEIN LIMITATION ACCOUNTS FOR VARIATION IN PRIMATE COLOR VISION PHENOTYPES: A COMPREHENSIVE FRAMEWORK FOR THE EVOLUTION OF PRIMATE TRICHROMATIC VISION

1. INTRODUCTION: AN OVERVIEW

The color vision of primates has captivated scientists for numerous decades and is expected to be a subject of interest for the foreseeable future. Primates represent the most visually adapted order of mammals, with a significant percentage of their substantial brain size allocated to the processing of visual information. (e.g. Barton, 2006). Most eutherian mammals have dichromatic (two-colour) vision, and chromatic distinctions are based on discriminating relatively shorter from relatively longer wavelengths within the visual spectrum (~400-700nm). These distinctions are established through brain comparisons of cone cells containing short (S) wavelength-sensitive photopigments, which exhibit peak sensitivity to blue light, and long (L) wavelength-sensitive pigments, which demonstrate peak sensitivity to green light. The photopigments are encoded by an autosomal S opsin gene and an X-linked L opsin gene, respectively. Primates possess an extra color channel that facilitates trichromatic vision through the duplication and divergence of the L opsin gene, leading to the formation of long and medium (L-M) wavelength-sensitive photopigments (Hunt et al., 2009; Regan et al., 2001). This configuration facilitates improved differentiation of light and the perception of various colors of green, yellow, orange, and red.

Old World monkeys, apes and humans are routinely trichromatic, having two loci for L-M opsin genes on each X-chromosome due to a gene duplication event (Jacobs, 2008), in addition to the autosomal S opsin locus. Alternatively, New World monkeys and some lemurs exhibit polymorphic trichromacy (Jacobs, 2007; Tan & Li, 1999; Tan et al., 2005; Veilleux & Bolnick, 2009). This alternate path to trichromacy results from a genetic polymorphism at a single locus of the X-chromosomal L-M opsin gene. Females with two different L-M opsin genes, combined with the common autosomal S opsin gene, possess trichromacy (Mollon et al., 1984). However, males, being hemizygotes, can inherit only one L-M opsin gene and are always dichromatic; homozygous females are also dichromatic. Variation in primate trichromacy and the selective pressures that led to trichromacy are under considerable debate (Caine et al., 2010).

To assess regular and polymorphic trichromacy in monkeys within an evolutionary context, it is essential to examine the behavioral, experimental, genetic, biogeographical, and ecological information related to primate trichromacy. This review concentrates on the evidence for and against foraging theories, as the majority of existing research assesses and endorses the efficacy of trichromacy in locating and picking food. We commence with an examination of the fruit foraging and young leaf hypotheses. We propose a novel hypothesis that both regular and polymorphism trichromacy provide a selective advantage to monkeys in the identification and selection of protein-rich foods, either leaves or fruit, and offer evidence supporting Ficus (Moraceae) as a protein source for neotropical primates. This idea is grounded in an examination of primate foraging behaviors on Ficus, alongside an analysis of the abundance and density of Ficus in African and neotropical forests.

2. THE HYPOTHESIS OF FRUGIVORY

Allen (1879) and Polyak (1957) argue that trichromacy is an adaptation to frugivory, which is the earliest hypothesis for the evolution of primate color vision. This hypothesis is based on the observation that the majority of primates significantly rely on dietary fruit (Chapman & Onderdonk, 1998; Fleagle, 1999). Trichromatic color vision should facilitate the detection and selection of edible fruits, as it enhances primates' abilities to detect red, orange, or yellowish ("colourful") fruits amidst green foliage and unripe fruits as a result of the fact that many tropical plants produce these colors (Regan et al., 2001). Primates may, in turn, offer these fruiting plants high-quality seed dispersal services (Garber & Lambert, 1998). Trichromatic advantages in these tasks are supported by compelling experimental and behavioral evidence. The capacity to differentiate between mature and unripe fruits, fruits from leaves, and edible from inedible species is a complex task that could be facilitated by trichromacy (Smith et al., 2003). The frugivory hypothesis was originally proposed and tested for catarrhine primates (Sumner & Mollon, 2000b) because until relatively recently platyrrhines (Mollon et al., 1984) and strepsirrhines (Tan & Li, 1999) were not known to possess trichromacy. The frugivory hypothesis has since been extended to platyrrhines and tested behaviourally (Caine et al., 2003). Theoretical studies have also assessed the conspicuity of dietary fruits to trichromatic versus dichromatic platyrrhines (Riba-Hernández et al., 2005; Stoner et al., 2005). Research evaluating the frugivory hypothesis is reviewed briefly below.

Tropical fruits exhibiting a "primate dispersal syndrome" (medium-sized red, orange or yellow fruits with succulent pulp, large seeds, tough exocarps; (Gautier-Hion et al., 1985)) have been found to occupy a narrow region of colour space that is detectable to trichromatic platyrrhines (Regan et al., 2001). Furthermore, the reflectance spectra of most primate-consumed fruits in Uganda were found to show chromatic changes as they ripen, and trichromatic catarrhines should be well adapted to discriminate fruit ripeness (Sumner and Mollon, 2000b). In the New World, the peak spectral sensitivities of trichromatic platyrrhine cone pigments were found to be well suited to the detection of fruits against a background of leaves under photopic conditions (Regan et al., 2001). The spectral tuning of the L-M cone pigments in the trichromatic platyrrhine *Alouatta seniculus* (Linnaeus, 1766) was found to be optimal for detecting ripe fruits against a background of leaf "noise" (Regan et al., 1998). Among six phenotypes in one white-faced capuchin (*Cebus capucinus*, Linnaeus, 1758) population, monkeys possessing the phenotype with the most spectrally separated L-M opsin alleles showed the highest acceptance index for red fruits (Melin et al., 2009). For this same population of capuchin monkeys, colourful fruits comprise 60% of the diet and are preferred to cryptically coloured fruits (Melin, 2011).

Trichromatic vision may also afford a selective advantage by allowing trichromats to forage on fruit under a greater range of light conditions than dichromats (Yamashita et al., 2005). In two species of *Saguinus* (Hoffmannsegg, 1807), captive trichromatic individuals were more efficient than dichromats at selecting ripe fruits, both in isolation and against a background of leaves (Smith et al., 2003). Riba-Hernandez et al. (2005) have provided further evidence to support this hypothesis in a study that connects fruit coloration to nutritional value. These authors established a positive correlation between the red-green color channel of spider monkeys (Ateles geoffreyi, Kuhl, 1820) and the glucose content of mature fruits, thereby arguing that sugar is the primary reward. In conclusion, the evidence supporting the foraging advantage of trichromats on ripe produce is compelling. The dietary significance of colorful fruits, the trichromat-accessible color space of tropical fruits, behavioral evidence for ripe fruit foraging advantages in wild and captive trichromats, and a positive correlation between nutritional value and fruit coloration all suggest that a ripe fruit foraging advantage may indeed contribute to the evolution and persistence of trichromacy. Nevertheless, the frugivory hypothesis is problematic on multiple levels. Alternate hypotheses regarding the evolution of trichromacy and issues with the frugivory hypothesis are discussed in the following sections.

2.1 Issues with the frugivory hypothesis

There is compelling empirical evidence that trichromacy is not necessary for detection of ripe fruits. Dominy and Lucas (2001) found that four routinely trichromatic catarrhine species in Uganda did not choose fruit on the basis of chromaticity, and that consumed fruits did not differ in chromaticity from unconsumed fruits. Comparing routinely trichromatic *Alouatta palliata* (Gray, 1849) fruit choices with polymorphically trichromatic *Ateles geoffroyi* in Costa Rica, Stoner et al. (2005) found that, contrary to expectations of a fruit foraging advantage for trichromats, the polymorphic *Ateles* consumed reddish fruits more often while routinely trichromatic *Alouatta* concentrated primarily on green fruits. Further studies of wild primates have also demonstrated that trichromats do not feed on brightly coloured fruits at faster rates than dichromats (Hiramatsu et al., 2008; Vogel et al., 2007). Additionally many fruits that are consumed and dispersed by primates are dull, and green or brown (Janson, 1983; Link & Stevenson, 2004), colours that can be detected through the blue-yellow colour channel of the dichromat (Dominy & Lucas, 2001). Intriguingly, the only genus of routinely trichromatic platyrrhine, *Alouatta*, is not a frugivore, but a folivore (Araújo et al., 2008; Jacobs et al., 1996).

3. THE HYPOTHESIS OF THE YOUNG LEAF

Many frugivorous primates, and specifically catarrhines, rely on leaves as fallback foods during periods of habitat-wide fruit scarcity (Lucas et al., 2003). Building on this observation, Dominy and Lucas (2001) argued that it is the ability to distinguish young leaves from mature leaves, rather than the ability to distinguish ripe fruits, that provided the selective force behind routine trichromacy. Measuring the luminance of young leaves and fruit consumed by catarrhines in Kibale, Uganda they found that while fruits could be reliably distinguished by the yellow-blue colour channels alone, the detection of young, proteinaceous leaves required the red-green colour channel of the trichromat. They additionally demonstrate that red-greenness of young leaves is positively correlated with the ratio of protein content to toughness. These observations formed the basis for the hypothesis that the evolution of both the routine trichromacy of catarrhines and the polymorphic trichromacy of platyrrhines results from the ability to detect fallback foods (Dominy et al., 2003).

Dominy et al. (2003) hypothesize that the evolution of routine trichromacy in catarrhines is the result of selection for colour vision that allows for the detection of fallback foods that sustain catarrhine populations during periods of low overall fruit availability[1]. In the case of catarrhines, trichromacy thus evolved as a means of detecting young, red leaves as fallback foods. Fallback foods have been operationally defined as "items assumed to be of relatively poor nutritional quality and high abundance, eaten particularly during periods when preferred foods are scarce" (Marshall & Wrangham, 2007, p. 1220). Dominy et al. (2003) posit that climatic cooling at the end of the Eocene resulted in the local extinction or decimation of figs and palms which led catarrhines to "fall back" on leaf resources in periods of low fruit availability. In the neotropics, where figs and palms remain abundant, platyrrhines have evolved mixed capabilities for the detection of these cryptic fallback resources.

Young leaves in the Old World are often red (Dominy et al., 2002). An estimated 50-62% of Old World species display red young leaf flush, compared to only 18-36% of New World species (Dominy & Lucas, 2001). The redness of young leaves results from delayed greening wherein plants postpone chloroplast function until full leaf expansion (Dominy & Lucas, 2001), coupled with the presence of anthocyanin pigment (Lee et al. 1987) which has several potential selective advantages to plants. Advantages to plants include fungicidal properties, photoprotection against UV damage of new leaves, the prevention of photoinhibition, and crypsis to protect against dichromatic herbivores (Coley & Aide, 1989; Gould et al., 1995; Stone, 1979). In a review of the evidence for and against these hypotheses, Dominy et al. (2003) conclude that crypsis is the most plausible, and that young leaf reddening is thus a plant strategy to reduce new leaf damage by herbivores. Dominy et al. (2002) argue that, based on the color space of young leaves, a dichromatic herbivore would see a young, red leaf as gloomy, lifeless, and unworthy of consumption. In Old World forests, where the emergence of new leaves aligns with limited fruit supply, the capacity to identify these crucial fallback meals would confer a distinct advantage, promoting trichromacy (Dominy & Lucas, 2001).

4. THE THEORY OF PROTEIN LIMITATION

Trichromacy facilitates the visual identification of long-wavelength colors of food items against a mature green leafy backdrop, regardless of whether the food items are young red leaves or ripe fruits. The discussion of whether frugivory or folivory has promoted trichromacy persists. Both hypotheses, frugivory and folivory, may be partially accurate and are supported by both sides. We propose a novel hypothesis that integrates the frugivory and young leaf ideas. We propose that trichromacy is advantageous for locating a vital resource essential to primates – protein. We propose that selective pressures may differ between catarrhine and platyrrhine monkeys due to variations in food availability and body size limitations. Protein sources for Old World primates are primarily red leaves, whereas for New World primates, they are predominantly red figs. We present the foundation of our unifying protein limitation hypothesis by examining the evidence that protein is a vital limiting resource and that red figs serve as a reliable and preferred protein source for New World primates, with trichromats possessing a foraging advantage for these fruits.

4.1 Protein as a limiting resource to frugivorous primates

Liebig's Law of the Minimum states that the functioning of a given organism is controlled (or limited) by an environmental factor or combination of factors present in the least favorable amount (Taylor, 1934). Since its definition, limiting resources have been shown to have profound effects on a wide array of species (Interlandi & Kilham, 2001). For primates that rely on ripe fruit as a dietary staple, that limiting factor is protein (Ganzhorn et al., 2009). Fruits are not considered to be sufficiently high in protein to meet the nutritional needs of primates (Kay, 1984; Milton, 1979; Oftedal, 1991), without supplementation from either leaves (in large bodied primates) or insects (in small bodied primates) (Fleagle, 1999). Obligate frugivores should feed on proteinaceous fruits whenever available, whereas facultative frugivores can supplement fruits with insects, seeds and leaves (Kunz & Diaz, 1995; Snow, 1981). Models of the evolution of primate diversity have identified the importance of the constraints of protein availability (Milton, 1979). The biomass of folivorous primates has been linked to nitrogen: fiber ratios of leaves in forests indicating a profound effect of protein availability on folivores (Chapman et al., 2004; Ganzhorn, 1992). Likewise in frugivores, protein as well as mineral content have been identified as factors influencing dietary selectivity (Barclay, 1995; Felton et al., 2009; Kunz & Diaz, 1995; O'Brien et al., 1998; Thomas, 1984; Wendeln et al., 2000). Further evidence for the importance of protein comes from nutritional analyses of fruits consumed by spider monkeys (*Ateles chamek*, Humboldt, 1812). Spider monkeys were found to regulate their daily protein intake much more tightly than either carbohydrates or fats, and their protein intake did not vary across seasons regardless of fluctuations in food availability (Felton et al., 2009).

Taken together, there is compelling evidence that protein is a limiting resource for primates. We suggest that acquisition of protein can provide a unified explanation of the adaptiveness of trichromacy. Protein acquisition by Old World monkeys and apes is facilitated by trichromacy via improved search efficiency for young, proteinaceous red leaves (after Dominy & Lucas, 2001). However, platyrrhines generally consume far fewer leaves than catarrhines (Dominy and Lucas, 2001). Given the general lack of leaf consumption by frugivorous platyrrhines and the abundance of proteinaceous figs in the neotropics, *Ficus* plants are a strong candidate for helping to meet the protein requirements of platyrrhines. For the smaller bodied, less folivorous neotropical primates, we suggest that improved foraging efficiency for red, proteinaceous figs favors trichromacy. In New World tropical forests primates prefer feeding on figs over other ripe fruits (Felton et al., 2009; Melin et al., 2009; Parr et al., In Press) and polymorphic trichromacy is the rule for all but one monochromatic nocturnal genus (*Aotus*, Gray, 1870), and one routinely trichromatic folivorous genus (*Alouatta*, Gray, 1849) (Jacobs, 2008). Following previous research, we submit that this may be maintained by heterozygote advantage (Hedrick, 2007) to trichromatic females via an increased ability to detect ripe fruits (Osorio et al., 2004; Osorio & Vorobyev, 1996; Smith et al., 2003; Sumner & Mollon, 2000a; Surridge et al., 2003). However, we emphasize that the primary advantage lies in the detection of conspicuous figs, which differ from other fruits in their high protein content. Our hypothesis differs from previous ideas concerning fruit foraging, which suppose fruit sugars are the primary reward. We suggest that protein, relatively abundant in figs, is the primary element favoring trichromacy.

In a comprehensive review of paleogeographic and biogeographic evidence for the dearth of *Ficus* plants in the Old World, Dominy et al. (2003) posit that climatic cooling at the end of the Eocene resulted in the local extinction or decimation of cryptically coloured (greenish- brownish), relatively nutritionally poor fruits of figs and palms in the Old World. They suggest that where figs and palms remain abundant in the neotropics, platyrrhine polymorphism (as opposed to routine trichromacy) is adaptive because dichromacy is suitable for the detection of these cryptic fallback resources. The hypothesis is compelling, elegant and has received empirical support. Additionally, recent studies have corroborated the importance of fallback foods in the evolution of primate traits (Lambert et al., 2004). However, we suggest that two components require a closer look: 1) the role of figs in the foraging decisions of platyrrhines, and 2) the nutritional value of figs as a key source of protein. We propose that accurately delineating the role of Ficus species facilitates an ecological enhancement of the young leaf hypothesis. Instead of figs being alternative resources, as they may be for Old World monkeys like orangutans (Marshall et al., 2009), we suggest that some species (namely those yielding red fruits) are favored and limiting resources for neotropical primates. The next sections include evidence regarding the nutritional value of Ficus fruits, behavioral evidence of primate feeding preferences for Ficus, and comparative abundance data for Ficus in Africa and the Neotropics. We also examine behavioral and genomic data on the adaptive significance of primate dichromacy and trichromacy, as well as the balancing selection that may sustain polymorphism trichromacy.

4.2 Ficus as protein

There are approximately 750 extant species of the genus *Ficus* which together constitute one of the most distinctive and peculiar genera of tropical flora in the world (Shanahan et al., 2001). The unique complex obligatory mutualism between *Ficus* species and pollinating agaonid fig wasps results in a universal pattern of asynchronous and unpredictable mast fruiting events (Janzen, 1979). Fruits of the genus *Ficus* are not in fact strictly fruits, but syconia, or inflorescences. Syconia are clusters of multiple flowers used in pollination, and bundled together in a single package with thousands of miniscule seeds (Janzen, 1979). These inflorescences are entered by one or more species-specific pollinating agaonid fig wasps through an ostiole in the pericarp. Wasps pollinate the florets inside the fruit as well as laying a single egg in each of the ovaries. Wasp larvae live inside the seed coat and eat developing seeds. A month or so later, wingless male wasps emerge from seed coats and mate with females through holes they have cut in the sides of ovaries. Developed female wasps are dependent upon males for this service, and where males fail to cut holes in the sides of *Ficus* ovaries, females remain trapped in their larval home. In addition to these mutualists, numerous species of parasitic Hymenoptera have been discovered. These species oviposit through the wall of the unripe fig without providing pollination services (Janzen, 1979).

The mutualism between figs and wasps, and its interruption by parasitic ovipositing Hymenoptera render the fig syconia (which we shall hereafter refer to as fruit) a bundle of protein-rich insects and larvae. Not only can grown female wasps become trapped inside of ovaries, but at some times of year parasitic larvae become so numerous that hundreds of fig fruits are eaten hollow and found completely filled with larvae (Janzen, 1979). In addition to wasp larvae, fig fruits can contain several other arthropod species, including mites, dipterans and nematodes (Frank, 1989). One analysis of a neotropical *Ficus* species found that animal matter from insects contributed 2.9% to fig dry matter, increasing protein content by 15.3% (Urquiza-Haas et al., 2008).

In addition to the presence of protein-rich insects, there is evidence of unique nutritional qualities of the actual fig fruits themselves. Many authors have referred to figs as low quality foods (Bronstein & Hoffman, 1987; Herbst, 1986; Jordano, 1983; Lambert, 1989). However, overall, nutritional analyses of tropical figs indicate that fig fruits contain moderate-to-high levels of protein relative to other fruits and relative to frugivore protein requirements (Conklin & Wrangham, 1994; Dierenfeld et al., 2002; Ganzhorn, 1988; Ganzhorn et al., 2009; Goodman et al., 1997; Kendrick et al., 2009; Molloy & Hart, 2002; O'Brien et al., 1998; Rogers et al., 1990; Wendeln et al., 2000). In one study, African fig fruits – without supplementation - were found to provide an acceptable baseline level of protein (Conklin & Wrangham, 1994). Neotropical figs may have nutritional value adequate to sustain at least some frugivores without additional food (Wendeln et al., 2000). A study of two neotropical *Ficus* species found they contained 4.5% - 6.1% (dry weight) protein, which is 2-3 times higher than that of a sympatric neotropical angiosperm, *Spondias mombin* (Anacardiaceae), and are more nutritionally balanced with higher protein:TNC and lipid ratios than sympatric species (Hladik et al., 1971). Evidence for the nutritionally balanced nature of figs is further indicated by the fact that some species of bat subsist entirely on certain *Ficus* species (Janzen, 1979; Wendeln et al., 2000) and capuchin monkeys decrease time spent foraging for insects, their primary protein source, on days they visit figs (Parr et al., In Press). Comparison of protein content (as proxied by percent nitrogen) of New World and Malagasy figs indicates that neotropical fruits contain enough protein to satisfy the needs of primates (7-11% protein, 1.1- 1.8% nitrogen), while those in Madagascar do not (Ganzhorn et al., 2009). Figs, then, might be well identified as important sources of protein for New World frugivorous primates. In order to determine the possible extent of this importance, it is necessary to review not only the nutritional value of figs, but also primate feeding patterns.

4.3 Who eats figs?

The concise answer to the question of who eats figs was offered by Daniel Janzen (1979): "Everybody." Indeed, figs are notable for the number of bird and mammal species that consume their fruits and they comprise a part of the diet for more species of animal than any other genus of wild tropical perennial fruit (Janzen, 1979). In this section, we focus on the role of figs in the diet and feeding ecology of platyrrhines.

A review of the literature on neotropical primates reveals that all diurnal frugivorous genera with a body mass exceeding Kay's 500g threshold (Fleagle, 1999) include figs in their diet to varying extents, with the exception of uakari monkeys of the genus *Cacajao* (Mivart, 1865) for which dietary information is scant (Boubli, 1999). Tamarins (*Saguinus* species) and marmosets (*Callithrix kuhlii*, Coimbra-Filho, 1985) have been found to include a variety of fig species in their diet across sites (Knogge et al., 2003; Raboy et al., 2008; Terborgh, 1983). Three of the four *Ficus* species present at a Brazilian site are included in the 55 "extremely valuable" species consumed by golden-lion tamarins (*Leontopithecus chrysomelas*, Kuhl, 1820) (Oliveira et al., 2009). The seventh most frequently consumed fruit of black titi monkeys (*Callicebus torquatus lugens*, Hoffmannsegg, 1807) in Columbia was the single species *Ficus mathewsii* (Palacios et al., 1997). At five different sites in South America (Yasuni, Caparu, Urucu, Tinigua and Pacaya) figs were one of the top seven genera consumed by woolly monkeys (*Lagothrix lagotricha*, Humboldt, 1812) at each site (di Fiore, 2004). In Cocha Cashu, Peru, brown capuchin monkeys (*Cebus apella*, Linnaeus, 1758) were found to eat figs whenever they were available, and South American squirrel monkeys (*Saimiri sciurius*, Linnaeus, 1758) and white-fronted capuchin monkeys (*Cebus albifrons*, Humboldt, 1812) concentrated feasting on figs whenever they were available (Terborgh, 1983). White-faced capuchins (*Cebus capucinus*) in Santa Rosa, Costa Rica prefer figs over all other ripe fruits, and concentrate their foraging efforts on *Ficus* whenever available, regardless of the presence and abundance of other ripe fruits in their habitat (Melin et al., 2009). Figs comprised nearly one-third of capuchin annual fruit foraging effort which represents over- selection relative to fig abundance (Melin et al., 2009). For black-faced spider monkeys (*Ateles chamek*) in the Guarayos Forest Reserve in Bolivia, the preference for figs relative to all other ripe fruits is striking, and when combined with data on habitat-wide fruit availability indicates that "spider monkeys consumed a diverse array of ripe fruits to overcome periods of fig scarcity rather than vice versa" (Felton et al., 2009). Even species that are considered to be folivorous show a preference for *Ficus* fruits. One study of Mexican howler monkeys (*Alouatta palliata mexicana*, Merriam, 1902) found that six *Ficus* species were amongst the eight most important species consumed at one site in Veracruz, Mexico (Serio- Silva et al., 2002). Diurnal frugivorous and folivorous neotropical primates thus all eat figs to a greater or lesser extent, with some species showing an unequivocal preference for them even during periods of high habitat wide fruit availability. The feeding ecology of platyrrhines, coupled with preliminary evidence for the protein content of neotropical figs indicate that rather than viewing figs as a fallback foods in times of habitat-wide fruit scarcity (Dominy et al., 2003; Shanahan et al., 2001) it may be more appropriate to view them as important, preferred and limiting resources.

4.4 The relative abundance of Ficus species in Africa and the Neotropics

In order for figs to be a more important source of protein for New World monkeys vis-à-vis catarrhines, it is necessary to demonstrate that figs are abundantly present in the neotropics relative to Old World forests. Fig abundance is especially salient given its asynchronous and unpredictable fruiting phenology (Janzen, 1979). We reviewed a worldwide database of plant species abundance maintained by the Center for Tropical Forest Science of the Smithsonian Tropical Research Institute (STRI) as well as our own data and published species abundance data to test the prediction that *Ficus* is more abundant in the neotropics than in Africa. Using the online database maintained by STRI, we counted the number of *Ficus* species and the number of individuals of each species recorded during the most recent census of each forest plot in Africa and in the neotropics for which data are currently available (N=3 African plots and 5 neotropical plots). A search of tree abundance literature uncovered three additional datasets containing information on *Ficus* abundance: one from Budongo Forest Reserve in Uganda (Tweheyo & Babweteera, 2007), one from Agaltepec Island in Mexico (Serio-Silva et al., 2002), and one from Santa Rosa National Park, Costa Rica (Parr et al., In Press).

In all plots censused by STRI all *Ficus* individuals >10cm diameter at breast height (DBH) were recorded. In the Budongo Forest Reserve in Uganda all *Ficus* trees >4cm DBH were recorded, and as such abundance data are overestimated. In the Agaltepec Island plot all *Ficus* trees >30cm DBH were recorded, and as such abundance data may be an underestimate. Parr et al. (In Press) recorded all *Ficus* trees >10cm DBH encountered in a transect survey of Sector Santa Rosa, Costa Rica.

Fig species are both more diverse and abundant in the New World relative to the Old World (Table 1). In the African plots *Ficus* species diversity ranges from two to four species per plot (N=70 trees, 114 hectares, 4 plots), with a mean number of 0.73 *Ficus* trees per hectare (range = 0.06 – 2.45). In the neotropical plots *Ficus* species diversity ranges from three to 12 species per plot (N = 458 trees, 145.96 hectares, 7 plots), with a mean number of 3.89 *Ficus* trees per hectare (range = 0.5 – 11.56). While the sample size is inappropriate for a statistical analysis of *Ficus* abundance data, the pattern in the data is clear. Figs are both more speciose and abundant in the neotropical plots relative to the African plots. This pattern holds despite one of the four African plot values representing an abundance overestimate (Budongo Forest Reserve, Uganda), and one of the six neotropical plot abundance values representing an underestimate (Agaltepec Island, Mexico). Removing the overestimated abundance data from Budongo Forest Reserve, the *Ficus* abundance in the remaining three African plots never meets or exceeds the abundance of the most *Ficus*-poor neotropical plot (Sherman, Panama, 0.5 individuals per hectare). This dataset supports the prediction of relatively high abundance of neotropical figs relative to African figs that has been noted elsewhere (Gautier-Hion & Michaloud, 1989; Shanahan et al., 2001).

4.5 Foraging for figs

Color vision polymorphism has endured in monkeys for as long as 14 million years (Surridge & Mundy, 2002). For polymorphic trichromacy in platyrrhines to be regarded as an adaptive strategy, it must be established that the preservation of diverse color vision phenotypes within a population confers some advantage (Melin et al., 2008). Substantial data supports a trichromat advantage in ripe fruit foraging (Melin et al., 2009; Regan et al., 1998; Riba-Hernández et al., 2005; Smith et al., 2003; Sumner & Mollon, 2000b; Yamashita et al., 2005). The sole published study on fig foraging by wild platyrrhines (Melin et al., 2009) indicates that trichromatic white-faced capuchin monkeys (*Cebus capucinus*) exhibit a foraging advantage when consuming ripe figs. Conspicuous figs were more prevalent than cryptic figs at the research site and were consumed significantly more often. Trichromatic monkeys had a superior acceptance index for prominent figs compared to dichromats, indicating an enhanced capacity to select riper figs based on visual stimuli. Dichromats employed extended and varied foraging sequences (e.g., frequent olfactory exploration) when evaluating figs. Although the differences in foraging behavior did not produce a net change in fig consumption rates, the authors propose that the enhanced discernment capabilities of trichromats may facilitate increased feeding rates in scenarios when ripe fruit is scarce and rapidly exhausted. Furthermore, trichromatic monkeys appeared more adept at selecting the ripest figs, but dichromats more frequently devoured mid-ripe (yellowish) figs (A.D Melin, pers. obs).

Table 1. Number of species and individuals of the genus Ficus in African and neotropical forests.

Site	Plot Size (hectares)	Census Date	N species	N Individuals	N Individuals per hectare	Source
Korup, Cameroon	50	1998	N/A	3	0.06	STRI
Edoro, Democratic Republic of Congo	20	2000	2	2	0.10	STRI
Lenda, Democratic Republic of Congo	20	2000	4	6	0.30	STRI
Budongo Forest Reserve, Uganda	24.13	2006	4*	59	2.45	Tweheyo and Babweteera 2007
La Planada, Columbia	25	2003	7	289	11.56	STRI
Yasuni, Ecuador	50	2003	12	30	0.60	STRI
BCI, Panama	50	2005	8	42	0.84	STRI
Cocoli, Panama	4	1998	4	9	2.25	STRI
Sherman, Panama	6	1999	3	3	0.50	STRI
Sector Santa Rosa, Costa Rica	2.66	2010	8*	4	1.5**	Parr et al, In Press
Agaltepec Island, Mexico	8.3	1996	6	81	9.76	Serio-Silva et al. 2002

* 8 species of *Ficus* are known in Sector Santa Rosa (Melin et al., 2009) however only 2 species have been encountered to date in transects (Parr et al., In Press).

** If only those fig trees with a DBH of 10cm or greater are included, then the number of individuals drops to three, and density drops to 1.13 (N.Parr, pers. comm.).

5. CONCISE SUMMARY

Although many hypotheses regarding the differences in color vision among catarrhines and platyrrhines emphasize the significance of detecting ripe fruit (Regan et al., 2001) or young leaves (Dominy et al., 2003), evidence suggests that this variation arises from visual adaptive strategies focused on identifying protein-rich foods. In continental Africa, frugivorous monkeys depend on young, reddish leaves for protein sources (Dominy et al., 2003), and routine trichromacy is prevalent. In New World tropical woods, effective foraging for vividly colored protein-rich figs may benefit trichromats and sustain color vision polymorphism through heterozygote advantage. It is important to note that the selection pressures influencing trichromacy may be diminished in the New World, as insects serve as a protein source for numerous neotropical monkeys, and dichromacy is advantageous for detecting surface-dwelling insects, due to the enhanced capacity of dichromats to disrupt camouflage. Instead of seeing figs as secondary resources (Dominy et al., 2003; Shanahan et al., 2001), we assert that they are primary and limiting resources. Furthermore, due to the timing of Old World phenophases, wherein young leaves are predominantly accessible during times of general fruit scarcity (Dominy et al., 2003), it is challenging to unequivocally classify young leaves as fallback foods, as the definition necessitates both spatial and temporal ubiquity (Marshall & Wrangham, 2007). The development and preservation of polymorphic trichromacy is a reaction to fluctuations in the detectability of not fallback foods, but rather the limiting resources for frugivorous primates: protein-rich diets. Our investigation has been confined to continental Africa and the Neotropics due to data availability from these regions. The recent identification of polymorphic trichromacy in several Malagasy strepsirrhines (Tan & Li, 1999; Veilleux & Bolnick, 2009) will offer a compelling comparison dataset for hypothesis testing. According to the protein limitation hypothesis, we anticipate that balancing selection will promote polymorphism trichromacy through a heterozygote advantage in frugivorous strepsirrhines with a conspicuous protein source. Considering the existing evidence regarding the limited presence of Ficus in Malagasy forests and the recent findings indicating a scarcity of protein in Malagasy Ficus fruits (Ganzhorn et al., 2009), it appears probable that Malagasy protein resources and their consequent impact on strepsirrhine color vision capabilities will be located elsewhere. Numerous hypotheses concerning primate morphological adaptations depend on the significance of fallback foods (Dominy & Lucas, 2001; Marshall & Wrangham, 2007) and foods that predominate in primate diets (Osorio & Vorobyev, 1996; Regan et al., 2001; Smith et al., 2003). This signifies an atypical divergence from monkey behavior research that typically emphasize the significance of resource limitations (e.g. Sterck et al., 1997; van Schaik, 1989; Wrangham, 1980). Research on monkey foraging adaptations could be enhanced by incorporating an analysis of morphological and sensory adaptations related to foraging on few resources.

6. REFERENCES

Allen, G. (1879). *The Colour-Sense: Its Origin and Development*, Trubner, ISBN 978-054-8102-93- 0, London

Araújo, A.C.; Didonet, J.J.; Araújo, C.S.; Saletti, P.G.; Borges. T.R. & Pessoa, V.F. (2008). Colour vision in the black howler monkey (*Alouatta caraya*). *Visual Neuroscience*, Vol.25, No.3, pp. 243-248, ISSN 0952-5238

Barclay, R.M.R. (1995). Does energy or calcium availability constrain reproduction by bats?

Symposia of the Zoological Society of London, Vol.67, pp. 245-258, ISSN 0084-5612 Barton, R.A. (2006). Olfactory evolution and behavioural ecology in primates. *American Journal of Primatology*, Vol.68, No.6, pp. 545-558, ISSN 0275-2565

Boubli, J.P. (1999). Feeding ecology of black-headed uacaris (*Cacajao melanocephalus melanocephalus*) in Pico da Neblina National Park, Brazil. *International Journal of Primatology*, Vol.20, No.5, pp. 719-749, ISSN 0164-0291

Bronstein, J.L. & Hoffman, K. (1987). Spatial and temporal variation in frugivory at a neotropical fig, *Ficus pertusa*. *Oikos*, Vol.49, pp. 261-268, ISSN 0030-1299

Caine, N.G.; Osorio, D. & Mundy, N.I. (2010). A foraging advantage for dichromatic marmosets (*Callithrix geoffroyi*) at low light intensity. *Biology Letters*, Vol.6, pp. 36- 38, ISSN 1744-957X

Caine, N.G.; Surridge, A.K. & Mundy, N.I. (2003). Dichromatic and trichromatic *Callithrix geoffroyi* differ in relative foraging ability for red-green colour-camouflaged and non-camouflaged food. *International Journal of Primatology*, Vol.24, No.6, pp. 1163- 1175, ISSN 0164-0291

Chapman, C.A.; Chapman, L.J.; Naughton-Treves, L.; Lawes, M.J. & McDowell, L.R. (2004). Predicting folivorous primate abundance: validation of a nutrition model. *American Journal of Primatology*, Vol.62, pp. 55-69, ISSN 0275-2565

Chapman, C.A. & Onderdonk, D.A. (1998). Forests without primates: primate/plant codependency. *American Journal of Primatology*, Vol.45, No.1, pp. 127-141, ISSN 0275-2565

Coley, P.D. & Aide, T.M. (1989). Red colouration of tropical leaves: A possible anti-fungal defence? *Journal of Tropical Ecology*, Vol.5, pp. 293-300, ISSN 0266-4674

Conklin, N.L. & Wrangham, R.W. (1994). The value of figs to a hind-gut fermenting frugivore: A nutritional analysis. *Biochemical Systematics and Ecology*, Vol.22, No.2, pp. 137-151, ISSN 0305-1978

di Fiore, A. (2004). Diet and feeding ecology of woolly monkeys in a western Amazonian rain forest. *International Journal of Primatology*, Vol.25, No.4, pp. 767-801, ISSN 0164- 0291

Dierenfeld, E.S.; Mueller, P.J. & Hall, M.B. (2002). Duikers: native food composition, micronutrient assessment, and implications for improving captive diets. *Zoo Biology*, Vol.21, pp. 185-196, ISSN 0733-3188

Dominy, N.J. & Lucas, P.W. (2001). Ecological importance of trichromatic vision to primates. *Nature*, Vol.410, No.6826, pp. 363-366, ISSN 0028-0836

Dominy, N.J.; Lucas, P.W.; Ramsden, L.W.; Riba-Hernandez, P.; Stoner, K.E. & Turner, I.M. (2002). Why are young leaves red? *Oikos* Vol.98, No.1, pp. 163-176, ISSN 0030-1299

Dominy, N.J.; Svenning, J.C. & Li, W.H. (2003). Historical contingency in the evolution of primate colour vision. *Journal of Human Evolution*, Vol.44, No.1. pp. 25-45, ISSN 0047-2484

Felton, A.M.; Felton, A.; Wood, J.T.; Foley, W.J.; Raubenheimer, D.; Wallis, I.R. & Lindenmayer, D.B. (2009). Nutritional Ecology of *Ateles chamek* in lowland Bolivia: how macronutrient balancing influences food choices. *International Journal of Primatology*, Vol.30, No.5, pp. 675-696, ISSN 0164-0291

Fleagle, J.G. (1999). *Primate Adaptation and Evolution*, Academic Press, ISBN 978-012-2603-41- 9, New York.

Frank, S.A. (1989). Ecological and evolutionary dynamics of fig communities. *Experientia*, Vol.45, pp. 674-680, ISSN 0014-4754

Ganzhorn, J.U. (1988). Food partitioning among Malagasy primates. *Oecologia*, Vol.75, pp. 436-450, ISSN 0029-8549

Ganzhorn, J.U. (1992). Leaf chemistry and the biomass of folivorous primates in tropical forests. *Oecologia*, Vol.91, No.4, pp. 540-547, ISSN 0029-8549

Ganzhorn, J.U.; Arrigo-Nelson, S.; Boinski, S.; Bollen, A.; Carrai, V.; Derby, A.; Donati, G.; Koenig, A.; Kowalewski, M. & Lahann, P. (2009). Possible fruit protein effects on primate communities in Madagascar and the neotropics. *PloS One*, Vol.4, No.12, pp. 1-8, ISSN 1932-6203.

Garber, P.A. & Lambert, J.E. (1998). Primates as seed dispersers: ecological processes and directions for future research. *American Journal of Primatology*, Vol.45, pp. 3-8, ISSN 0275-2565.

Gautier-Hion, A.; Duplantier, J.M.; Quris, R.; Feer, F.; Sourd, C.; Decoux, J.P.; Dubost, G.; Emmons, L.; Erard, C. & Hecketsweiler, P. (1985). Fruit characters as a basis of fruit choice and seed dispersal in a tropical forest vertebrate community. *Oecologia*, Vol.65, No.3, pp. 324-337, ISSN 0029-8549

Gautier-Hion, A. & Michaloud, G. (1989). Are figs always keystone resources for tropical frugivorous vertebrates? A test in Gabon. *Ecology*, Vol.70, No.6, pp. 1826-1833, ISSN 0012-9658

Goodman, S.M.; Ganzhorn, J.U. & Wilme, L. (1997). Observations at a *Ficus* tree in Malagasy humid forest. *Biotropica*, Vol.29, No.4, pp. 480-488, ISSN 0006-3606

Gould, K.S.; Kuhn, D.N.; Lee, D.W. & Oberbauer, S.F. (1995). Why leaves are sometimes red. *Nature*, Vol.378, pp. 241-242, ISSN 0028-0836

Hedrick, P.W. (2007). Balancing selection. *Current Biology*, Vol.17, No.7, pp. R230-R231, ISSN 0960-9822

Herbst, L.H. (1986). The role of nitrogen from fruit pulp in the nutrition of a frugivorous bat, *Carollia perspicillata*. *Biotropica*, Vol.18, pp. 39-44, ISSN 0006-3606

Hiramatsu, C.; Melin, A.D.; Aureli, F.; Schaffner, C.M.; Vorobyev, M.; Matsumoto, Y. & Kawamura, S. (2008). Importance of achromatic contrast in short-range fruit foraging of primates. *PLoS ONE*, Vol.3, No.10, pp. 3356, ISSN 1932-6203

Hladik, C.M.; Hladik, A.; Bousset, J.; Valdebouze, P.; Viroben, G. & Delort-Laval, J. (1971). Le regime alimentaire de primate de l'il de Barro Colourado (Panama). *Folia Primatologica*, Vol.16, pp. 85-122, ISSN 0015-5713

Hunt, D.M.; Carvalho, L.S.; Cowing, J.A. & Davies, W.L. (2009). Evolution and spectral tuning of visual pigments in birds and mammals. *Philosophical Transactions of the Royal Society of London B*, Vol.364, No.1531, pp. 2941-2955, ISSN 1364–503X

Interlandi, S.J. & Kilham, S.S. (2001). Limiting resources and the regulation of diversity in phytoplankton communities. *Ecology*, Vol.82, No.5, pp. 1270-1282, ISSN 0012-9658

Jacobs, G.H. (2007). New world monkeys and colour. *International Journal of Primatology*, Vol.28, pp. 729-759, ISSN 0164-0291

Jacobs, G.H. (2008). Primate colour vision: a comparative perspective. *Visual Neuroscience*, Vol.25, pp. 1-15, ISSN 0952-5238

Jacobs, G.H.; Neitz, M.; Deegan, J.F. & Neitz, J. (1996). Trichromatic colour vision in New World monkeys. *Nature*, Vol.382, No.6587, pp. 156-158, ISSN 0028-0836

Janson, C.H. (1983). Adaptation of fruit morphology to dispersal agents in a neotropical forest. *Science*, Vol.219, No.4581, pp. 187-189, ISSN 0036-8075

Janzen, D.H. (1979). How to be a fig. *Annual Review of Ecology and Systematics*, Vol.10, No.1, pp. 13-51, ISSN 0066-4162

Jordano, P. (1983). Fig-seed predation and dispersal by birds. *Biotropica*, Vol.15, No.1, pp. 38- 41, ISSN 0006-3606

Kay, R.F. (1984). On the use of anatomical features to infer foraging behaviour in extinct primates. In: *Adaptations for Foraging in Nonhuman Primates*, Rodman, P.S. & Cant, J.G.H. (Eds.), 21-53, Columbia University Press, New York, ISBN 978-023-1052-27-6

Kendrick, E.L.; Shipley, L.A.; Hagerman, A.E. & Kelley, L.M. 2009. Fruit and fibre: the nutritional value of figs for a small tropical ruminant, the blue duiker (*Cephalophus monticola*). *African Journal of Ecology*, Vol.47, pp. 556-566, ISSN 0141-6707

Knogge, C.; Herrera, E.R.T. & Heymann, E.W. (2003). Effects of passage through tamarin guts on the germination potential of dispersed seeds. *International Journal of Primatology*, Vol.24, No.5, pp. 1121-1128, ISSN 0164-0291

Kunz, T.H. & Diaz, C.A. (1995). Folivory in fruit-eating bats, with new evidence from *Artibeus jamaicensis* (Chiroptera: Phyllostomidae). *Biotropica*, Vol.27, pp. 106-120, ISSN 0006-3606

Lambert, F.R. (1989). Fig-eating by birds in a Malaysian lowland rain forest. *Journal of Tropical Ecology*, Vol.5, pp. 401-412, ISSN 0266-4674

Lambert, J.E.; Chapman, C.A.; Wrangham, R.W. & Conklin-Brittain, N.L. (2004). The hardness of cercopithecine foods: Implications for the critical function of enamel thickness in exploiting fallback foods. *American Journal of Physical Anthropology* Vol.125, pp. 363-368, ISSN 0002-9483

Lee, D.W.; Brammeier, S. & Smith, A.P. (1987). The selective advantages of anthocyanins in developing leaves of mango and cacao. *Biotropica* Vol.19, No.1, pp. 40-49, ISSN 0006-3606

Link, A. & Stevenson, P.R. (2004). Fruit dispersal syndromes in animal disseminated plants at Tinigua National Park, Colombia. *Revista Chilena de Historia Natural* Vol.77, pp. 319-334, ISSN 0716-078X

Lucas, P.W.; Dominy, N.J.; Riba-Hernandez, P.; Stoner, K.E.; Yamashita, N.; Petersen-Pereira, W.; Salas-Pena, R.; Solis-Madrigal, S.; Osorio, D. & Darvell, B.W. (2003). Evolution and function of routine trichromatic vision in primates. *Evolution*, Vol.57, No.11, pp. 2636-2643, ISSN 1558-5646

Marshall, A.; Boyko, C.M.; Feilen, K.L.; Boyko, R.H. & Leighton, M. (2009). Defining fallback foods and assessing their importance in primate ecology and evolution. *American Journal of Physical Anthropology*, Vol.140, pp. 603-614, ISSN 0002-9483

Marshall, A.J. & Wrangham, R.W. (2007). Evolutionary consequences of fallback foods. *International Journal of Primatology*, Vol.28, pp. 1219-1235, ISSN 0164-0291

Melin, A.D. (2011). Polymorphic Colour Vision and Foraging in White-faced Capuchins: Insights from Field Research and Simulations of Monkey Vision. PhD Dissertation. Calgary: University of Calgary.

Melin, A.D.; Fedigan, L.M.; Hiramatsu, C.; Hiwatashi, T.; Parr, N. & Kawamura, S. (2009). Fig foraging by dichromatic and trichromatic *Cebus capucinus* in a tropical dry forest. *International Journal of Primatology*, Vol.30, No.6, pp. 753-775, ISSN 0164-0291

Melin, A.D.; Fedigan, L.M.; Hiramatsu, C. & Kawamura, S. (2008). Polymorphic colour vision in white-faced capuchins (*Cebus capucinus*): is there foraging niche divergence among phenotypes? *Behavioural Ecology and Sociobiology*, Vol.62, No.5, pp. 659-670, ISSN 1045-2249

Melin, A.D.; Fedigan, L.M.; Hiramatsu, C.; Sendall, C.L. & Kawamura, S. (2007). Effects of colour vision phenotype on insect capture by a free-ranging population of white- faced capuchins, *Cebus capucinus*. *Animal Behaviour*, Vol.73, No.1, pp. 205-214, ISSN 0003-3472

Melin, A.D.; Fedigan, L.M.; Young, H.C. & Kawamura, S. (2010). Can colour vision variation explain sex differences in invertebrate foraging by capuchin monkeys? *Current Zoology*, Vol.56, pp. 300-312, ISSN 1674-5507

Milton, K. (1979). Factors influencing leaf choice by howler monkeys: a test of some hypotheses of food selection by generalist herbivores. *American Naturalist*, Vol.114, No.3, pp. 362-378, ISSN 0003-0147

Mollon, J.D.; Bowmaker, J.K. & Jacobs, G.H. (1984). Variations of colour vision in a New World primate can be explained by polymorphism of retinal photopigments. *Proceedings of the Royal Society of London Series B, Biological Sciences*, Vol.222, pp. 373- 399, ISSN 1364-5021

Molloy, L. & Hart, J.A. (2002). Duiker food selection: palatability trials using natural foods in the Ituri Forest, Democratic Republic of Congo. *Zoo Biology*, Vol.21, pp. 149-159, ISSN 0733-3188

Morgan, M.J.; Adam, A. &Mollon, J.D. (1992). Dichromats detect colour-camouflaged objects that are not detected by trichromats. *Proceedings of the Royal Society B, Biological Sciences*, Vol.248, No.1323, pp. 291-295, ISSN 0962-8452

O'Brien, T.G.; Kinnaird, M. & Dierenfeld, E.S. (1998). What's so special about figs? *Nature*, Vol.392, pp. 668, ISSN 0028-0836

Oftedal, O.T. (1991). The nutritional consequences of foraging in primates: the relationship of nutrient intakes to nutrient requirements. *Philosophical Transactions of the Royal Society of London Series B*, Vol.334, pp. 161-170, ISSN 0962-8436

Oliveira, L.C.; Hankerson, S.J.; Dietz, J.M. & Raboy, B.E. (2009). Key tree species for the golden-headed lion tamarin and implications for shade-cocoa management in southern Bahia, Brazil. *Animal Conservation*, Vol.13, pp. 60-70, ISSN 1367-9430

Osorio, D.; Smith, A.C.; Vorobyev, M. & Buchanan-Smith, H.M. (2004). Detection of fruit and the selection of primate visual pigments for colour vision. *American Naturalist*, Vol.164, No.6, pp. 696-708, ISSN 0003-0147

Osorio, D. & Vorobyev, M. (1996). Colour vision as an adaptation to frugivory in primates. *Proceedings of the Royal Society B*, Vol.263, pp. 593-599, ISSN 0962-8452

Palacios, E.; Rodriguez, A. & Defler, T.R. (1997). Diet of a group of *Callicebus torquatus lugens* (Humboldt, 1812) during the annual resource bottleneck in Amazonian Colombia. *International Journal of Primatology*, Vol.18, No.4, pp. 503-522, ISSN 0164-0291

Parr, N.; Melin, A.D. & Fedigan, L. In Press. Figs are not always fallback foods: The relationship between *Ficus* and *Cebus* in a tropical dry forest. *International Journal of Zoology*, ISSN 2231-3516

Polyak, S. (1957). *The Vertebrate Visual System*, University of Chicago Press, ISBN 978-022- 6674-94-0, Chicago

Raboy, B.E.; Canale, G.R. & Dietz, J.M. (2008). Ecology of *Callithrix kuhlii* and a review of eastern Brazilian marmosets. *International Journal of Primatology*, Vol.29, No.2, pp. 449-467, ISSN 0164-0291

Regan, B.C.; Julliot, C.; Simmen, B.; Vienot, F; Charles-Dominique, P. & Mollon, J.D. (1998). Frugivory and colour vision in *Alouatta seniculus*, a trichromatic platyrrhine monkey. *Vision Research*, Vol.38, No.21, pp. 3321-3327, ISSN 0042-6989

Regan, B.C.; Julliot, C.; Simmen, B.; Vienot, F.; Charles-Dominique, P. & Mollon, J.D. (2001). Fruits, foliage and the evolution of colour vision. *Philosophical Transactions of the Royal Society of London B*, Vol. 356, pp. 229-283, ISSN 0962-8436

Riba-Hernández, P.; Stoner, K.E. & Lucas, P.W. (2005). Sugar concentration of fruits and their detection via colour in the Central American spider monkey (*Ateles geoffroyi*). *American Journal of Primatology*, Vol.67, No.4, pp. 411-423, ISSN 0275-2565

Rogers, M.E.; Maisels, F.; Williamson, E.A.; Fernandez, M. & Tutin, C.E.G. (1990). Gorilla diet in the Lope Reserve, Gabon: A nutritional analysis. *Oecologia*, Vol.84, pp. 326- 339, ISSN 0029-8549

Serio-Silva, J.C.; Rico-Gray, V.; Hernández-Salazar, L.T. & Espinosa-Gómez, R. (2002). The role of *Ficus* (Moraceae) in the diet and nutrition of a troop of Mexican howler monkeys, *Alouatta palliata mexicana*, released on an island in southern Veracruz, Mexico. *Journal of Tropical Ecology*, Vol.18, No.6, pp. 913-928, ISSN 0266-4674

Shanahan, M.; So, S.; Compton, S.G. & Corlett, R. (2001). Fig-eating by vertebrate frugivores: a global review. *Biological Reviews of the Cambridge Philosophical Society*, Vol.76, No.4, pp. 529-572, ISSN 1464-7931

Smith, A.C.; Buchanan-Smith, H.M.; Surridge, A.K.; Osorio, D. & Mundy, N.I. (2003). The effect of colour vision on the detection and selection of fruits by tamarins (*Saguinus* spp.). *Journal of Experimental Biology*, Vol.206, pp. 3159-3165, ISSN 0022-0949

Snow, D.W. (1981). Tropical frugivorous birds and their food plants: a world survey. *Biotropica*, Vol.13, pp. 1-14, ISSN 0006-3606

Sterck, E.H.M.; Watts, D.P. & van Schaik, C.P. (1997). The evolution of female social relationships in nonhuman primates. *Behavioural Ecology and Sociobiology*, Vol.41, No.5, pp. 291-309, ISSN 1045-2249

Stone, B.C. (1979). Protective colouration of young leaves in certain Malaysian palms. *Biotropica*, Vol.11, pp. 126, ISSN 0006-3606

Stoner, K.E.; Riba-Hernández, P. & Lucas, P.W. (2005). Comparative use of colour vision for frugivory by sympatric species of platyrrhines. *American Journal of Primatology*, Vol.67, No.4, pp. 399-409, ISSN 0275-2565

Sumner, P. & Mollon, J.D. (2000a). Catarrhine photopigments are optimized for detecting targets against a foliage background. *Journal of Experimental Biology*, Vol.203, pp. 1963-1986, ISSN 0022-0949

Sumner, P. & Mollon, J.D. (2000b). Chromaticity as a signal of ripeness in fruits taken by primates. *The Journal of Experimental Biology*, Vol.203, pp. 1987-2000, ISSN 0022-0949

Surridge, A.K. & Mundy, N.I. (2002). Trans-specific evolution of opsin alleles and the maintenance of trichromatic colour vision in Callitrichine primates. *Molecular Ecology*, Vol.11, No.10, pp. 2157-2169, ISSN 0962-1083

Surridge, A.K.; Osorio, D. & Mundy, N.I. (2003). Evolution and selection of trichromatic vision in primates. *Trends in Ecology & Evolution*, Vol.51, pp. 198-205, ISSN 0169-5347

Tan, Y. & Li, W.H. (1999). Trichromatic vision in prosimians. *Nature*, Vol. 402, pp. 36, ISSN 0028-0836:

Tan, Y.; Yoder, A.D.; Yamashita, N. & Li, W.H. (2005). Evidence from opsin genes rejects nocturnality in ancestral primates. *Proceedings of the National Academy of Sciences*, Vol.102, No.41, pp. 14712-14716, ISSN 0027-8424

Taylor, W.P. (1934). Significance of extreme or intermittent conditions in distribution of species and management of natural resources, with a restatement of Liebig's law of minimum. *Ecology*, Vol.15, No.4, pp. 374-379, ISSN 0012-9658

Terborgh, J. (1983). *Five New World Primates: A Study in Comparative Ecology*, Princeton University Press, ISBN 978-069-1083-38-4, New York

Terborgh, J. (1986). Community aspects of frugivory in tropical forests. In: *Frugivores and Seed Dispersal*, Estrada, A. & Fleming, T.H. (Eds.), pp. 370-384, Kluwer Academy Publishers, ISBN 978-906-1935-43-8, New York

Thomas, D.W. (1984). Fruit intake and energy budgets of frugivorous bats. *Physiological Zoology*, Vol.57, pp. 457-467, ISSN 0031-935X

Tweheyo, M. & Babweteera, F. (2007). Production, seasonality and management of chimpanzee food trees in Budongo Forest, Uganda. *African Journal of Ecology*, Vol.45, No.4, pp. 535-544, ISSN 1365-2028

Urquiza-Haas, T.; Serio-Silva, J.C. & Hernandez-Salazar, L.T. (2008). Traditional nutritional analyses of figs overestimates intake of most nutrient fractions: A study of *Ficus perforata* consumed by howler monkeys (*Alouatta palliata mexicana*). *American Journal of Primatology*, Vol.70, pp. 432-438, ISSN 0275-2565

van Schaik, C.P. (1989). The ecology of social relationships amongst female primates. In: Comparative Socioecology: The Behavioural Ecology of Humans and Other Mammals, Standen, V. & Foley, R.A. (Eds.), Blackwell, Oxford

Veilleux, C.C. & Bolnick, D.A. (2009). Opsin gene polymorphism predicts trichromacy in a cathemeral lemur. *American Journal of Primatology*, Vol.71, pp. 86-90, ISSN 0275-2565

Vogel, E.; Neitz, M. & Dominy, N.J. (2007). Effect of colour vision phenotype in the foraging of white-faced capuchins, *Cebus capucinus*. *Behavioural Ecology*, Vol.18, pp. 292-297, ISSN 1045-2249

Wendeln, M.C.; Runkle, J.R. & Kalko, E.K.V. (2000). Nutritional values of 14 fig species and bat feeding preferences in Panama. *Biotropica*, Vol.32, No.3, pp. 489-501, ISSN 0006- 3606

Wrangham, R.W. (1980). An ecological model of female-bonded primate groups. *Behaviour*, Vol.75, pp. 262-299, ISSN 0005-7959

Yamashita, N.; Stoner, K.E.; Riba-Hernández, P.; Dominy, N.J. & Lucas, P.W. (2005). Light levels used during feeding by primate species with different colour vision phenotypes. *Behavioural Ecology and Sociobiology*, Vol.58, No.6, pp. 618-629, ISSN 1045-2249

CHAPTER 3

THE ACOUSTIC BEHAVIOR AS A TOOL FOR BIODIVERSITY AND PHYLOGENETIC RESEARCH: CASE OF THE *RHAMMATOCERUS* SPECIES INHABITING URUGUAY (ORTHOPTERA, ACRIDIDAE, GOMPHOCERINAE)

1. INTRODUCTION: AN OVERVIEW

Species variety plays a crucial role in the examination and understanding of biodiversity (Boero, 2010). A primary objective of Zoology is the examination of animal diversity, specifically animal species, with Taxonomy serving as a fundamental discipline to facilitate this study. Biodiversity study presently necessitates a multidisciplinary approach (Boero, 2009), incorporating several fields like as morphology, molecular biology, ecology, and ethology, which contribute novel traits for consideration. The investigation of biodiversity must advance through the incorporation of integrative taxonomy (Boero, 2010), including not only classical taxonomy but also other valuable fields, such as behavioral studies. In this context, behavior and vocalizations are pertinent attributes for identifying novel taxa (Valdecasas, 2011). Sound generation in insects is prevalent and has been documented across various orders. It encompasses various behaviors, primarily defense against predators, aggression, and mating or sexual conduct. Insects utilize four sensory modalities in their reproductive behavior: tactile, visual, chemosensory, and auditory. The auditory sense has generally garnered more attention; this stimulus is frequently perceived by humans, and its creation entails the movement of specialized structures that are typically visible. The sorts of sounds generated and their methods can be classified into three primary categories: stridulation, vibration, and percussion. Insect noises have been categorized for behavior into many types: call or announcement, courtship, aggregation, aggression, mating, and interaction sounds (Lewis, 1984; Ragge & Reynolds, 1998). The auditory signals in Orthoptera facilitate social interactions in a comprehensive manner. This form of acoustic behavior cannot be examined in isolation and can only be comprehended within the context of the species' whole behavior, which extends beyond mere sound creation. From this perspective, we demonstrate that precise understanding of the physical characteristics of sounds, along with the anatomy of the sound-producing organs, is crucial (Busnel, 1954). The sounds generated by Orthoptera are significant for taxonomic classification and serve as a crucial method for identification. They are essential for determining the actual state of local populations that may exhibit minimal morphological variations (Blondheim, 1990; García et al., 1995). The phenomenon of acoustic communication during reproduction is well-documented in Ensiferan insects (crickets and katydids): male singing Ensifera (Orthoptera) produce distinct calling songs for species identification, whereas courtship songs tend to be less specific and are primarily influenced by sexual selection. In contrast, singing Caelifera (Orthoptera) exhibit a greater variety of behaviors before mating. Within the subfamily Gomphocerinae (Acrididae), numerous species produce calling songs that are distinct enough for field identification. Additionally, they engage in intricate, frequently multimodal courtship behaviors that encompass sequences of acoustic, vibrational, and/or visual signals (Nattier et al., 2011; Ragge & Reynolds, 1998). The Neotropical Gomphocerinae comprise a group of grasshoppers with insufficiently understood taxonomy, systematics, and biology (Otte, 1979; Otte & Jago, 1979). Rhammatocerus Saussure, 1861 is extensively spread within this group, spanning from the southern United States to central Argentina. The species possess significant economic value, with several serving as notable pests for crops and grazing (Carbonell et al., 2006; Cigliano & Lange, 1998; Salto et al., 2003), particularly in Brazil and Colombia (Lecoq & Assis-Pujol, 1998). The genus Rhammatocerus is associated with several genera, including Parapellopedon Jago, 1971, and Cauratettix Roberts, 1937. Since its first description, the most significant investigations on this genus have been conducted by Jago (1971) and Carbonell (1995); nonetheless, no comprehensive taxonomic revisions of all its species have been undertaken. The genus now comprises 13 species (Assis-Pujol, 1997a 1997b, 1998; Carbonell, 1995), some of which remain inadequately delineated (Carbonell pers. comm.). The group exhibits significant intraspecific diversity and notable variability in its outward shape. Additional characters beyond physical traits have been investigated to distinctly differentiate the species. In the Scyllinini tribe, similar to other Gomphocerinae, genitalia facilitate species distinction only in exceptional instances (Carbonell, 1995). The phallic complex is a significant physical trait among Acrididae; yet, in Gomphocerinae, it lacks practical utility (Carbonell, 1995). The female genitalia, particularly the spermatheca, have demonstrated its efficacy in species identification (Assis-Pujol & Lecoq, 2000). Moreover, molecular investigations have failed to yield information conducive to species identification (Loreto et al., 2008). The sounds generated by Orthoptera hold significant taxonomic importance. They provide a significant function as an identification system during mating, thus holding considerable importance for determining the actual status of local populations exhibiting minor morphological variations (Blondheim, 1990; García et al., 1995). Research on the sounds generated by Gomphocerinae has consistently shown its effectiveness in addressing species identification challenges (Ragge & Reynolds, 1998); nonetheless, the acoustic behavior of Neotropical Gomphocerinae has been investigated to a minimal degree (COPR, 1982; García et al., 2003; Lorier, 1996; Lorier et al., 2010; Riede, 1987). Rhammatocerus pictus (Brunner, 1900) and Rhammatocerus brunneri (Giglio-Tos, 1895) are strong species residing in moist high pastures in northern and central South America, approximately between latitudes 15 and 40 South. Rhammatocerus pictus is found in central and southern Brazil (Rio Grande do Sul), Paraguay, Chile (Malleco), northern and central Argentina, Bolivia, and Uruguay (Assis-Pujol, 1998; Carbonell et al., 2006), while Rhammatocerus brunneri is located in central-southern Brazil, Paraguay, Bolivia, and Uruguay (Assis-Pujol, 1997a). In Uruguay, these species inhabit analogous environments characterized by low-humidity regions adorned with long, lush grass and gentle hill inclines. They are dispersed throughout the country (Assis-Pujol, 1998) and are predominantly located north of Río Negro. Both species coexist in the Artigas, Rivera, and Lavalleja Departments (Assis-Pujol, 1997a, 1998). The two species are systematically proximate, having been distinguished by Assis–Pujol (1998) solely on morphological criteria: the coloration of hind femora and tibiae, as well as the form of the spermatheca. Their acoustic behavior remains currently unknown. Our aim is to ascertain the true status of these two taxa in relation to their sound production and acoustic behavior, detailing the songs generated in various behavioral contexts, the discovered behavioral units, and the sound-morphological structures.

2. MATERIALS AND METHODS

This research on the sound production and stridulatory structures of Rhammatocerus pictus and Rhammatocerus brunneri involved males and females sourced from the "Colección de Entomología de la Facultad de Ciencias de la Universidad de la República," Uruguay, as well as specimens collected in Uruguay.

2.1 Microscopy Techniques and Characteristics Evaluated in Stridulatory File Analysis

To study the structure of the stridulatory apparatus, 10 specimens of each sex and species (Table 1) were observed under a binocular microscope (Olympus SZH provided with 10X ocular lenses, 0.66-4X zoom objective, 2X lens and graduated eye piece) as well as with a Jeol 6100 scanning microscope, equipped with SEI (secondary electron images), working at an acceleration voltage of 15 kV and at 21 mm (working distance). Images were captured with the LINK ISIS program. Because the samples are hard and have not risk of being dehydrated, they only had to be cleaned as proposed by Clemente et al. (1989) and, then, coated with pure gold.

Measurements were taken by means of a sliding stage mounted on the stereoscopic binocular microscope, the displacement of which was measured by an attached dial calliper in combination with a graduated eye piece. The accuracy of the dial calliper was 0.05 mm.

For this study, the shape of the file and pegs have been taken into account, as well as the number of pegs in the file, femur length (HFL), file length (FL), peg density all along the file (PD) and in its middle area (PDM) and file length / femur length ratio (FLx100/HFL). Measurements are expressed in mm (Table 2).

Table 1. Summary of the information concerning origin, collection date and number of specimens used to study the sound producing organs of *Rhammatocerus*.

	Specimens	Locality	Date	Collector
Rhammatocerus brunneri	4 males (nº 61, 62, 63, 65)	Cerro Chato Dorado. Rivera. Uruguay	12-II-2000	Clemente, García, Lorier
	1 male (nº 64) 2 females (nº 93, 94)	Sierra de la Aurora. Rivera. Uruguay	14-III-1961	C.S.C, A. Mesa, P. San Martín
	2 males (nº 67, 68) 1 female (nº 91)	Cuchilla de Cuñapirú. Rivera. Uruguay	21-I-1956	C.S.C.
	1 males (nº 69) 3 female (nº 86, 90, 95)	Ronda Alta. Río Grande do Sul. Brasil	24-II-1964	A. Mesa, M.A. Monné
	1 males (nº 70)	Bom Jesús.Río Grande do Sul. Brasil	26-II-1964	A. Mesa, M.A. Monné
	1 male (nº 71) 4 females (nº 87, 88, 89, 92)	40 km. N de Caaguazú. Paraguay	13-III-1965	C.S.C, A. Mesa, M.A. Monné
Rhammatocerus pictus	2 males (nº 56, 59)	Cerro Chato Dorado. Rivera. Uruguay	17-18-III-2001	E. Lorier
	2 males (nº 60, 66)	Lunarejo. Rivera. Uruguay	1-IV-1999	E. Lorier
	1 male (nº 57)	V.Lunarejo Rivera. Uruguay	13-II-2000	Clemente, García, Lorier

	Specimens	Locality	Date	Collector
	1 male (nº 72) 1 female (nº 79)	Sª de la Aurora. Rivera. Uruguay	14-III-1961 11-III-1961	C.S.C, A. Mesa, R. San Martín
	1 male (nº 73) 1 female (nº 76) 1 female (nº 77)	Las Piedras. Canelones. Uruguay	5-II-1966 5-XI-1966 20-III-1964	A. Cármenes
	1 male (nº 74)	Pto. Pepeají. Paysandú. Uruguay	IV-1954	C.S.C.
	1 male (nº 58)	Buena Vista Agraciada. Soriano. Uruguay	8-II-2001	García, Lorier, Presa
	1 male (nº 75) 1 female (nº 83)	Lagoa Vérmelha. Rio Grande do Sul. Brasil	18-II-1964	C.S.C, A. Mesa, M.A. Monné
	1 female (nº78) 1 female (nº 80)	Ayo. Tres Cruces. Artigas. Uruguay	12-XI-1955 14-II-1955	Fac. Humanidades y Ciencias
	1 female (nº 81)	Tartagal. Salta. Argentina	29,31-I-1965	A. Mesa, R. Sandulski
	2 females (nº 82, 84)	San Lorenzo. Salta. Argentina	3-II-1965	A. Mesa, R. Sandulski
	1 female (nº 85)	Nonoai. Río Grande do Sul. Brasil	20-II-1964	A. Mesa, M.A. Monné

Two-ways multivariate analysis of variance (MANOVA) was used to assess whether there were overall differences between study species regarding the stridulatory file, considering as dependent variables the abovementioned morphological traits. In this case separated analyses were run for each sex. For these and all other statistical analyses, software SPSS (v. 15.0) was used. All variables were log-transformed. The signification value, in all the cases, was $P \leq 0.05$.

Table 2. Summarized data related to stridulatory apparatus of *Rhammatocerus brunneri* and *R. pictus*. HFL: hind femur length; FL: whole stridulatory file length; Pegs: total number of pegs in the whole file; PD: density of pegs in the whole file; PDM: density of pegs in the middle of the file. Values are expressed as mean ± SD and range in parenthesis.

		HFL (mm)	FL (mm)	Pegs	PD (mm^{-1})	PDM (mm^{-1})	FL x 100/ HFL
R. brunneri	Males n=10	16.96 ± 0.76 (15.80-17.90)	5.09 ± 0.76 (4.5-5.70)	74.4 ± 12.4 (58-99)	14.60 ± 1.98 (11.60-18.68)	19.5 ± 3.2 (13-24)	29.96 ± 0.97 (28.61-31.84)
	Females n=10	23.86± 0.86 (22.70-25.25)	11.61 ± 0.73 (10.70-12.90)	84 ± 7.1 (72-95)	7.26 ± 0.70 (5.81-8.12)	12.6 ± 1.9 (10-16)	49.02 ± 3 (44.95-53.69)
R. pictus	Males n=10	17.92 ± 0.94 (16.50-19.40)	5.21 ± 0.62 (4.40-6.00)	73 ± 10.70 (58-90)	14.18 ± 2.61 (11.05-18.88)	18.6 ± 2.8 (14-22)	29.04 ± 2.78 (24.71-33.33)
	Females n=10	22.78 ± 1.32 (21.10-24.65)	10.78 ± 1.05 (8.95-12.05)	74.2 ± 3.29 (69-79)	6.94 ± 0.69 (6.09-8.16)	9.8 ± 1.8 (8-12)	47.57 ± 6.10 (36.30-54.50)

2.1 Examination of Behavior and Acoustic Emission

This study was conducted with 9 males (Table 3), 6 males of *R. pictus* and 3 males of *R. brunneri*, and 312 different songs (126 from *R. pictus* and 186 from *R. brunneri*), registered in 5 different tapes (Table 3). The specimens were held in cages in the laboratory and fed mainly with grasses that were changed daily. Humidity was provided by daily watering and by cotton imbibed in water. Two types of cages were used: a glass cage, with net top, 20 x 11 x 14 cm, and a wooden cage with metal mesh top and glass front, 35 x 35 x 55 cm, both exposed to natural light or artificial light provided by a 40W bulb 12 hours per day (Table 3).

The sounds produced were recorded under different conditions, such as isolated specimens, a male together a female of the same species, a species together other different species, both species together.

Sound recordings were made in the laboratory (Table 3) using a Uher 4000 and a Uher 6000 analogical tape recorders (Uher Werke München, Barmseestrasse 11, 8000 München 71, Germany), at a tape speed of 9.5 cm/s, with a frequency response (in Hz) of 20-25000 and signal-to-noise ratio better than 66 dB A remote-control Uher M655 and a Uher M518 dynamic microphones, were located 10-20 cm from the specimens.

Table 3. Summary of the information concerning specimens used to study the acoustic behaviour of *Rhammatocerus* and studied recordings.

		Locality of capture	Date and collector	Recording date	Recording conditions	Tape	Type of sound and recording number
R. pictus	1 male	Cerro Batoví. Tacuaremb. Uruguay	1/IV/1999 E.Lorier	18/IV/1999	Wood cage, (bulb 25W) 20ºC	EL6/1999	1 calling 4 courtship Rec.3/6/99
	1 male	Valle del Lunarejo. Rivera. Uruguay	13/II/2000 Clemente, García, Lorier	14/II/2000	Wood cage, (bulb 40W) 33ºC	EL3/2000	1 calling Rec. 2/3/2000 Rec.3/3/2000
	1 male	Cerro Chato Dorado. Rivera. Uruguay	11/II/2000 Clemente, García, Lorier	22/II/2000	Wood cage, (2 bulbs 40W) 31-32ºC	EL4/2000	4 calling 14 courtship Rec. 2/4/2000
	3 males	Rivera. Uruguay	18/III/2001 Lorier	20/III/2001 23/III/2001	Glass cage (bulb 40W) 27-28 ºC	EL3/2001	96 disturbance 6 courtship Rec. 1/3/2002 Rec. 2/3/2001
R. brunneri	3 males	Cerro Chato Dorado. Rivera. Uruguay	12/II/2000 Clemente, García, Lorier	14/II/2000	Wood cage, (bulb 40W) 33ºC	EL2/2000	36 disturbance 2 calling Rec. 8/2/2000
					Wood cage, (bulb 40W) 32ºC	EL3/2000	118 disturbance 10 calling 7 courtship Rec. 1/3/2000 Rec. 3/3/2000
				22/II/2000	Wood cage, (2 bulbs 40W) 32 ºC	EL4/2000	1 courtship 12 disturbance Rec. 1/4/2000

Observations of communicative and interactive behaviour and of the general activity of individuals were made in the laboratory and recorded with a JVC GR-AXM23 video-camera for subsequent analysis. Specimens were observed throughout the recordings, including mute periods, and the behaviour of the specimens in each situation was noted.

Sound recordings were analysed using a Mingograph 420 System attached to a digital oscilloscope (Tektronix 2211) and to a Krohn-Hite 3550 filter. To study the physical characteristics of the sound, the analogical signal was digitized with a Sound Blaster® AWE64 Gold at 8 bits and at a 44 kHz sampling frequency, and then studied using the Avisoft® SAS Lab Pro 3.8. PC software for MS-Windows. Oscillograms were obtained using the option One-dimensional functions, selecting the function Time-signal. The spectral characteristics were obtained using the same option than before, selecting, in the function Amplitude spectrum (linear), the Hamming evaluation window, bandwidth 0.526 Hz and resolution 0.336 Hz.

Males of both species produce different types of sound in different behavioural situations fitting with the categories from the literature (García et al., 2003; García et al., 2005; Ragge & Reynolds, 1998). The terminology used to describe songs follows that of Ragge & Reynolds (1998). The variables used for the study of the sounds are: echeme length, number of syllables, syllables length, rate of emission of syllables, peak of maximum amplitude, low quartile, middle quartile, upper quartile, minimum frequency, maximum frequency and band width. The following types of sounds were identified: (1) the calling song, produced spontaneously by a male; (2) the courtship song, produced by males when close to a female and (3) the disturbance song, produced by males when interacting with other individuals.

To explore the relationship among different types of sounds produced by the studied species, a principal component analysis (PCA) was performed considering all the variables used to study the sounds (see before). Since for disturbance songs not all response variables could be recorded, they were not included in the PCA. Two-ways MANOVA was used to formally assess the differences among type of sound (calling song and courtship song) and species (*R. pictus* and *R. brunneri*), considering the variables used for the study of sounds as dependent variables. When cases corresponding to different groups overlapped after running the PCA, a separate MANOVA was performed for these groups considering the same dependent variables and factors as for the overall MANOVA. Variables were log- transformed for both PCA and MANOVA and a significance level of 5% was selected.

Specimens and recordings are kept in the "Colección de Entomología de la Facultad de Ciencias de la Universidad de la República", Uruguay, and in the "Colección del Área de Zoología de la Universidad de Murcia", Spain. Recordings can be reviewed using the recording number (Table 3). Sample songs are also available at OSF (http://osf2.orthoptera.org) (Eades 2001).

3. RESULTS

The recorded sounds, generated by males in various behavioral contexts, were all produced through the femoro-tegminal mechanism, which involves the friction of the stridulatory file pegs of the hind femora against particular tegminal veins. No disparities among songs have been noted concerning the recording conditions. No auditory signals from females have been documented in any instance, despite their hind leg motions. These movements are of moderate amplitude, starting with the leg folded at about 45º respect the vertical and displacing about 45º from the starting position. This movement can be isolated, from resting position to the vertical, or be repeated several times, when close to other individuals, males or females.

3.1 Stridulatory file

In *Rhammatocerus pictus* the hind femora of both sexes have, along their inner surface, a stridulatory file that is almost linear and long in relation to femur length. The stridulatory pegs are well developed and regularly spread except at ends, where they are more irregularly and more separately distributed (Fig. 1A-D).

Male file, although shorter than that of the female, contains more pegs than that of the female. The peg density on the whole file is greater in males than in females (Table 2, Figs. 1A and C). The male pegs are conic shaped. They are inserted in the alveolus, which have a raised margin, and have a short peduncle (Fig. 1B). The female pegs, well developed, are also conic shaped with rounded apex, and have a short peduncle. They are also inserted in alveolus with raised margins (Fig. 1D).

In both sexes of *Rhammatocerus brunneri* the stridulatory file is linear and long in relation to femur length. Pegs are well developed and regularly spread except at ends, where they are more separated and irregularly disposed (Fig. 1 E-H).

Male file is half the length than that of female, but it has almost the same number of pegs and, so, the peg density is greater (Table 2, Fig. 1E and G). The male pegs are conic shaped and have a short peduncle. They are inserted by in the alveolus, which have a raised margin (Fig. 1F). The female pegs are also conic shaped, slightly irregularly spread and more separated than in males. Pegs are also inserted by a short peduncle in alveolus with raised margins (Fig. 1H).

The MANOVA revealed the lack of overall differences among species regarding the stridulatory file in the case of males ($F_{5, 14}$= 1.563; P= 0.234), although significant differences were detected for femur length ($F_{1, 18}$= 6.205; P=0.023). Females of different species showed significant overall differences ($F_{5, 14}$= 3.724; P= 0.024), as well as for the responses variables number of pegs in the file ($F_{1, 18}$= 15.947; P= 0.001) and peg density all along the file ($F_{1, 18}$= 11.920; P= 0.003).

3.2 Calling and courtship songs

The PCA performed extracted two components explaining most of the variance of the original data (79.5%). Component 1 was positively correlated with those variables dealing with frequency and negatively with duration of the echeme. Component 2 was positively correlated with both the number of syllables and the rate of emission of syllables and negatively with the duration of syllables (Fig. 2).

When scores for each case were plotted against component 1 and 2 three major groups were identified: 1) courtship and calling songs of *R. brunneri* (closely overlapped); 2) calling song of *R. pictus* and one of the types of courtship song of this species (pictus 1); 3) other type of courtship song of *R. pictus* (pictus 2) (Fig. 3). The MANOVA performed for both the courtship and calling songs of the studied species indicated the existence of overall significant differences among species (F11, 23= 18.042; P<0.0001) and type of song (F33, 68.47= 6.318; P<0.0001).

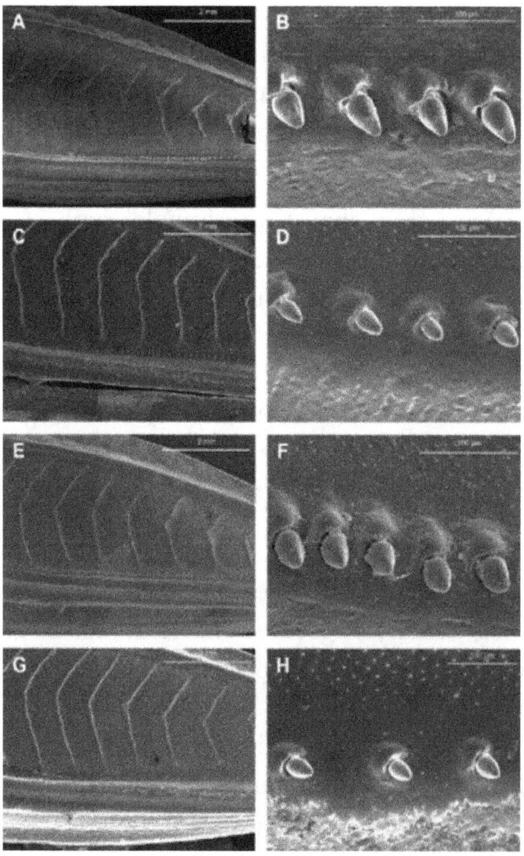

Fig. 1. Stridulatory files. A-D: *Rhammatocerus pictus*. A: male file general appearance, B: detail of male pegs of middle zone, C: female file general appearance, D: detail of female pegs of middle zone. E-H: *Rhammatocerus brunneri*. E: male file general appearance, F: detail of male pegs of middle zone, G: female file general appearance, H: detail of female pegs of middle zone.

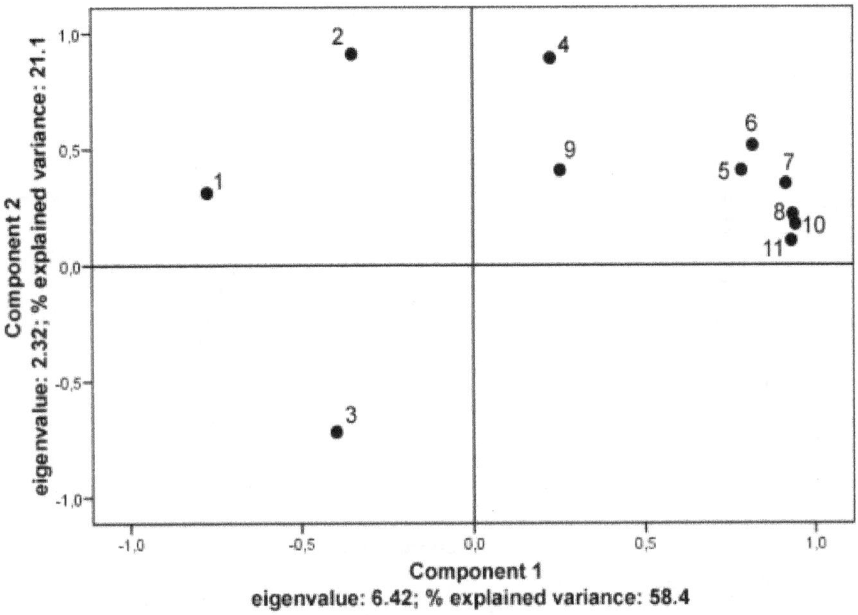

Fig. 2. Results of PCA performed to explore relationships among types of sound considering all the variables used: 1 echeme length; 2 number of syllables; 3 syllables length; 4 rate of emission of syllables; 5 peak of maximum amplitude; 6 low quartile; 7 middle quartile; 8 maximum frequency; 9 minimum frequency; 10 upper quartile;

3.2.1 *Rhammatocerus pictus*

Calling song was a spontaneous song, consisting of echemes composed of a variable number of syllables (Table 4).

Its frequency spectrum occupied a broad band, with the peak of maximum amplitude at around 7 kHz (Table 4, Fig. 4C).

The echeme started with short and almost inaudible syllables. The sound increased in intensity after 1/3 of the echeme had been emitted. Syllables were clearly separated one from other by evident gaps (Figs. 4A and B).

It was produced by synchronously rubbing the two hind femora against the tegminae, both hind legs moving rapidly. At the beginning of the echeme the legs movement was hardly visible.

Courtship songs were composed of echemes of a variable number of syllables (Table 4). Two different types of song have been observed, the here called pictus 1 (Fig. 4D-F) emitted at a broad band, with the maximum amplitude at around 5 kHz, and the called pictus 2 (Fig. 4G- I) which occupied a much narrower band, with the maximum amplitude at around 2 kHz; both are significantly different (Fig. 3).

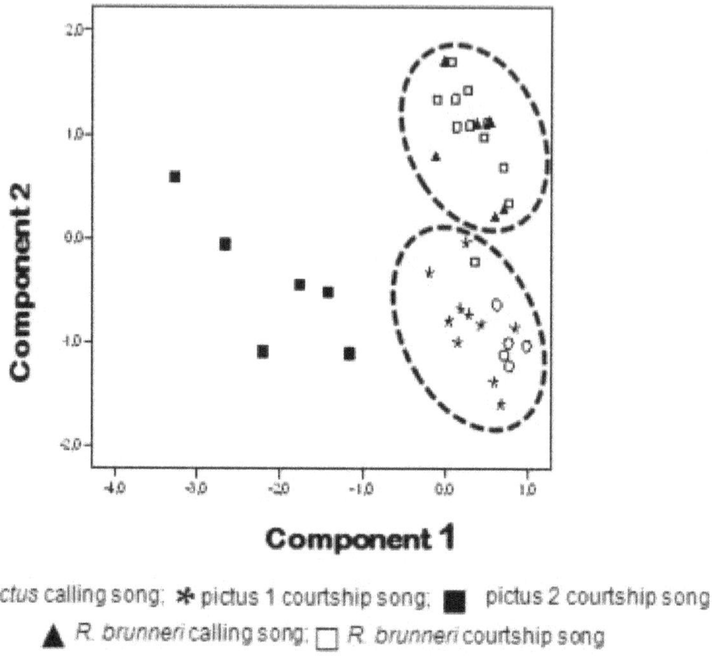

Fig. 3. Graphic showing the results of plotting scores for each cases against components 1 and 2 of PCA.

When the courtship started, the male moved walking towards the female; stopping when close and perpendicular to her. Then he moved his antennae up and down, synchronously at the beginning and, then, alternately, and started to sing. While singing, he directed his antennae towards the female, forming an around 150º angle. After the song, the male usually jumped suddenly on the female, trying to mate while touching her with his antennae. In most cases, the mating was not effective, the male being rudely rejected by the female by kicking the male, raising the hind legs or simply going away him.

While the courtship song, males performed two kinds of movement: 1) Mute movement: one to three slow and wide up and down movements of hind legs. They rose synchronously from the rest position (around 30º in relation to the main corporal axis) to around 75º, and then bent slightly asynchronously. 2). Stridulatory movement: a series of synchronous, quick and small up and down movements, of little amplitude (30-55º) respect to the corporal axis. In some cases, the movement 1 lacked, the courtship starting directly with the quick movement.

The songs started as imperceptible, almost inaudible, and sound increased in intensity until the end. At the final section, syllables were intense, similar in structure, with gaps between them (Fig. 4 D-E and G-H).

The MANOVA run separately for the group 2 identified by the PCA revealed that for this species, overall, calling song and pictus 1 courtship song were not significantly different ($F_{11,\,3}$= 0.702; P= 0.712), existing differences only in the case of some the response variables dealing with frequency (P≤0.037): peak maximum amplitude ($F_{1,\,13}$= 5.401; P= 0.037), low quartile ($F_{1,\,13}$= 11.494; P= 0.005), upper quartile ($F_{1,\,13}$= 6.026; P= 0.029), middle quartile ($F_{1,\,13}$=7.279; P= 0.018) and minimum frequency ($F_{1,\,13}$=11.527; P= 0.005).

3.2.2 *Rhammatocerus brunneri*

The calling song consisted of an echeme composed of syllables (Table 4, Fig. 5A and B), short and almost inaudible at the beginning and increasing in intensity further on. At the end, the echeme had a high intensity and the syllables are very close with almost no gap between them. The frequency spectrum of the sound occupies a broad band, the main peak being at around 8500 Hz (Table 4, Fig. 5C).

The leg movements to produce the sound are almost imperceptible at the beginning of the echeme. They became wider as the echeme went on and at the end legs seemed to move much quickly than before.

The courtship song was composed of echemes of a variable number of syllables (Table 4, Fig. 5D and E). As regards the spectral characteristics (Fig. 5F) they were similar to that of calling song there having not been found any statistically significant difference between this song and calling song.

The leg movements to produce the sound were also similar to that to produce calling song. When a male was performing a courtship, could be interrupted by other males, who started to perform the disturbance song.

During courtship the male follows the female, standing behind very close to her.

The MANOVA run separately for the group 1 identified by the PCA revealed that for *R. brunneri* there were not significant overall differences between courtship and calling songs ($F_{11,\,5}$= 0.607; P=0.773), although the duration of the echeme (Table 4) was significantly ($F_{1,\,15}$= 4.786; P= 0.045) different.

3.3 Disturbance song

3.3.1 *Rhammatocerus pictus*

Songs were composed of isolated syllables of variable duration (Table 4) irregularly emitted (Fig. 6A and B). Its frequency spectrum showed a quite broad band of emission; the maximum amplitude peak being at below 3 kHz (Fig. 6C). To produce song, the males moved rapidly and almost synchronously either hind legs or only one hind leg.

Males produced these sounds when the specimens were at near or in contact with each other. The signals were sometimes emitted in alternation. Some individuals were observed singing while walking. When a male was courting a female, other males near the couple started to sing the disturbance song that, in this case, played the role of rivalry song.

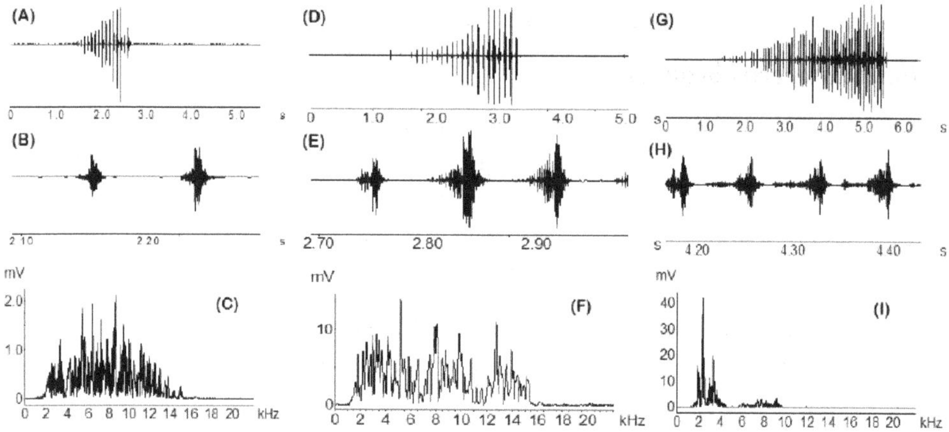

Fig. 4. ***Rhammatocerus pictus*****. Calling (A-C) and courtship (pictus 1 D-E and pictus 2 G-I) songs. Echeme (A, D and G); syllable detail (B, E and H), and frequency spectra (C, F and I).**

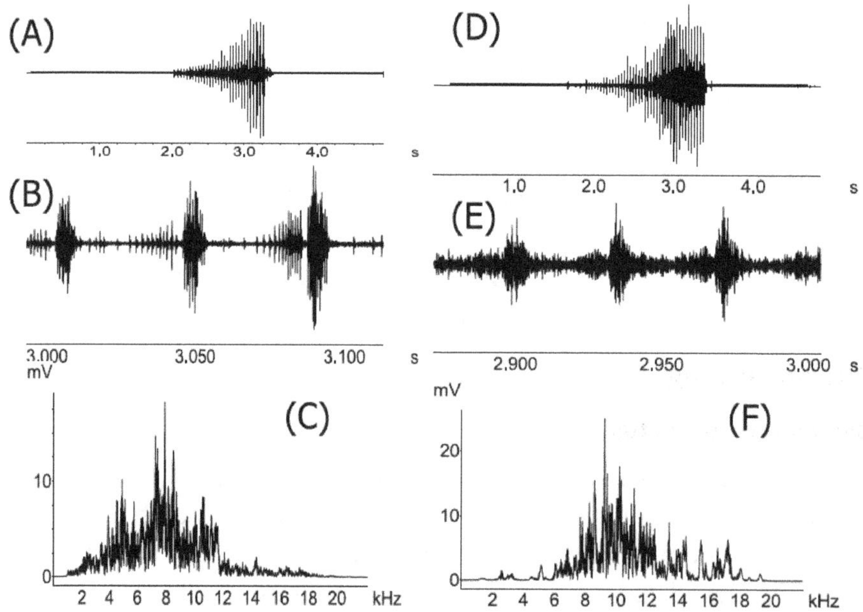

Fig. 5. ***Rhammatocerus brunneri*****. Calling song (A-C) and courtship song (D-F). Echeme (A and D); syllable detail (B and E) and frequency spectra (C and F).**

3.3.2 *Rhammatocerus brunneri*

Songs were composed of isolated syllables (Table 4, Fig. 6D and E) emitted alternately by different individuals. The frequency spectrum occupied a broad band, with the peak of maximum amplitude at around 10 kHz (Fig. 6F). The leg movements to produce this song were similar to that performed by *R. pictus*, usually one or two quick asynchronous up and down movements.

This song was emitted in different circumstances, such as coinciding with or at the end of a calling song of other male or as rivalry song.

For disturbance songs, indicated the existence of overall significant differences among species (MANOVA: F8, 15= 53.073; P<0.001).

No disturbance songs produced by females have been recorded. One or two mute up and down movements of their hind legs were observed after a male interaction or courtship song or when interacting with other female. These movements have been also observed performed by males when close to other individuals or interacting with them.

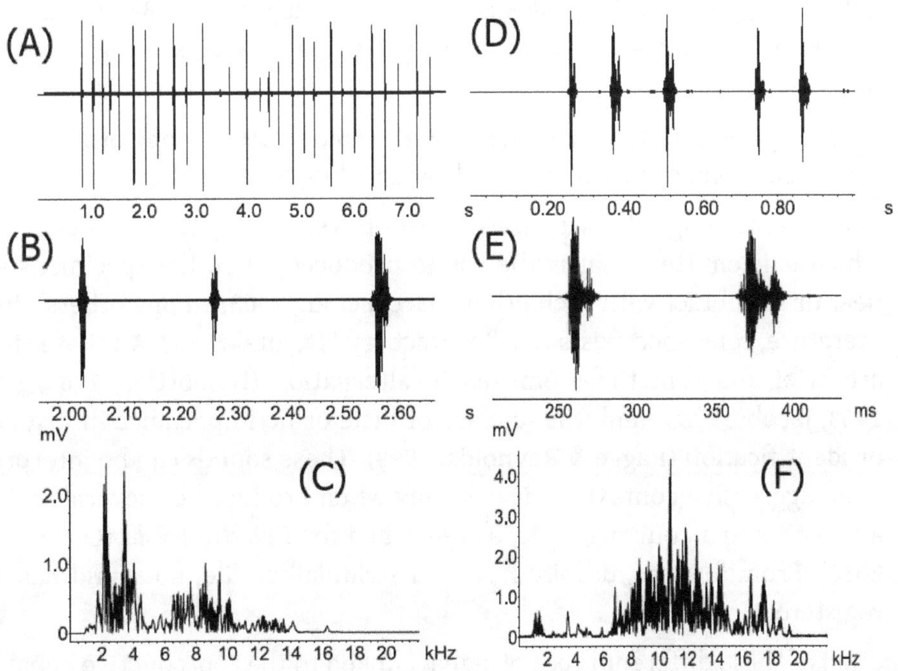

Fig. 6. Disturbance song. *Rhammatocerus pictus* (A-C) and *Rhammatocerus brunneri* (D-F). Sequence of syllables (A and D); syllable detail (B and E) and frequency spectra (C and F).

4. DISCUSSION

The stridulatory file of *R. pictus* and *R. brunneri* are well developed in both males and females. The peg spread, regular all along the file except at the ends, fits with that of other Gomphocerinae (Clemente et al., 1989; Pitkin, 1976). It can be pointed that the metrical characteristics referred to the stridulatory file are not species specific for males of *R. pictus* and *R. brunneri* but can serve to differentiate the females of both species. Nevertheless, there seem to be some morphological differences between stridulatory pegs of males. In both cases pegs are conic shaped, but *R. pictus* have pegs clearly more elongated, with a more acute apex, and the alveolus have the raised margin more irregular than *R. brunneri*.

The acoustic repertoire of *R. pictus* and *R. brunneri* is similar to that of other neotropical Gomphocerinae species, such as *Parapellopedon instabilis* (Rehn, 1906), *Euplectrotettix ferrugineus* Bruner, 1900 and *Fenestra bohlsi* Giglio-Tos, 1895 (García et al., 2003; Lorier, 1996; Riede, 1987), being made up of:

- Calling song, as defined by Bailey (1991), Dumortier (1963), Ewing (1989), García et al. (1995), Green (1995), Helversen & Helversen (1983), Lorier et al. (2010), Ragge (1987), Ragge & Reynolds (1998), Reynolds (1986), Riede (1987), among others, since the songs were produced by males being apart from other specimens.
- Courtship song, as defined by Ragge & Reynolds (1998) among others, that is, the special song produced by a male when close to a female.
- Disturbance song, characterized by the physical structure and context in which it is emitted, that is the sound produced when the specimens were near or in contact with each other (García et al., 2003, among others). In the literature, reference exists to a "contact cry" (Kontaktlaut). All these signals are brief and sometimes emitted in alternation (Dumortier, 1963; Faber, 1953; Jacobs, 1953) and it is said are of little or no importance in taxonomy or identification (Ragge & Reynolds, 1998). These sounds can be interpreted in an aggressive context as rivalry song when produced by several males as an answer to a courting male, as described for *Fenestra bolshi* (Lorier et al., 2010). From females, despite having a stridulatory file, no sound has been registered.

The function of these different types of signals linked to the reproductive behaviour in Orthoptera and other insects has been analysed and interpreted in the context of the sexual selection theory. Sound can evolve, among other possibilities, through sexual selection. Evolutionary pressures have to act towards a minor energetic cost and a minimum risk of predation in the mating (Bailey, 1991). The significance of the male calling songs of Orthoptera lies in the fact that they are believed to provide the main means of mater recognition and hence of reproductive isolation of sympatric species (Ragge, 1987; Ragge & Reynolds, 1998). Acoustic signals provide enough relevant information on species identification and sexual selection. So, sound is considered essential to solve taxonomical problems at the specific level. Closely related, sympatric species often use strikingly different acoustic criteria to discriminate the same set of conspecific and heterospecific signals (Gehardt & Huber, 2002). A marked difference between songs is a strong indicative of different species (Ragge & Reynolds, 1998). When two morphologically similar populations of Orthoptera have consistently different calling song, it is likely that they belong to different species. Conversely, when two populations show small morphological differences but have exactly the same calling song, they are probably forms of a single species (Ragge, 1987).

In most Gomphocerinae some acoustic signals identify the species and can avoid the crossing between sympatric close species. This is the case of *Rhammatocerus* species here studied, which show differences in physical characteristics of all types of song. Both species have overall different songs, as shown in Table 4 and Figures 4, 5 and 6. Differences are more or less evident depending on the type of song.

The calling song of *R. brunneri* is very similar to the courtship song (Fig. 5). Thus, this species uses a similar song, with little differences (Table 4), in two different behavioural situations. In *R. pictus* a similar situation exists between the calling song and one of the courtship songs recorded (pictus 1) (Fig. 4A-F). At the light of the recorded songs, this species may start courting with a song similar to the calling song (pictus 1) changing afterwards to a particular courtship song (pictus 2), accompanied by legs and antennae movements. Although the whole sequence has not been registered, it would be not surprising since the use of a courtship song similar to calling song has been observed in other Gomphocerinae species, such as *Chorthippus binotatus binotatus* (Charpentier, 1924) (García et al., 1995) and *Omocestus antigai* (Bolívar, 1987) (Clemente et al., 1999). In the last case, it could be observed that the courtship started with the calling song and, later, the courtship song was emitted.

This kind of behaviour fits with that observed, for example, in genus *Stenobothrus* Fischer, 1853, the ancestor of which should have had produced simple and likely identical calling and courtship songs. The complex courtship songs observed in many species resulted from the addition of new acoustic traits derived from the common calling song pattern and from the addition of visual signals produced by the movement of legs and antennae (Berger, 2008 as cited in Nattier et al., 2011). In *R. pictus*, during the courtship new traits related to the spectral characteristics of the sound and, on the other hand, visual signals by legs and antennae movements would have been added.

These similarity of calling and courtship songs in *R. brunneri* and, partly, in *R. pictus* could be explained by the signals flexibility observed in acridids by which the same signal can be used in different contexts, as here occurred. This would imply following the opinion of Otte (1977), a further stage in an evolutionary process not ended still, probably accompanying a sympatric speciation process. Probably, sounds should not be the unique barrier to avoid hybridizing but there could also take part other kind of signals, such as those visual or chemical (Otte, 1977), although behavioural traits are often evolutionary labile and hence communication systems often diverge more rapidly than do morphological and molecular characteristics (Gehardt & Huber, 2002). Nevertheless, the question of which of the two song types calling or courtship song is the phylogenetic older one remains open (Berger & Gottsberger, 2010) but a simplification of courtship repertoires in some species suggests that the evolution of courtship song, and that of mating behaviour as a whole, could also result from a dynamic process (Berger, 2008 as cited in Nattier et. al., 2011).

As regards the disturbance song, in some species phonotaxis is reduced when other males interfere, diminishing the efficiency of calling or courtship song. It could be pointed, for male disturbance songs – either spacing or answering other male courtship –, that the answering male, when alternating its song, uses its interference potential of acoustic channel (Cade, 1985; Greenfield, 1997). Thus, the emitters compete for the receipt's attention, in this case the silent female. Thereby it could be understood, for *R. brunneri*, the courtship interruption provoked by rivalry songs (McGregor & Peake, 2000). Differences between disturbance songs of the two species, *R. pictus* and *R. brunneri*, support the specific communicative value even of this type of song.

Acoustic communication necessitates a distinct signal recognizable by conspecifics to prevent interspecific mating; thus, its properties must be specific. The analyzed data demonstrate that acoustic properties, in both temporal and frequency domains, facilitate the obvious differentiation of cryptic species and provide valuable insights for phylogenetic analysis. Until now, the sole distinguishing characteristics between R. pictus and R. brunneri were the coloration of the hind femora and tibiae, as well as the morphology of the spermatheca (Assis-Pujol, 1998); nonetheless, the true status of both species has remained ambiguous. Our findings may impact taxonomy by introducing new factors for the genus revision, hence substantiating the classification of R. pictus and R. brunneri as distinct valid species (Carbonell pers. comm.).

Table 4. Data summary concerning the physical characteristics of the songs emitted by the males of *Rhammatocerus pictus* and *R. brunneri* on the time and frequency domains (data of which are expressed in Hz)

	Calling song		Courtship song			Disturbance song	
	R. pictus	R. *brumteri*	*R pictus 1*	*R pictus 2*	R. *brunner*	*R pictus*	R. *brumteri*
Echeme length (sec)	1.509 ± 0.164 (1.241-1.658)	2.130 ± 0.480 (1.618-3.22)	2.022 ± 0.588 (1.21-3.23)	3.02 ± 0.968 (2.151-4.807)	1.651 ± 0.380 (1.233-2.291)	–	–
Number of syllables	20.166 ± 1.602 (18-23)	49.416 ± 12.101 (34-67)	24.38± 7.65 (14– 10)	38.85± 12.33 (28-61)	38± 12.23 (25–26)	–	–
Syllables length (sec)	0.049 s ±0.012 (0.028-0.064	0.028 ± 0.002 (0.024-0.032)	0.041± 0.008 (0.031-0.070)	0.052 ± 0.005 (0.043-0.059)	0.027 ± 0.003 (0.023-0.032)	0.035 ±0.008 (0.020-0.066)	0.058 ± 0.023 (0.023-0.127)
Rate of emission (syll/sec)	13.471 ± 1.772 (12.06-16.61)	22.684 ± 3.542 (13.63-27.47)	12.07 ± 1.682 (9.14-15.39)	13 ± 0.328 (12.68-13.56)	22.9 ± 4.67 (13.63-29.29)	–	–
N	6	12	18	7	8	96	166
Peak of maximum amplitude	7198.08 ± 2962.745 (2790-11040)	8570.63 ± 2201.574 (1571-11864)	5014 ± 2239.707 (1297-10680)	2196.71 ±499,754 (1884-3474)	7772.93 ± 2307.929 (3466-11170)	2752.14 ± 644.780 (2210-3830)	10065.38 ± 2145.442 (7550-14443)
Low quartile	5316.54 ± 796.125 (3760-6540)	6687.78 ± 1343.213 (3334-8645)	4340.39 ± 771.349 (2530-5351)	2236.79 ± 330.158 (1913-3213)	6129.47 ± 1524.788 (3639-9000)	2844.29 ± 332.883 (2150-3550)	8388.63 ± 1043.882 (6610-9619)

	Calling song		Courtship song			Disturbance song	
	R. pictus	R. *brumteri*	*R pictus 1*	*R pictus 2*	*R. brunner*	*R pictus*	R. *brumteri*
Middle quartile	8419.23 ± 803.845 (6990-9880)	9081.74 ± 1357.565 (6193-11148)	6879. 75 ± 1361.498 (3924-8893)	3416.43 ± 1426.617 (2330-7380)	8720.00 ± 1293.614 (6341-10675)	4696.43 ± 1097.841 (3570-6930)	10899.75 ± 1054.661 (9560-12758)
Upper quartile	11354.62 ± 801.969 (8910-12270)	12040.63 ± 1507.873 (9146-14499)	10159.11 ± 1635277 (6346-12917)	7058.50 ± 3157.601 (2853-14136)	11790.47 ± 1228.933 (8694-13339)	9090.00 ± 1350.504 (7080-11640)	13516.00 ± 1555.696 (10740-15500)
Minimum frequency	1864.62 ± 281.769 (940-2340)	1865.52 ± 1207.976 (963-7068)	1349.04 ± 327.541 (931-1920)	1301.29 ± 292.857 (1014-1838)	2078.73 ± 1178.164 (807-5307)	1542.14 ± 200.851 (1030-1930)	3237.38 ± 603.467 (2540-4430)
Maximum frequency	15971.92 ± 1312.951 (13130-18820)	17174.26 ± 2293.155 (12220-20916)	14593.64 ± 2035.698 (9735-17520)	7315.57 ± 3018.844 (3441-12973)	16105.13 ± 2529.865 (10874-18809)	11807.86 ± 1944.448 (8220-13970)	17744.50 ± 1357.726 (14470-18680)
Band width	14102.69 ± 1365.625 (11140-17070)	15306.85 ± 2635.109 (10150-19902)	13243.75 ± 1932.220 (8780-15993)	6013.64 ± 3030.971 (1606-11929	14025.87 ± 3022.831 (8640-17463)	10259.29 ± 1980.810 (6710-12460)	14248.38 ± 1372.988 (11920-15860
N	26	27	28	14	15	14	8

5. CONCISE SUMMARY

The exploration of biodiversity is essential (Boero, 2009), particularly for the identification and differentiation of cryptic species. In addition to well-defined and easily observable anatomical traits, other qualities, such as genetic and behavioral attributes, are also available. Insects' sound production has remarkable qualities that provide species discrimination based on sound characteristics in both temporal and frequency domains, as well as behavioral factors relevant to the environment of sound generation. Rammathocerus brunneri and R. pictus are sympatric Neotropical Gomphocerinae species exhibiting highly analogous physical characteristics, including those pertaining to the stridulatory file, the organ responsible for sound production. Nonetheless, they create songs that are species-specific, introducing new characteristics to be evaluated for their current unique state. Furthermore, the noises and their associated behaviors offer additional information to consider in the phylogenetic history of the genus. Consequently, the auditory behavior, when observed, merits consideration in taxonomy revisions due to its established utility. The task of naming animals remains incomplete (Boero, 2009), and sound can assist in fulfilling this task.

6. REFERENCES

Assis-Pujol, C.V. de (1997a). Notas Sinonímicas e Redescrições de Duas Espécies de *Rhammatocerus* Saussure, 1861 (Orthoptera, Acrididae, Gomphocerinae, Scyllinini). *Boletim do Museu Nacional* (N.S.), Río de Janeiro, Vol. 376, pp. 1-12, ISSN 0080-312X.

Assis-Pujol, C.V. de (1997b). Duas Novas Espécies Brasileiras de *Rhammatocerus* Saussure, 1861. (Acrididae, Gomphocerinae, Scyllinini). *Boletim do Museu Nacional* (N.S.), Río de Janeiro, Vol. 380, pp. 1-10, ISSN 0080-312X.

Assis-Pujol, C.V. de (1998). Aspectos Morfológicos, Taxonómicos e Distribuição Geográfica de Cinco Espécies de *Rhammatocerus* Saussure, 1861 (Acrididae, Gomphocerinae, Scyllinini). *Boletim do Museu Nacional* (N.S, Río de Janeiro, Vol. 387, pp. 1-27, ISSN 0080-312X.

Assis-Pujol, C.V. de & Lecoq, M. (2000). Comparative study of spermathecae in eleven *Rhammatocerus* Saussure, 1861 grasshopper species (Orthoptera: Acrididae: Gomphocerinae: Scyllini). *Proceedings of the Entomological Society of Washington,* Vol. 120, No. 1, pp. 120-128, ISSN 0013-8797.

Bailey, W.J. (1991). *Acoustic behaviour of insects. An evolutionary perspective,* Chapman and Hall, ISBN 0-412-31980-2, London.

Berger, D. & Gottsberger, B. (2010). Analysis of the courtship of *Myrmeleotettix antennatus* (Fieber, 1853) - with general remarks on multimodal courtship behaviour in gomphocerine grasshoppers. *Articulata,* Vol. 25, No. 1, pp. 1-21, ISSN 0171-4090.

Blondheim, S.A. (1990). Patterns of reproductive isolation between two sibling grasshopper species *Dociostaurus curvicercus* and *D. jagoi jagoi* (Orthoptera: Acrididae: Gomphocerinae). *Transactions of the American Entomological Society,* Vol. 116, No. 1, pp. 1-64, ISSN 0002-8320.

Boero, F. (2009). Zoology in the era of biodiversity. *Italian Journal of Zoology*, Vol. 76, No. 3, pp. 239, ISSN 1125-0003

Boero, F. (2010). The study of species in the era of biodiversity: a tale of stupidity. *Diversity*, Vol. 2, pp. 115-126, ISSN 1424-2818.

Busnel, R.G. (Ed.) (1954). *Colloque sur l´Acoustique des Orthoptères,* Institut National de la Recherche Agronomique, Paris.

Cade, W. (1985). Insect mating and courtship behaviour. In: *Comprehensive insect physiology biochemistry and pharmacology,* Kerkut, G.A. & Gilbert, L.I. (Eds.), pp. 591-619, Pergamon Press, ISBN 0-08-030812-0, Oxford and New York.

Carbonell, C.S. (1995). Revision of the tribe Scyllinini, nov. (Acrididae: Gomphocerinae), with descriptions of new genera and species. *Transactions of the American Entomological Society,* vol. 121, No. 3, pp.87-152, ISSN 0002-8320.

Carbonell, C.S.; Cigliano, M.M. & Lange, C.E. (2006). *Especies de Acridomorfos (Orthoptera) de Argentina y Uruguay*, CD ROM, Publications on Orthopteran Diversity, The Orthopterist´s Society at the Museo de La Plata, ISBN 987-05-0546-5, Argentina.

Cigliano, M.M. & Lange, C.E. (1998). Orthoptera. In: *Biodiversidad de Artrópodos,* Morrone, J.J. & Coscarón, S. (Eds.), pp. 67-83,. A. Ediciones Sur, ISBN 950-9715-42-5, La Plata, Argentina.

Clemente, M.E.; García, M.D. & Presa, J.J. (1989). Estudio comparativo de la fila estriduladora de las especies de los géneros *Stenobothrus* Fischer, 1853, *Omocestus* Bolívar, 1878 y *Myrmeleotettix* Bolívar, 1914 presentes en la Península Ibérica (Orthoptera, Caelifera, Gomphocerinae). *Boletín de la Real Sociedad Española de Historia Natural* (Sección Biológica), Vol. 84, No. 3-4, pp. 343-361, ISSN 0366-3272.

Clemente, M.E.; García, M.D.; Arnaldos, M.I.; Romera, E. & Presa, J.J. (1999). Confirmación de las posiciones taxonómicas específicas de *Omocestus antigai* (Bolívar, 1897) y *Omocestus navasi* Bolívar, 1908 (Orthoptera, Acrididae). *Boletín de la Real Sociedad Española de Historia Natural* (Sección Biológica), Vol. 95, no. 3-4, pp. 27-50, ISSN 0366-3272.

C.O.P.R. (1982). *The locust and grasshopper agricultural manual,* Centre for Overseas Pest Research, ISBN 9780851351209, London.

Dumortier, B. (1963). Ethological and physiological study of sound emissions in Arthropoda. In: *Acoustic Behaviour of Animals*, Busnel, R.G. (Ed.), pp. 583-654, Elsevier, New York, Amsterdam.

Eades, D. (2001). Version 2 of the Orthoptera species file online. *Journal of Orthoptera Research*, Vol. 10, pp. 153-163. ISSN 1082-646.

Ewing, A.W. (1989). *Arthropod Bioacoustics. Neurobiology and Behaviour*, Comstock Publishing Associates, Cornell University Press, ISBN 08-014-2478X, Ithaca, New York.

Faber, A. (1953). *Laut-und Gebärdensprache bei Insekten. Orthoptera (Geradflügler). I*, Mitteilungen Staatliches Museum für Naturkunde in Stuttgart, Stuttgart.

García, M.D.; Clemente, M.E. & Presa, J.J. (1995). Manifestaciones acústicas de *Chorthippus binotatus binotatus* (Charpentier, 1825) (Orthoptera: Acrididae). Su estatus taxonómico y su distribución en la Península Ibérica. *Boletín de la Asociación española de Entomología*, Vol. 19, pp. 229-242, ISSN 0210-8984.

García, M.D.; Lorier, E.; Clemente, M.E. & Presa, J.J. (2003). Sound production in *Parapellopedon instabilis* (Rehn, 1906) (Orthoptera: Gomphocerinae). *Annales de la Société entomologique de France* (n.s), Vol. 39, No. 4, pp. 335-342, ISSN 0037-9271.

García, M.D.; Larrosa, E.; Clemente, M.E. & Presa, J.J. (2005). Contribution to the knowledge of genus *Dociostaurus* Fieber, 1853 in the Iberian Peninsula, with special reference to its sound production (Orthoptera: Acridoidea). *Anales de Biología*, Vol. 27, pp. 155- 189, ISSN 0213-3997.

Gehardt, H.C. & Huber, F, (2002). *Acoustic communication in insects and anurans*, The University of Chicago Press, ISBN 0-226-28833-1, Chicago.

Green, S.V. (1995). Song characteristics of certain Namibian grasshoppers (Orthoptera: Acrididae: Gomphocerinae). *African Entomology*, Vol. 3, pp. 1-6, ISSN 1021-3589.

Greenfield, M.D. (1997). Acoustic communication in Orthoptera. In: *The bionomics of grasshoppers, katydids and their kin*, Gangwere, S.K., Muralirangan, M.C. & Muralirangan, M. (Eds.), pp. 197-230, Cab International, ISBN 0-85199-141-6, Wallingford, United Kingdom.

Helversen, D.v. & Helversen, O.v. (1983). II.3. Species recognition and acoustic localization in acridid grasshoppers: a behavioural approach. In: *Neuroethology and behavioral physiology*, Huber, F. & Markl, H. (Eds.), pp. 95-107, Springer – Verlag, ISBN 0387126449, Berlin, Heidelberg.

Jacobs, W. (1953). Vergleichende Verhaltensstudien an Feldheuschrecken (Orthoptera, Acrididae) und einigen anderen Insekten. *Zoologische Anzeiger*, Suppl. 17, pp.115- 138, ISSN 0044-5231.

Jago, N.D. (1971). A review of the Gomphocerinae of the world, with a key to the genera (Orthoptera, Acrididae). *Proceedings of the Academy of Natural Sciences of Philadelphia*, Vol. 123, No. 8, pp. 205-343, ISSN 0097-3157.

Lecoq, M. & Assis-Pujol, C.V. (1998). Identity of *Rhammatocerus schistocercoides* (Rehn, 1906) from South and North of the Amazonian rain forest and new hypotheses on the outbreaks determinism and dynamics. *Transactions of the American Entomological Society*, Vol. 124, No. 1, pp. 13-23, ISSN 0002-8320.

Lewis, T. (Ed.) (1984). *Insect Communication*, 12 th. Symposium of the Royal Entomological Society of London, Academic Press Inc, ISBN 0-12-447175-7, London.

Loreto, V.; Cabrero, J.; López-León, M.D.; Camacho, J.P.M. & de Souza, M.J. (2008). Comparative analysis of r-DNA location in five neotropical Gomphocerinae grasshopper species. *Genetica*, Vol. 132, pp. 95-101, ISSN 0016-6707.

Lorier, E. (1996). *Estudio de la comunicación acústica en especies de Euplectrottettix (Acrididae, Gomphocerinae),* Repertorios de tesis de maestría y doctorado 1986-1994, Ministerio de Educación y Cultura, PEDECIBA, UDELAR, PNUD, Uruguay, Oficina del libro AEM, Montevideo, Uruguay.

Lorier, E.; Clemente, M.E.; García, M.D. & Presa, J.J. (2010). El comportamiento acústico de *Fenestra bohlsii* Giglio-Tos, 1895 (Orthoptera, Acrididae, Gomphocerinae). *Neotropical Entomology,* Vol. 39, No. 6, pp. 839-853, ISSN 1519-566X.

McGregor, P.K.& Peake, T.M. (2000). Comunication networks: social environments for receiving and signaling behaviour. Acta Ethologica, Vol. 2, pp. 71-81, ISSN 0873- 9749.

Nattier, R.; Robillard, T.; Amedegnato, C.; Couloux, A.; Cruaud, C. & Desutter-Grandcolas, L. (2011). Evolution of acoustic communication in the Gomphocerinae (Orthoptera: Caelifera: Acrididae). *Zoologica Scripta*, Vol. 40, No. 5, pp. 479-497, ISSN 0300-3256.

Otte, D. (1977). Comunication in Orthoptera. In: *How animals communicate,* Sebeok, T. (Ed.), pp. 334-361, Indiana University Press, ISBN 0-253-32855-1, Bloomington, Indiana.

Otte, D. (1979). Revision of the grasshopper tribe Orphulellini (Gomphocerinae, Acrididae). *Proceedings of the Academy of Natural Sciences of Philadelphia*, Vol. 131, pp. 52-58, ISSN 0097-3157.

Otte, D. & Jago, N.D. (1979). Revision of the grasshopper genera *Silvitettix* and *Compsacris* (Gomphocerinae, Acrididae). *Proceedings of the Academy of Natural Sciences of Philadelphia*, Vol. 131, pp. 257-288, ISSN 0097-3157.

Pitkin, L.M. (1976). A comparative study of the stridulatory files of the British Gomphocerinae (Orthoptera: Acrididae). *Journal of Natural History,* Vol. 10, pp.17- 28, ISSN 0022-2933.

Ragge, D.R. (1987). The songs of the western European grasshoppers of the genus *Stenobothrus* in relation to their taxonomy (Orthoptera: Acrididae). *Bulletin of the British Museum (Natural History)* Entomology series, Vol. 55, pp. 393-424, ISSN 0524- 6431.

Ragge, D.R. & Reynolds, W.J. (1998). *The song of the Grasshoppers and Crickets of Western Europe*, Harley Books, ISBN 0-946589-49-6, Colchester.

Reynolds, W.J. (1986). A description of the song of *Omocestus broelemannnni* (Orthoptera: Acrididae) with notes on its taxonomic position. *Journal of Natural History,* Vol. 20, pp. 111-116, ISSN 0022-2933.

Riede, K. (1987). A comparative study of mating behaviour in some Neotropical grasshoppers (Acridoidea). *Ethology,* Vol. 76, pp. 265-296, ISSN 0179-1613.

Salto, C.; Primo, J. & Luiselli, S. (2003). Preferencias alimenticias de *Rhammatocerus pictus* (Bruner) y *Aleuas lineatus* Stål (Orthoptera: Acrididae) en condiciones semicontroladas. INTA Rafaela. Anuario 2003. Available from: http://www.inta.gov.ar/rafaela/info/documentos/anuario2003/a2003_p148.pdf (accessed November 9, 2010).

Valdecasas, A.G. (2011). Una disciplina científica en la encrucijada: la Taxonomía. *Memorias de la Real Sociedad Española de Historia Natural, Segunda época,* Vol. 9, pp. 9-17. ISSN 1132-0869.

CHAPTER

IDENTIFICATION OF NON-LOCAL JAPANESE PINE SAWYERS IN YUNNAN, SOUTHWESTERN CHINA USING CONTEMPORARY MOLECULAR METHODS

4

1. INTRODUCTION: AN OVERVIEW

The term "biological invasion" encompasses two dimensions: it denotes the introduction of an exotic species or non-native populations to a specific geographical region, and it also pertains to the ecological and economic repercussions of such actions (Perrings et al., 2002). As socioeconomic growth and globalization progress, the concerns over biological invasions are intensifying, as certain invasive alien species can inflict significant damage on local ecosystems and economy. (Xu et al., 2006; Meyerson & Mooney, 2007).

Human beings consistently bring creatures to new ecosystems, whether intentionally or inadvertently. In instances of deliberate introduction, several species, including potatoes, maize, peanuts, and sunflowers, have been demonstrated to be advantageous for agronomic purposes (Pope et al., 2001; S.M. Wang, 2004). Nonetheless, species such as Eichhornia crassipes (Martius) have engendered a variety of issues (Harley et al., 1996; Julien, 2001; T.J. Hu et al., 2009). The repercussions of inadvertent introduction are frequently detrimental, as these species typically remain undiscovered until their populations have significantly proliferated. The incursion and proliferation of Eupatorium adenophorum Sprengel in western China exemplifies a typical case (R. Wang & Y.Z. Wang, 2006; R. Wang et al., 2011).

2. BACKGROUND INFORMATION

2.1 The Japanese pine sawyer

The Japanese pine sawyer, *Monochamus alternatus* Hope (Coleoptera: Cerambycidae), is a stem-boring beetle widely distributed in eastern Asia and the northern portion of the Indochinese peninsula (Davis et al., 2008). This polyphagous beetle feeds on conifers throughout the life history, including many unrelated species of *Pinus*, *Picea*, *Abies*, *Cedrus*, *Larix*, and *Cupressus* (Ning et al., 2004; Davis et al., 2008). *M. alternatus* is a univoltine species, which produces only one generation per year (Togashi, 1989; L.P. Wang, 2004; Zhao et al., 2004). After copulation, the female adults gnaw ovipositing wounds in the bark of host trees, and lay eggs in the space between the phloem and the xylem (Anbutsu & Togashi, 2000). The newly hatched larvae ingest wood tissue from both the phloem and the xylem, and start to excavate "U" shaped tunnels from the third instar. The final instar larvae build oval pupal chambers at the end of the tunnels to pupate. The newly eclosed adults feed on shoots, needles, and bark to obtain nutrients for maturation (Shibata, 1987). Copulation usually takes place five to ten days after emergence, each adult is able to mate more than once (L.P. Wang, 2004; H. Yang et al., 2006). As a typical secondary stem-boring species, the female adults tend to select stressed hosts for oviposition (S.J. Hu et al., 2009), which is induced by the volatile chemicals (i.e., α-pinene, β-pinene, 3-carnine, and ethanol) emitted from the hosts (Ikeda et al., 1986; Yamasaki et al., 1989), but often influenced and deterred by bark thickness, branch diameter, ovipositing scars from other female adults, and larval frass (Nakamura et al., 1995a, 1995b; Anbutsu & Togashi, 2000; Li & Z.N. Zhang, 2006; S.J. Hu et al., 2009; Z.X. Yang et al., 2010).

2.2 Association with the pine wood nematodes

The pine wood nematode, *Bursaphelenchus xylophilus* (Steiner et Buhrer) (Nematoda: Aphelenchoididae), is a quarantine phytopathogenic organism of conifers indigenous to northern America but casually spread to Eurasia and southern Japan in last century (CABI & EPPO, 1997). The nematode, like other species of genus *Bursaphelenchus*, lives in the vascular tissue of their coniferous hosts, which decreases the transportation of water and resin, and subsequently weakens the hosts and causes a syndrome named the pine wilt disease (Mamiya, 1983). The pathogenicity of *B. xylophilus* varies significantly. In North America, reports on its damages are rarely seen (Wingfield et al., 1986). However, in the vast majority of Japan and China, where *B. xylophilus* is considered as an invasive species, the infestation is often fatal, and has caused catastrophic timber loss (Mamiya, 1988; X.B. Hu et al., 1997). In nature, the relocation of *Bursaphelenchus* nematodes depends on coleopterous vectors, but the association between nematodes and vectors differs with taxa (Linit, 1988). For *B. xylophilus*, the long- horned beetles of genus *Monochamus* are the major vector (Linit et al., 1983;

Evans et al., 1996; Kulinich & Orlinskii, 1998), and research has confirmed that *M. alternatus* is the key vector of *B. xylophilus* in eastern Asia (Mamiya & Enda, 1972; Kobayashi et al., 1984). In infested hosts, the immature pine wood nematodes are able to locate the pupae of *M. alternatus* and board into the tracheae, and the newly eclosed adults carry the nematodes when seeking food and new hosts. During the course of feeding and ovipositing, the nematodes detach from the beetles and enter the new hosts through the feeding and ovipositing wounds to initiate a new round of infestation (Edwards & Linit, 1992; Naves et al., 2007).

2.3 Economic importance of *M. alternatus*

There are fundamentally four dimensions of the economic significance of M. alternatus. The stem-boring coleopteran of conifers directly destroys the tender shoots, needles, and bark of the host by adult eating and ovipositing, potentially weakening the host when a substantial population is present in a stand. The tunnels constructed by the larvae within the host frequently inflict greater harm, ultimately leading to the host's demise and significantly diminishing the timber's economic worth (L.P. Wang, 2004). M. alternatus remains in its immature stage within the host, rendering it challenging to identify within lumber and wood packaging materials, and it can be readily transported to other ecosystems through trade (Haack, 2006). Timber and wood packaging materials containing B. xylophilus can be transported (Braasch et al., 2001; Gu et al., 2006), allowing the disguised M. alternatus to subsequently disseminate the nematodes into new environments through their dispersal. Consequently, *M. alternatus* has been identified as a significant forest pest by forestry authorities globally.

2.4 The pine wilt disease in China

In mainland China, the pine wilt disease was reported for the first time in a small patch of *Pinus thunbergii* near the Sun Yat-sen Mausoleum in Nanjing, Jiangsu Province in 1982 (Y.Y. Wang et al., 1991). Afterwards, the infested area expanded quickly throughout eastern China. From the bulletins released by the SFA (State Forestry Administration) of China, a clear tendency of westward expansion can be observed (SFA, 2004 ~ 2007). During the past two years, some counties in Henan, Shaanxi, and Sichuan provinces in northern, northwestern, and southwestern China also reported infested areas of the pine wilt disease (SFA, 2011) (Fig. 1).

With the expansion of pine wilt disease, an interesting distribution pattern was revealed. In eastern and southern China, the distribution concentrated near harbors, like the deltas of the Yangtze River and the Pearl River, while in the central portion of China, the infested areas are mostly along the Yangtze River. Such a distribution pattern suggested that the dispersal of *B. xylophilus* was closely related to transportation.

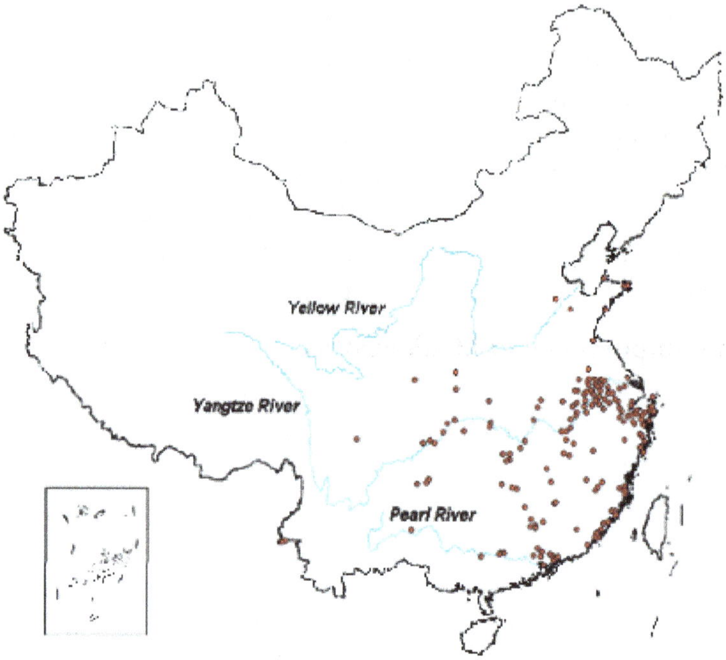

Fig. 1. The distribution pattern of the pine wilt disease (red dots) in mainland China based on the historical data from 2004 ~ 2011, data source: SFA (2004 ~ 2011).

The occurrence of the pine wilt disease in Yunnan Province, southwestern China was distinct. Not only because it was one of the latest infested areas after the massive expansion, but also due to its unique geographical position. Unlike eastern China where many harbors are distributed, the first, single, and isolated pine stand (*Pinus kesiya* var. *langbianensis*) in Yunnan infested by *B. xylophilus* was located in the westernmost corner, a remote township named Wanding near the Sino-Burmese border (SFA, 2007).

3. THE HYPOTHESIS

Considering the absence of reports of B. xylophilus in Burma (CABI & EPPO, 1997), it is illogical to conjecture that the pine wood nematode in Wanding originated from any Burmese source. Furthermore, Wanding, the sole site of infestation, is situated in the western section of the LRGR, which is considered a natural barrier impeding the expansion of organisms by natural processes (He et al., 2005). Consequently, it was illogical to conjecture that the nematodes originated from neighboring regions in Yunnan. After a thorough inspection of the initially contaminated pine stand, investigators found that the infested trees were concentrated around a telecommunication facility under construction, which utilized a significant quantity of electronics produced in eastern China (Z.Q. Li, pers. comm.). Consequently, a plausible hypothesis was formulated suggesting that the nematodes in Wanding were inadvertently introduced alongside non-native M. alternatus populations concealed under wood-packaging materials. If *M. alternatus* from various localities exhibit distinct genetic profiles, non-local individuals can be identified by examining genetic differences at the population level, and the origin can be determined by comparing the genetic profiles with samples obtained from the suspected area.

4. VALIDATING THE HYPOTHESIS – THEORY AND EXPERIMENT

4.1 Molecular achievements on *Monochamus*

Although *Monochamus* species are important forest pests across the Eurasia, research employing molecular techniques were quite limited compared to other arthropod pests. Cesari et al. (2005) started the first molecular taxonomy research on seven *Monochamus* species by using the combined mtDNA data of *cox1* and *12s* genes. Afterward, such research were quickly developed in Japan, where *M. alternatus* had become a threat to forestry. Kawai et al. (2006) published a paper on the genetic structure of 27 populations of *M. alternatus* from Japan and China (with 25 populations from Japan and two populations from China). Shoda-Kagaya (2007) published paper on microsatellite markers. Then, Koutroumpa et al. (2009) conducted a criticizing research which revealed *Numts* resembling the *cox1* and *cox2* genes of *M. galloprovincialis* and *M. sutor*, alerting scientists that precautions are necessary when applying the gene markers from the mitogenome. There was no such research on *M. alternatus* in China published by the end of 2009.

4.2 Designing the experiment

Appropriate sampling strategy requires samples to be taken from populations scattered in the entire research range, and for each population, sufficient individuals should be collected to represent it (S.Y. Chen & Y.P. Zhang, 2006). Based on this, seven populations representing

M. alternatus from central, southern, southwestern, and northern Yunnan were selected, in which three populations were chosen from the southwestern Yunnan where the pine wood nematodes were reported. In an attempt to determine the source of the pine wood nematode, two reference populations from outside Yunnan were set, with one from Hubei Province representing central China and the other from Zhejiang Province representing eastern China (Fu et al., 2010; Fig. 2). Adult *M. alternatus* were collected by flight traps baited with barkborer lure (Chinese Academy of Forestry, Zhejiang, China).

Choosing the correct genetic marker can minimize problems in data analysis (Simon et al., 1994) and maximize the future potential of cross references with data published by other researchers. Population phylogeny requires gene markers with moderate evolutionary rate, free of recombination, which means that mtDNA, a maternal heritage genome without recombination, is the ideal choice. Considering that there has been no report on *Numts* in *M. alternatus*, which can also be detected and excluded by strict data examination when encountered. The prefunding research by Kawai et al. (2006) resolved a clear population phylogeny using a gene fragment of *cox2*; the research discussed here chose the same *cox2* marker.

The same PCR primers developed by Roehrdanz (1993) and Kawai et al. (2006) were applied, and the PCR reaction was performed by the protocols described by Kawai et al. (2006). Bioinformatic and statistic software like DAMBE 5.0.7 (Xia & Xie, 2001), MEGA 4.0 (Tamura et al., 2007), SAMOVA 1.0 (Dupanloup et al., 2002), AMOVA 3.1 (Excoffier et al., 2005), and SSPS 13.0 (SPSS Inc., Illinois, US) were applied to analyze haplotype assemblage, Kimura two-parameter (K2P) distance (Kimura, 1980), NJ phylogenetic reconstruction (Saitou & Nei, 1987), population grouping, and multidimensional scaling (MDS) (Lessa, 1990). The mapping of haplotype distribution was performed in AcrView 3.3 (ESRI, USA).

Fig. 2. The sampling sites of *M. alternatus* in Yunnan, Hubei, and Zhejiang. Dots on the map of China represent the capital cities of the three provinces. After Fu et al. (2010).

5. VALIDATING THE HYPOTHESIS – MOLECULAR EVIDENCES

5.1 Haplotype assemblage

Eighteen haplotypes were defined by 22 polymorphic sites in the 565 bp *cox2* sequence, with haplotypes 1, 2, and 5 showing much higher frequencies. The distribution pattern of these 18 haplotypes was locality related: haplotypes 1 to 4 were widely distributed, dominating most of the central, northern, and southern Yunnan; haplotypes 6 to 18 were found within single localities. Haplotype 5 was shared by populations from southwestern Yunnan and Zhejiang; however, haplotypes 1 to 4 were neither shared by population from Hubei nor population from Zhejiang; and haplotypes 12 to 15 were only found in Hubei population. Further analysis of the haplotypes derived from the three populations in southwestern Yunnan discovered a high ratio of haplotype 5: with 69.2%, 70.0%, and 71.4% in Ruili, Wanding, and Lianghe populations, respectively. The matrix of shared haplotypes showed a boundary between the three populations from southwestern Yunnan and the remaining populations, with the shared haplotypes being confined within each population group and the unique haplotypes being presented only in two of the populations from southwestern Yunnan (Table 1). When mapping haplotype frequencies into pie charts, a more obvious population boundary as well as the existence of non-local individuals in three populations from southwestern Yunnan can be observed (Fig. 3).

Table 1. Matrix of shared haplotypes of the nine *M. alternatus* populations (below diagonal) and the numbers of unique haplotypes (last row). Population codes correspond to those in Fig. 2.

Population	PE	RL	WD	HN	SF	LH	YS	ZJ	HB
Pu'er									
Ruili	0								
Wanding	0	1							
Huaning	1	0	0						
Stone Forest	2	0	0	2					
Lianghe	2	1	1	1	2				
Yongsheng	1	0	0	1	1	1			
Zhejiang	0	1	1	0	0	1	0		
Hubei	0	0	0	0	0	0	0	0	
Unique haplotypes	0	3	3	0	0	0	0	3	4

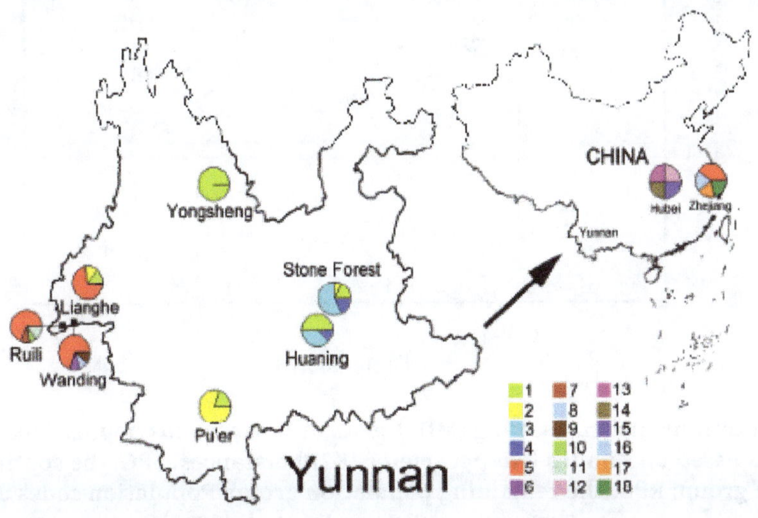

Fig. 3. Mapping of haplotype frequencies of all nine populations of *M. alternatus*.

5.2 Genetic distance

The Kimura two-parameter (K2P) distance varied from 0.0014 to 0.0132, with the distance between Pu'er (southern Yunnan) and Yongsheng (northern Yunnan) being the minimum and that between Stone Forest (central Yunnan) and Hubei being the maximum (Table 2). The K2P distances among populations from Pu'er, Huaning, Stone Forest, and Yongsheng and among populations from Ruili, Wanding, and Lianghe showed close genetic connection; but the K2P distances between any given population from Pu'er, Huanian, Stone Forest, or Yongsheng and that from Ruili, Wanding, or Lianghe suggested otherwise (Table 2). Hence, the seven populations sampled from Yunnan could be divided into two groups, hereafter designated as the SPG, the southwestern population group containing populations from Ruili, Wanding, and Lianghe, and the RPG, the remaining population group containing populations from Pu'er, Huaning, Stone Forest, and Yongsheng. The grouping result was then supported by the multidimensional scaling (MDS) analysis (Fig. 4). Notably, both the K2P distances and the MDS analysis showed a close relationship between the population from Zhejiang and all populations from the SPG, however, the relationship between the population from Hubei and any population from both the SPG and the RPG was much greater (Table 2; Fig. 4).

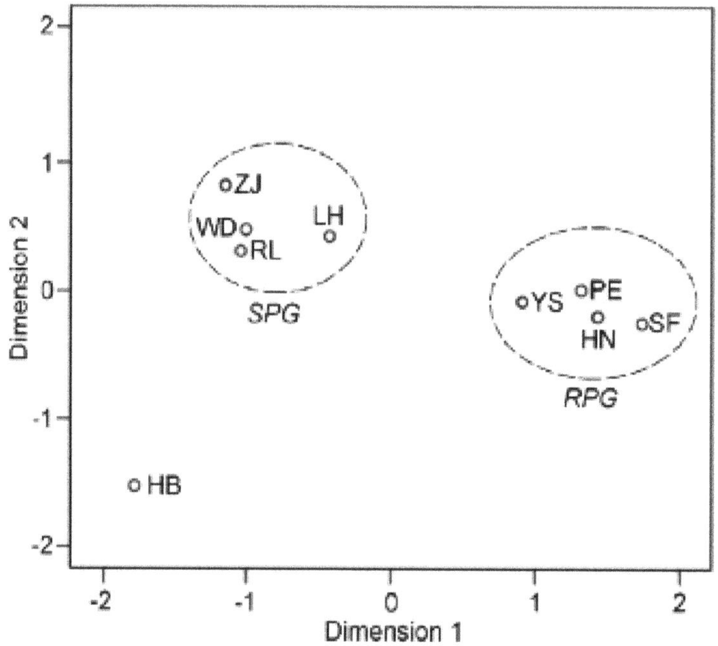

Fig. 4. The multidimensional scaling (MDS) plots of *M. alternatus* populations at different locations based on Kimura two-parameter (K2P) distances. SPG, the southwestern population group; RPG, the remaining population group. Population codes correspond to those in Fig. 2. After Fu et al. (2010).

Table 2. Matrix of the Kimura two-parameter (K2P) distance (below diagonal) and standard errors (above diagonal) of between populations of *M. alternatus*. Population codes correspond to those in Fig. 2. Data source: Fu et al. (2010).

Site	PE	RL	WD	HN	SF	LH	YS	ZJ	HB
PE		0.0037	0.0035	0.0012	0.0015	0.0028	0.0013	0.0037	0.0040
RL	0.0084		0.0007	0.0036	0.0040	0.0012	0.0034	0.0008	0.0024
WD	0.0084	0.0018		0.0035	0.0038	0.0012	0.0032	0.0009	0.0024
HN	0.0020	0.0090	0.0090		0.0013	0.0029	0.0012	0.0037	0.0040
SF	0.0022	0.0101	0.0101	0.0022		0.0032	0.0020	0.0040	0.0042
LH	0.0064	0.0027	0.0031	0.0071	0.0081		0.0026	0.0013	0.0025
YS	0.0014	0.0070	0.0070	0.0020	0.0031	0.0053		0.0034	0.0038
ZJ	0.0093	0.0021	0.0026	0.0098	0.0109	0.0035	0.0079		0.0027
HB	0.0122	0.0070	0.0077	0.0123	0.0132	0.0084	0.0107	0.0086	

5.2 Phylogenetic cladistics

The neighbor-joining (NJ) phylogenetic tree based on the K2P distance of the 18 haplotypes resolved three major clades designated from A to C (Fig. 5). Clade A contained six populations from Yunnan, with four from northern, central, and southern portion, and only one sample from Wanding, southwestern portion was included. Clade B contained the three populations from southwestern Yunnan and the population from Zhejiang, eastern China.

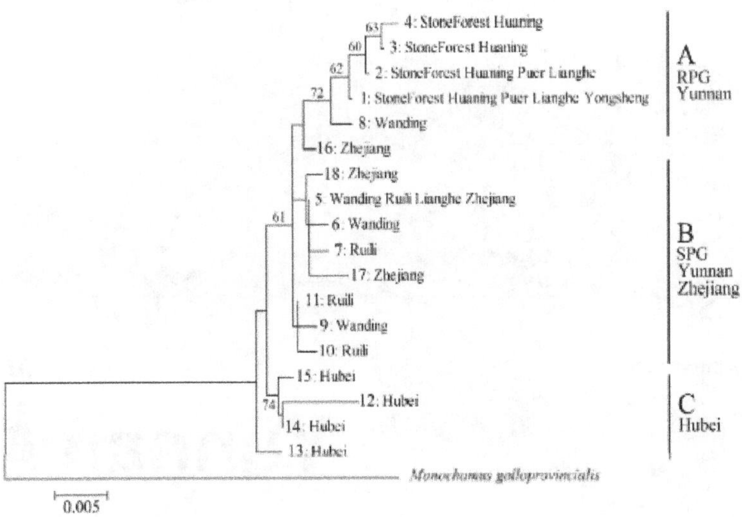

Fig. 5. The neighbor-joining (NJ) phylogenetic tree of the 18 haplotypes of *M. alternatus*, numbers above the branches represent the bootstrap values (> 50). After Fu et al. (2010).

However, no sample from other portions of Yunnan grouped in clade B. Clade C only contained the population from Hubei, central China. The topological structure of the

NJ phylogenetic tree suggests that the seven populations consisted of two groups, the RPG and the SPG, regardless of the connection between them by a few samples from Wanding and Lianghe. It also suggested that the genetic profile of the population from Hubei is different from the remaining ones involved in the research.

5.3 Population genetic structure

SAMOVA and AMOVA analyses were applied to verify the genetic structure of the seven populations from Yunnan. When the number of groups (K) grows and the F_{CT} value reaches the plateau, the SAMOVA analysis yields the optimal result. In this research, the F_{CT} value reached the maximum (F_{CT} = ٠,٧٣) when $K = 2$, with populations from Pu'er, Huaning, Stone Forest, and Yongsheng being one group and populations from Ruili, Wanding, and Lianghe being the other (Fig. 6), supporting the grouping result mentioned previously.

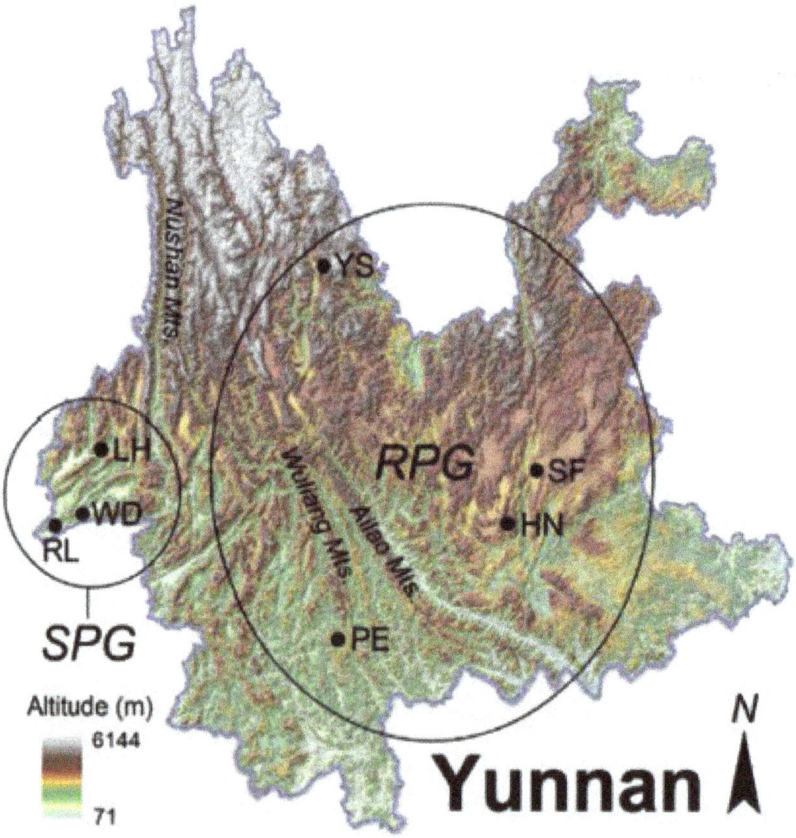

Fig. 6. The terrain of Yunnan and the grouping result of seven populations of *M. alternatus* determined by SAMOVA and AMOVA analyses with major geographical barriers marked (Ailao Mountains, Wuliang Mountains, and Nushan Mountains). SPG, the southwestern population group; RPG, the remaining population group. Population codes correspond to those in Fig. 2.

The grouping was then tested by the AMOVA analysis, which indicated that 72.54% of the variation occurred between the two groups, 21.86% occurred within populations, and only 5.60% occurred among populations within groups (Table 3). The relatively high variation within populations was also mirrored by the haplotype assemblage that most of the populations possessed multiple haplotypes, especially the three populations from southwestern Yunnan rich in unique haplotypes (Table 1; Fig. 3).

Table 3. AMOVA analysis results. Data source: Fu et al. (2010).

Source of variation	d.f.	Variance components	Percentage of variation	F	P
Between groups	1	1.725	72.54	FCT = 0.73	0.025
Among populations within groups	5	0.133	5.60	FSC = 0.21	0.001
Within populations	54	0.520	21.86	FST = 0.78	0.000
Total	60	2.378	100	--	--

6. IDENTIFICATION OF NON-LOCAL *M. ALTERNATUS*

The experiment provided molecular evidence strongly supporting the following two hypotheses: 1) the genetic structure divides the seven *M. alternatus* populations from Yunnan into two groups, the southwestern population group (SPG) and the remaining population group (RPG), and 2) it is most likely that there were non-local individuals of *M. alternatus* in the three populations sampled from southwestern Yunnan, as these populations shared one of the dominant haplotypes of the reference population from Zhejiang. However, in an attempt to finally determine the identity of non-local *M. alternatus*, the relationship between such genetic divergence and the natural geographical characteristics of Yunnan should be considered thoroughly.

Genetic divergence has been frequently discussed in terms of natural geographical barriers (Yagi et al., 2001; Shoda-Kagaya, 2007; Shi et al., 2010). Yunnan is a typical LLPR with complex terrain, where numerous population phylogenetic studies on different insects have revealed the role that geographical barriers play in obstructing the gene flow among populations from various localities (Shi & Ye, 2004; Liu et al., 2007; P. Chen & Ye, 2008; Shi et al., 2010). Basically, the mountain ranges in the LRGR portion of Yunnan are considered effective geographical barriers, which restrict gene flow in the latitudinal direction.

Since the natural environment of Yunnan is able to cause genetic divergence, the separation of two population groups could be either the result of natural geographical barriers or the consequence of introduced non-local sources. The seven sampling sites in Yunnan were separated by great mountain ranges. Ruili, Wanding, and Lianghe were separated from other populations by Nushan Mountains, Pu'er was separated from Stone Forest, Huaning, and Yongsheng by Wuliang and Ailao Mountains (Fig. 6). But it is interesting to note that the genetic distances among populations from the RPG were limited to 0.0014 to 0.0031 (Table 2), the while genetic variation among them comprised only 5.6% of the total divergence (Table 3), and they shared all of the haplotypes defined by the samples taken from them (Table 1; Fig. 3). Hence, the geographical barriers such as Wuliang and Ailao Mountains had not caused sufficient genetic divergence among these populations. Judging from the phylogenetic tree (Fig. 5), these four populations were latest diverged, which may represent a short inhabitation history of *M. alternatus* in Yunnan under natural condition. To the contrast, the genetic distances between populations from the SPG and the RPG were much higher than that within each group, varied from 0.0064 to 0.0101 (Table 2), the genetic variation reached 72.54% (Table 3), and most of the unique haplotypes were defined (Table 1). Given the discussion above, the geographical barriers can not explain such divergence.

Assuming that the SPG and the RPG were two genetically independent sources, there must be gene flows among these populations in the course of evolution, which will be detected by shared haplotypes. It is noticeable that population from Lianghe shared two haplotypes with populations from the RPG; however, no haplotype from the SPG was shared by any other populations. Given that the individuals from the RPG were able to "migrate" into the SPG, the same phenomenon should have happened the other way around. Another assumption was that the genetic divergence between the SGP and the RPG was caused by introduction of non-local *M. alternatus*. The haplotype distribution demonstrated that about 90% individuals of the southwestern populations shared haplotype 5 with population from Zhejiang (Table 1; Fig. 3), and the phylogenetic tree indicated a close genetic connection between haplotype 5 and those unique haplotypes (Fig. 5), which had been regarded as a symbol of establishment of non-local populations (Slatkin & Hudson, 1991; J. Hu et al., 2008).

Pine wilt disease in Wanding was observed four years following the introduction of electronics with wood packaging materials from eastern China, where the disease was most prevalent. Given the traits of the immature stages of M. alternatus and its relationship with B. xylophilus, it is plausible that the untreated, infested wood packaging materials discarded carelessly near the building site served as the source of infestation. This will not only elucidate the considerable genetic diversity between the SPG and the RPG in Yunnan, but also clarify the genetic relationship between the SPG and the population from Zhejiang rather than Hubei.

7. CONCISE SUMMARY

The study demonstrated the presence of non-local M. alternatus in southwestern Yunnan by the application of contemporary molecular techniques. This study is the first, if not the exclusive, investigation of the population phylogeny of Monochamus longhorn beetles in China, and serves as an additional example of employing DNA tools in the study of biological invasion. In addition to illustrating the viability of identifying non-local populations, it revealed that, unlike the dispersal of other invasive alien species, such as fruit flies, the proliferation of pine wood nematodes in mainland China primarily resulted from human-assisted transport of infested, untreated wood packaging materials. Consequently, restricting the movement or distribution of contaminated wood components may emerge as the paramount measure to inhibit the proliferation of pine wilt disease.

8. PROSPECTIVE INVESTIGATIONS AND AIMS

The research presented herein utilized contemporary molecular techniques that yielded compelling evidence demonstrating that the dispersal of pine wood nematodes in southwestern Yunnan was instigated by the incidental introduction of the non-native Japanese pine sawyer, Monochamus alternatus. Furthermore, it facilitated the identification of the potential origin of these non-local M. alternatus by integrating molecular data with historical information regarding telecommunication infrastructure development in the affected region. Contemporary technology enables scientists to directly detect pine wood nematodes from wood samples or insects (X.R. Wang et al., 2010; X.R. Wang et al., 2010; Y.Q. Hu et al., 2011). Through the application of modern techniques, foresters and plant quarantine personnel can transform the identification of invasive organisms from mere conjecture to concrete evidence. Advocating for this concept in conjunction with contemporary molecular tools would not only aid future law enforcement but also effectively address regional and worldwide trade embargoes related to quarantine and invasive organisms. In addition to identifying non-local populations, population phylogeny can be utilized to examine the historical dispersal of insects, encompassing both natural and anthropogenic migration, as indicated by prior research (Kawai et al., 2006; Shoda-Kagaya, 2007; Shi et al., 2010). Given China's 28-year history of pine wilt disease propagation, the dispersal patterns of the pine wood nematode and its principal vector, M. alternatus, warrant thorough investigation. To date, there exists only one publication regarding the dispersal pattern of nematodes in China (Sun et al., 2008). The senior author and his colleagues are presently undertaking a comprehensive study on the population phylogeny of *M. alternatus* collected from several locations in mainland China. This research partially addresses the longstanding inquiry concerning the distribution pattern of *M. alternatus* in relation to the proliferation of pine wilt disease in China. With the capability to concurrently analyze the genetic profiles of B. xylophilus and M. alternatus, forthcoming research can elucidate the dispersal patterns and originating sources of pine wood nematodes in China with more precision.

9. REFERENCES

Anbutsu, H. & Togashi, K. (2000). Deterred ovipostion response of *Monochamus alternatus* (Coleoptera: Cerambycidae) to oviposition scars occupied by eggs. *Agricultural and Forest Entomology*, 2 (3): 217-223.

Braasch, H., Tomiczek, C., Metge, K., Hoyer, U., Burgermeister, W., Wulfert, I., & Schönfeld, U. (2001). Records of *Bursaphelenchus* spp. (Nematoda, Parasitaphelenchidae) in coniferous timber imported from the Asian part of Russia. *Forest Pathology*, 31 (3): 129-140.

CABI & EPPO. (1997). *Quarantine Pests for Europe (2nd edition).* Wallingford: CAB International.

Cesari, M., Marescalchi, O., Francardi, V., & Mantovani, B. (2005). Taxonomy and phylogeny of European *Monochamus* species: first molecular and karyological data. *Journal of Zoological Systematics and Evolutionary Research*, 43 (1): 1-7.

Chen, P. & Ye, H. (2008). Relationship among five populations of *Bactrocera dorsalis* based on mitochondrial DNA sequences in western Yunnan, China. *Journal of Applied Entomology*, 132 (7): 530-537.

Chen, S.Y. & Zhang, Y.P. (2006). The genetic approaches and applications in the research of the origin of domesticated animals. *Chinese Science Bulletin*, 51 (21): 2469-2475. [in Chinese]

Chen, Z.Y. (2001). *The Pandect of Yunnan Climate.* Beijing: China Meteorological Press. [in Chinese]

Davis, E.E., Albercht, E.M., & Venette, R.C. (2008). *Monochamus alternatus.* In: Venette, R.C. *Exotic Pine Pests: Survey Reference*, USDA Forest Service.

Dupanloup, I., Schneider, S., & Excoffier, L. (2002). A simulated annealing approach to define the genetic structure of populations. *Molecular Ecology*, 11 (12): 2571-2581.

Edwards, O.R. & Linit, M.J. (1992). Transmission of *Bursaphelenchus xylophilus* through oviposition wounds of *Monochamus carolinensis* (Coleoptera: Cerambycidae). *Journal of Nematology*, 24 (1): 133-139.

Evans, H.F., McNamara, D.G., Braasch, H., Chadoeuf, J., & Magnusson, C. (1996). Pest risk analysis for the territories of the European Union (as the PRA area) on *Bursaphelenchus xylophilus* and its vectors in the genus *Monochamus. EPPO Bulletin*, 26: 199-249.

Excoffier, L., Lava, G., & Schneider, S. (2005). Arlequin ver. 3.0: an integrated software package for population genetics data analysis. *Evolutionary Bioinformatics Online*, 1: 47-50.

Fu, D.Y., Hu, S.J., Ye, H., Haack, R.A., & Zhou, P.Y. (2010). Pine wilt disease in Yunnan, China: evidence of non-local pine sawyer *Monochamus alternatus* (Coleoptera: Cerambycidae) population revealed by mitochondrial DNA. *Insect Science*, 17 (5): 439-447.

Gu, J., Braasch, H., Burgermeister, W., & Zhang, J. (2006). Records of *Bursaphelenchus* spp. intercepted in imported packaging wood at Ningbo, China. *Forest Pathology*, 36 (5): 323-333.

Guo, H.J. and Long, C.L. (1998). *The Biodiversity of Yunnan*. Kunming: Yunnan Science & Technology Press: 1-10. [in Chinese]

Haack, R.A. (2006). Exotic bark- and wood-boring Coleoptera in the United States: recent establishments and interceptions. *Canadian Journal of Forest Research*, 36 (2): 269-288.

Harley, K.L.S., Julien, M.H., & Wright, A.D. (1996). Water hyacinth: a tropical worldwide problem and methods for its control. In: Cussans, B.H., Devine, G.W., Duke, M.D., Fernandez-Quintanilla, S.O., Helweg, C., Labrada, A., Landes, R.E., Kudsk, M., & Streibig, P. (eds.) *Proceedings of the 2nd International Weed Control Congress*. Copenhagen: Slagelse: 630-644.

He, D.M., Wu, S.H., Peng, H., Yang, Z.F., Ou, X.K., and Cui, B.S. (2005). A study of ecosystem changes in Longitudinal Range-Gorge Region and transboundary eco- security in Southwest China. *Advances in Earth Science*, 20 (3): 932-943. [in Chinese with English abstract]

Hu, J., Zhang, J.L., Nardi, F., & Zhang, R.J. (2008). Population genetic structure of the melon fly, *Bactrocera cucurbitae* (Diptera: Tephritidae), from China and Southeast Asia. *Genetica*, 134 (3): 319-324.

Hu, S.J., Fu, D.Y., Li, Z.Q. & Ye, H. (2009). Factors effecting the distribution of the oviposition scars of *Monochamus alternatus* in *Pinus kesiya* var. *langbianensis*. *Forest Pest and Disease*, 6 (4): 1-2, 11. [in Chinese with English abstract]

Hu, T.J., Wang, Y.C., & Lian, Q.P. (2009). Advances on hazards and controls of common water hyacinth, *Eichhornia crassipes*. *Fisheries Science and Technology*, 25 (6): 27-30. [in Chinese with English abstract]

Hu, X.B., Qu, T., & Zheng, H. (1997). On the control strategies of pine wilt disease in China. *Forest Pest and Disease*, 16 (3): 30-32. [in Chinese]

Hu, Y.Q., Kong, X.C., Wang, X.R., Zhong, T.K., Zhu, X.W., Mota, M.M., Ren, L.L., Liu, S., &

Cai, M. (2011) Direct PCR-based method for detecting *Bursaphelenchus xylophilus*, the pine wood nematode in wood tissue of *Pinus massoniana*. *Forest Pathology*, 41 (2): 165-168.

Ikeda, T., Yamane, A., Enda, N., Oda, K., Makihara, H., Ito, K., & Ôkochi, I. (1986). Attractiveness of volatile components of felled pine trees for *Monochamus alternatus* (Coleoptera: Cerambycidae). *Journal of the Japanese Forestry Society*, 68 (1): 15-19.

Julien, M.H. (2001). Biological control of water hyacinth with arthropods: a review to 2000. In: Julien, M.H., Hill, M.P., Center, T.D., and Ding, J.P. (eds). *ACIAR Proceedings*, 102: 8-20.

Kawai, M., Shoda-Kagaya, E., Maehara, T., Zhou, Z.L., Lian, C.L., Iwata, R., Yamane, A., & Hogetsu, T. (2006). Genetic structure of pine sawyer *Monochamus alternatus* (Coleoptera: Cerambycidae) populations in Northeast Asia: consequences of the spread of pine wilt disease. *Environmental Entomology*, 35 (2): 569-579.

Kimura, M. (1980). A simple method for estimating evolutionary rates of base substitutions through comparative studies of nucleotide sequences. *Journal of Molecular Evolution*, 16 (2): 111-120.

Kobayashi, F., Yamane, A., & Ikeda, T. (1984). The Japanese pine sawyer beetle as the vector of pine wilt disease. *Annual Review of Entomology*, 29: 115-135.

Koutroumpa, F.A., Lieutier, F., & Roux-Morabito, G. (2009). Incorporation of mitochondrial fragments in the nuclear genome (*Numts*) of the longhorned beetle *Monochamus galloprovincialis* (Coleoptera, Cerambycidae). *Journal of Zoological Systematics and Evolution Research*, 47 (2): 141-148.

Kulinich, O.A. & Orlinskii, P.D. (1998). Distribution of conifer beetles (Scolytidae, Curculionidae, Cerambycidae) and wood nematodes (*Bursaphelenchus* spp.) in European and Asian Russia. *EPPO Bulletin*, 28: 39-52.

Lessa, E.P. (1990). Multidimensional analysis of geographical genetic structure. *Systematic Zoology*, 39 (3): 242-252.

Li, S.Q. & Zhang, Z.N. (2006). Influence of larval frass extracts on the oviposition behaviour of *Monochamus alternatus* (Col., Cerambycidae). *Journal of Applied Entomology*, 130 (3): 177-182.

Linit, M.J. (1988). Nematode-vector relationships in the pine wilt disease system. *Journal of Nematology*, 20 (2): 227-235.

Linit, M.J., Kondo, E., & Smith, M.T. (1983). Insects associated with the pinewood nematode, *Bursaphelenchus xylophilus* (Nematoda: Aphelenchoididae), in Missouri. *Environmental Entomology*, 12 (2): 467-470.

Liu, J.H., Shi, W., and Ye, H. (2007). Population genetics analysis of the origin of the oriental fruit fly, *Bactrocera dorsalis* Hendel (Diptera: Tephritidae), in northern Yunnan Province, China. *Entomological Science*, 10 (1): 11-19.

Mamiya, Y. (1983). Pathology of pine wilt disease caused by *Bursaphelenchus xylophilus*. *Annual Review of Phytopathology*, 21: 201-220.

Mamiya, Y. (1988). History of pine wilt disease in Japan. *Journal of Nematology*, 20 (2): 219- 226.

Mamiya, Y. & Enda, N. (1972). Transmission of *Bursaphelenchus lignicolus* (Nematoda: Aphelenchoididae) by *Monochamus alternatus* (Coleoptera: Cerambycidae). *Nematologica*, 18 (2): 159-162.

Meyerson, L.A. & Mooney, H.A. (2007). Invasive alien species in an era of globalization. *Frontiers in Ecology and the Environment*, 5 (4): 199-208.

Nakamura, H., Tsutsui, N., & Okamoto, H. (1995a). Oviposition habit of the Japanese pine sawyer, *Monochamus alternatus* Hope (Coleoptera, Cerambycidae) I. Factors affecting the vertical distribution of oviposition scars in a pine tree. *Japanese Journal of Entomology*, 63 (3): 633-640.

Nakamura, H., Tsutsui, N., & Okamoto, H. (1995b). Oviposition habit of the Japanese pine sawyer, *Monochamus alternatus* Hope (Coleoptera, Cerambycidae) II. Effect of bark thickness on making oviposition scars. *Japanese Journal of Entomology*, 63 (4): 739- 745.

Naves, P.M., Camacho, S., de Sousa, E.M., & Quartau, J.A. (2007). Transmission of the pine wood nematode *Bursaphelenchus xylophilus* through feeding activity of *Monochamus galloprovincialis* (Col., Cerambycidae). *Journal of Applied Entomology*, 131 (1): 21-25.

Ning, T., Fang, L.Y., Tang, J., & Sun, J.H. (2004). Advances in research on *Bursaphelenchus xylophilus* and its key vector *Monochamus* spp. *Entomological Knowledge*, 41 (2): 97-104. [in Chinese with English abstract]

Perrings, C., Williamson, M., Barbier, E.B., Delfino, D., Dalmazzone, S., Shogren, J., Simmons, P., & Watkinson, A. (2002). Biological invasion risks and the public good: an economic perspective. *Conservation Ecology*, 6 (1): 1-7.

Pope, K.O., Pohl, M.E.D., Jones, J.G., Lentz, D.L., von Nagy, C., Vega, F.J., & Quitmyer, I.R. (2001). Origin and environmental setting of ancient agriculture in the lowlands of Mesoamerica. *Science*, 292 (5520): 1370-1373.

Roehrdanz, R.L. (1993). An improved primer for PCR amplification of mitochondrial DNA in a variety of insect species. *Insect Molecular Biology*, 2 (2): 89-91.

Saitou, N. & Nei, M. (1987). The neighbor-joining method: a new method for reconstructing phylogenetic trees. *Molecular Biology and Evolution*, 4 (4): 406-425.

SFA (State Forestry Administration). (2004). *The Bulletin of State Forestry Administration of P.R. China, (2004-2).* http://www.forestry.gov.cn/distribution/2004/04/15/zwgk- 2004-04-15-5965.html. [in Chinese]

SFA (State Forestry Administration). (2005). *The Bulletin of State Forestry Administration of P. R. China, (2005-2).* http://www.forestry.gov.cn/distribution/2005/01/31/zwgk- 2005-01-31-5962.html. [in Chinese]

SFA (State Forestry Administration). (2006). *The Bulletin of State Forestry Administration of P. R. China, (2006-1).* http://www.forestry.gov.cn/distribution/2006/02/22/zwgk- 2006-02-22-5960.html. [in Chinese]

SFA (State Forestry Administration). (2007). *The Bulletin of State Forestry Administration of P. R. China, (2007-4).* http://www.forestry.gov.cn/distribution/2007/07/11/zwgk- 2007-07-11-5949.html. [in Chinese]

SFA (State Forestry Administration). (2009). *The Bulletin of State Forestry Administration of P. R. China, (2009-3).* http://www.forestry.gov.cn/portal/main/govfile/13/govfile_ 1635.html [in Chinese]

SFA (State Forestry Administration). (2011). *The Bulletin of State Forestry Administration of P.R. China. (2011-2).* http://www.forestry.gov.cn/portal/main/govfile/13/govfile_ 1792.htm. [in Chinese]

Shi, W., Kerdelhué, C., & Ye, H. (2010). Population genetic structure of the oriental fruit fly, *Bactrocera dorsalis* (Hendel) (Diptera: Tephritidae) from Yunnan province (China) and nearby sites across the border. *Genetica*, 138: 377-385.

Shi, W. & Ye, H. (2004). Genetic differentiation in five geographic populations of oriental fruit fly, *Bactrocera dorsalis* (Hendel) (Diptera: Tephritidae) in Yunnan province. *Acta Entomologica Sinica*, 47 (3): 384-388. [in Chinese with English abstract]

Shibata, E. (1987). Oviposition schedules, survivorship curves, and mortality factors within trees of two cerambycid beetles (Coleoptera: Cerambycidae), the Japanese pine sawyer, *Monochamus alternatus* Hope, and sugi bark borer, *Semanotus japonicus* Lacordaire. *Researches on Population Ecology*, 29 (2): 347-367.

Shoda-Kagaya, E. (2007). Genetic differentiation of the pine wilt disease vector *Monochamus alternatus* (Coloeptera: Cerambycidae) over a mountain range --- revealed from microsatellite DNA markers. *Bulletin of Entomological Research*, 97 (2): 167-174.

Simon, C., Francesco, F., Beckenbach, A., Crespi, B., Liu, H., & Flook, P. (1994). Evolution, weighting, and phylogenetic utility of mitochondrial gene sequences and a compilation of conserved polymerase chain reaction primers. *Annals of the Entomological Society of America*, 87 (6): 651-701.

Slatkin, M. & Hudson, R.R. (1991). Pairwise comparisons of mitochondrial DNA sequences in stable and exponentially growing populations. Genetics, 129 (2): 555-562.

Sun, J., Yang, S.Y., Cui, C.L., Zhang, C.X., Lin, M.S., & Zhang, K.Y. (2008). Possible transmission routes of *Bursaphelenchus xylophilus* in China based on molecular data. *Journal of Nanjing Agricultural University*, 31 (2): 55-60. [in Chinese with English abstract]

Tamura, K., Dudley, J., Nei, M., & Kumar, S. (2007). MEGA 4: Molecular Evolutionary Genetics Analysis (MEGA) Software Version 4.0. *Molecular Biology and Evolution*, 24 (8): 1596-1599.

Togashi, K. (1989). Development of *Monochamus alternatus* Hope (Coleoptera: Cerambycidae) in relation to oviposition time. *Japanese Journal of Applied Entomology and Zoology*, 33 (1): 1-8.

Wang, L.P. (2004). Study on the biological characteristic of *Monochamus alternatus* Hope. *Journal of Fujian Forestry Science and Technology*, 31 (3): 23-26. [in Chinese with English abstract]

Wang, R., Wang, J.F., Qiu, Z.J., Meng, B., Wan, F.H., & Wang, Y.Z. (2011). Multiple mechanisms underlie rapid expansion of an invasive alien plant. *New Phytologist*, 191 (3): 828-839.

Wang, R. & Wang, Y.Z. (2006). Invasion dynamics and potential spread of the invasive alien plant species *Ageratina adenophora* (Asteraceae) in China. Diversities and Distributions 12: 397–408.

Wang, S.M. (2004). Introduction of the American-originated crops and its influence on the Chinese agricultural production structure. *Agricultural History of China*, 24 (2): 16-27. [in Chinese with English abstract]

Wang, S.Y. & Zhang, W. (2005). *Yunnan Geography*. Kunming: The Nationalities Publishing House of Yunnan. [in Chinese]

Wang, X.R., Kong, X.C., Jia, W.H., Zhu, X.W., Ren, L.L., & Mota, M.M. (2010). A rapid staining-assisted wood sampling method for PCR-based detection of pine wood nematode *Bursaphelenchus xylophilus* in *Pinus massoniana* wood tissue. *Forest Pathology*, 40 (6): 510-520.

Wang, X.R., Zhu, X.W., Kong, X.C., & Mota, M.M. (2010). A rapid detection of the pinewood nematode, *Bursaphelenchus xylophilus* in stored *Monochamus alternatus* by rDNA amplification. *Journal of Applied Entomology*, 135 (1/2): 156-159.

Wang, Y.Y., Song, Y.S., Zang, X.Q., & Liu, Y. (1991). The ten-year review and future control measures of the pine wilt disease in China. *Forest Pest and Disease*, 10 (3): 39-42. [in Chinese]

Wingfield, M.J., Bedker, P.J., & Blanchette, R.A. (1986). Pathogenicity of *Bursaphelenchus xylophilus* on pines in Minnesota, Wisconsin. *Journal of Nematology*, 18 (1): 44-49.

Xia, X.H. & Xie, Z. (2001). DAMBE: Data analysis in molecular biology and evolution. *Journal of Heredity*, 92 (4): 371-373.

Xu, H.G., Ding, H., Li, M.Y., Qiang, S., Guo, J.Y., Han, Z.M., Huang, Z.G., Sun, H.Y., He, S.P., Wu, H.R., & Wan, F.H. (2006). The distribution and economic losses of alien species invasion to China. *Biological Invasions*, 8: 1495-1500.

Yagi, K., Katoh, T., Chichvarkhin, A., Shinkawa, T., & Omoto, T. (2001). Molecular phylogeny of butterflies *Parnassius glacialis* and *P. stubbendorfii* at various localities in East Asia. *Genes & Genetic Systems*, 76 (4): 229-234.

Yamasaki, T., Sakai, M., & Miyawaki, S. (1989). Oviposition stimulants for the beetle, *Monochamus alternatus* Hope, in inner bark of pine. *Journal of Chemical Ecology*, 15 (2): 507-516.

Yang, H., Wang, J.J., Zhao, Z.M., Yang, D.M., & Zhang, H. (2006). Effects of multiple mating on quantitative depletion of spermatozoa, fecundity, and hatchability in *Monochamus alternatus*. *Zoological Research*, 27 (3): 286-290. [in Chinese with English abstract]

Yang, Z.X., Wang, J.M., Chen, X.M., Duan, Z.Y., & Ye, S.D. (2010). The vertical distribution characteristics of *Monochamus alternatus* on *Pinus yunnanensis* trunks. *Forest Research*, 23 (4): 607-611. [in Chinese with English abstract]

Zhao, Y.X., Dong, Y., & Xu, Z.H. (2004). Bionomics and geographical distribution of *Monochamus alternatus* Hope (Coleoptera: Cerambycidae) in Yunnan Province. *Forest Pest and Disease*, 23 (5): 13-16. [in Chinese with English abstract]

CHAPTER 5

DEVELOPMENT OF A SIMULATION MODEL FOR THE SPREAD OF CITRUS GREENING DISEASE BY THE VECTOR INSECT DIAPHORINA CITRI ON AN INDIVIDUAL SCALE

1. INTRODUCTION: AN OVERVIEW

In order to implement an Integrated nuisance Management (IPM) system for crop nuisance insects and disease, it is imperative to conduct a risk assessment of pertinent obstacles. The data for this type of risk assessment analysis is typically obtained through field surveys or experiments. Nevertheless, the evaluation of the interaction between significant factors frequently necessitates clarification. Modelling serves as a foundation for the theoretical description of this interaction, which is based on field data and relevant simulations. We have created a simulation model that estimates the transmission of citrus greening disease (Huanglongbing, HLB) by the vector insect Diaphorina citri Kuwayama. Subsequently, we utilized the model to inform our risk assessment by analyzing the parameters that influence the disease's dissemination dynamics. We focused on the Mekong Delta region of Vietnam, which has been significantly impacted by the spread of HLB. At the outset of this chapter, we describe the vector insect and the disease. Subsequently, we delineate the methodology for the development of the HLB disease-spread estimation model. Lastly, we offer recommendations to mitigate the transmission of HLB, which are predicated on the results of the disease-spread estimation in a variety of scenarios.

1.1 Citrus greening disease

Citrus greening disease is a limiting factor in citrus production worldwide (Bové, 2006). The symptoms of HLB are similar to those of nutritional stress (Halbert & Manjunath, 2004). A survey conducted over an eight-year period in Réunion Island, for example, indicated that 65% of the trees were badly damaged and rendered unproductive within seven years of planting (Aubert et al., 1996). In Thailand, citrus trees generally decline within 5-8 years after planting due to HLB (Roistacher, 1996). In his compilation of global infection statistics, Toorawa (1998) estimated that 50 million trees were infected in South and Southeast Asia, and 3 million were infected in Africa. In India and Saudi Arabia, there has been a marked decline in the citrus industry as a result of HLB.

The pathogens are phloem-inhabiting bacteria in the generous *Candidatus* Liberibacter (Halbert et al., 2004). Although these bacteria have hitherto not been sufficiently cultured for the application of Koch's postulates, some experimental results have strongly suggested that they are the pathogen in HLB (Su et al., 1986; Buitendag & von Broembsen, 1993). There are two principal means of transmission for a healthy tree: graft transmission, whose frequency has been estimated in many previous studies (e.g., Lin & Lin, 1990; van Vuuren, 1993), but with widely varying values; and transmission through vector insects. In Asian countries, HLB is borne by the Asian citrus psyllid, *D. citri*.

Detection of the HLB pathogen has been achieved by several methods. DNA identification through the PCR method was used to detect the bacteria both in citrus plants and vector insects (Bové et al., 1993; Tian et al., 1996). Wang et al. (2006) conducted a study which developed a reproducible conventional PCR method with several primer sets, and two quantitative real-time PCR methods for detection and monitoring of the pathogen. The HLB pathogen also can be detected with an electron microscope, ELISA (Garnier & Bové, 1993).

1.2 Life cycles of the vector insect and the disease

The *D. citri* is a Hemipteran insect, measuring 3 to 4 mm in length, with piercing-sucking mouthparts that allow this pest to feed on the phloem of citrus spp. and other related rutaceous plants. The eggs of *D. citri* are laid on the new leaf growth of expanding terminals, in the folds of unfurled leaves, and behind developing leaf buds (Chavan & Summanwar, 1993). There are five nymphal instars (Aubert, 1987). Adults may live several months, and females may lay as many as 800 eggs in a lifetime, under artificial rearing conditions (Mead, 1977); however, the longevity and fecundity in actual field conditions are not well known. Temperature-dependent development of *D. citri* has been estimated (Liu & Tsai, 2000; Yang et al., 2006; Nakata, 2006), and the results reveal a relatively consistent trend, even though the host plants used in the experiments varied. In general, nymphs grew faster at higher temperatures, except for 32.5 °C. Nakata (2006) estimated the developmental zero and effective accumulative temperature of the egg as 13.7°C and 46.9 degree-days, respectively. The developmental zero and effective accumulative temperature of the nymphs were 11.6°C and 192.3 degree-days, respectively (Nakata, 2006).

Fourth and fifth instar nymphs and adults of *D. citri* can acquire the pathogen, and emerged adults that have fed on infected plants as nymphs can transmit the pathogen to healthy plants (Capoor et al., 1974; Xu et al., 1988). Once the bacterium is acquired, the psyllid will retain and transmit the bacterium throughout its life. Multiplication of the pathogen in individual *D. citri* was investigated by Inoue et al. (2009). The efficiency of transmission to healthy plants by the virulent adult *D. citri* was estimated in several studies. The virulency of vectors grown on infected trees, and the transmission rate of virulent vectors to healthy host trees, were different in each report. Inoue et al. (2009) estimated a successful transmission of 67% (for test plants) by inoculative adult *D. citri* that were given acquisition feeding in the nymphal developmental period, suggesting that the pathogenic bacteria was present in the salivary glands of these psyllids. In another study, the transmission rate of virulent *D. citri* to healthy plants was estimated 1% (Huang et al., 1984). In our previous study, the virulence of vectors grown on infected trees, and the transmission rate of virulent vectors to healthy host trees, were almost 90% and almost 25%, respectively (Ohto & Kobori, unpublished data).

2. THE MODEL

Until recently, the majority of pathogen and host models were constructed using mean-field models, which presupposed that the individuals were uniformly distributed and that the interactions were uniform. However, in recent years, the accelerated calculation speed of computers has facilitated the development of an Individual-Based Model (IBM). In this model, each pathogen and vector operate independently, adhering to a predetermined set of rules. The model is capable of analyzing the cumulative outcomes of the individual behaviors in a simulation field to investigate the dynamics of disease transmission. The IBM provides a novel and potent instrument for the investigation of insect-borne plant diseases due to its ability to address the vector and host separately. Consequently, we implemented this methodology to create an HLB disease-spread model that was based on the D. citri vector, with a reference to the pine wilt disease-spread model developed by Takasu (2009). The C language employed for the models source code was derived from the C language technology of simulated entities, as published electronically by Takasu (2008).

2.1 Framework of the model and estimation of the parameters

In this simulation, we targeted the region of the Mekong Delta, Vietnam, in order to establish the basic parameters. Some parameters were provided by previous reports and some by our own observations in the target area. The model is summarized in Fig. 1.

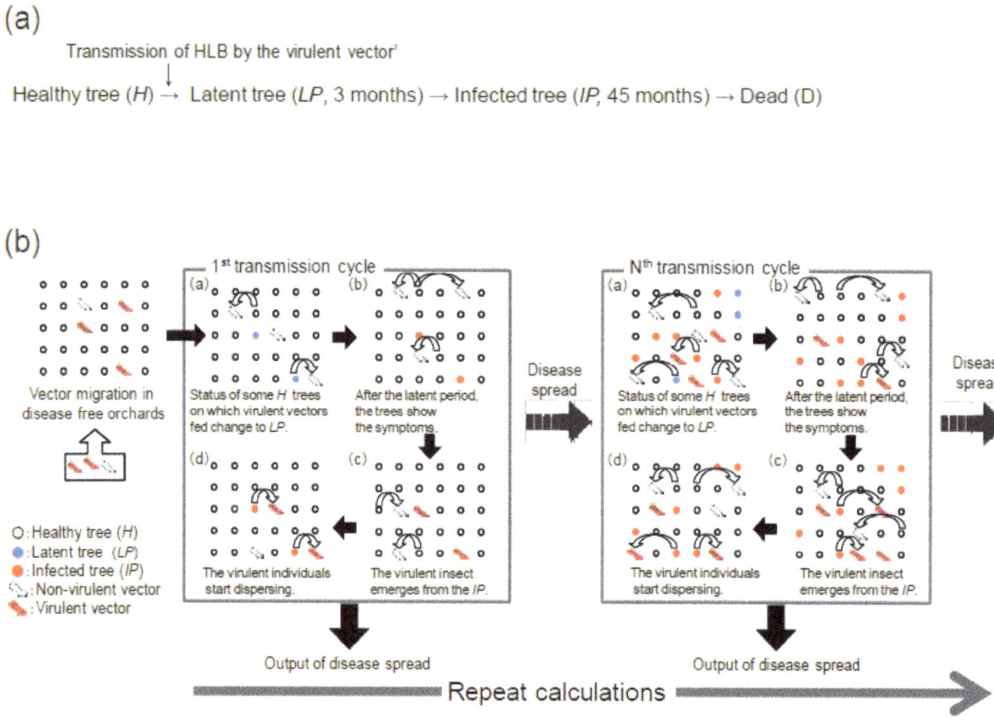

Fig. 1. (a) Change in tree status over time. (b) Summary of citrus greening disease-spread simulation by our model.

From Kobori et al. (2010).

We assumed that the citrus trees were arranged in a two-dimensional regular lattice, with the position of a given tree as '(x, y)', where x and y are integers, and $-L \leq x \leq L$ and $-M \leq y \leq M$ determines the spatial dimensions of the lattice. The distance between trees was assumed to be 2.5 m, in accordance with the common planting regime in the region. We assumed a discrete time progression defined by one month (t).

Tree status was defined as one of the following: (1) healthy period (*H*), when the tree is healthy and has the potential to be infected with HLB by the feeding of virulent *D. citri*; (2) latent period (*LP*), when the tree has been infected with HLB by *D. citri* but still does not have infective ability; (3) infectious period (*IP*), when the *D. citri* growing on the tree have the potential to transmit HLB with a probability of *V*; and (4) dead period (*D*), when the tree is dead or there is no tree at the given point in the lattice. The value *V* was estimated at 0.9 from our aforementioned previous study (Ohto & Kobori, unpublished data). Transition through these tree statuses is irreversible: *H*→*LP*→*IP*→*D*. There have been few previous reports that estimated the latent and infectious periods of HLB. The growth stage, cultivar and environmental differences of the host plant may affect these, but we do not know the details. Hence, in this model, we estimated the default values, based on our own field observation and a survey questionnaire conducted in Vietnam, as the following: the *LP* tree essentially changes status to *IP* in 3*t*, and the *IP* to *D* in 45*t* (Fig. 1 (a)).

The vector insects are assumed to move around according to a given dispersal kernel. If the insect is virulent, it transmits HLB with a probability of *Tr* to tree *H*. In this report, we set *Tr* at 0.3, in accordance with our aforementioned previous study (Ohto & Kobori, unpublished data). A vector insect individual produces an average of *Rp* next-generation individuals according to Gaussian distribution. In this report we assumed an *Rp*-value of 10, in accordance with our life-table study in Japan (Kobori, un-published data) and similar observations in Vietnam. We also established a threshold number, of 100 individuals of the next generation on a given tree, from our observations in Vietnam.

To exclude edge and corner effects associated with a rectangular two-dimensional space, the model space was assumed to be a torus. In addition, we distributed enough *D* trees around the simulation to prevent adults from arriving at the edge.

2.2 Determination of the dispersal kernel

To determine the dispersal kernel of the *D. citri*, we carried out choice and no-choice tests in a glass house, and artificial release experiments in the experimental field in Japan.

In the glass house, we estimated factors affecting the dispersal activity of *D. citri*, first by means of a no-choice test. Movement from a 'growth' tree (that is, one on which the *D. citri* has grown from egg to adult) to another tree was observed, in time, for each situation (Kobori et al., 2011a). We examined high – low insect density situations, and new-bud – no- new-bud situations, for the growth tree. The dispersal movement of the adult *D. citri* was greater for trees with new buds than for trees without new buds (Fig. 2). The effects of *D. citri* density were not detected in this experiment. Additionally, we investigated host-plant selection by *D. citri* through choice tests (Fig. 3). Choice testing between trees with and without buds revealed that released adults preferred trees with buds during the testing period. This preference disappeared when both trees were treated with a sticky spray to prevent adult movement between trees. These results suggest that adult *D. citri* do not select trees with new buds from a distance but recognize them by means of random movement between trees.

Through artificial release experiments, we estimated mean dispersal distance, frequency of movement and dispersal direction (Kobori et al., 2011b). We artificially released, in separate experiments, 10,000 and 1,000 adult *D. citri*, marked in pink, at the center of the experimental field, in which 33 X 33 potted host plants were set out in a grid at 2.5 m intervals (for the marking method, see Nakata, 2008).

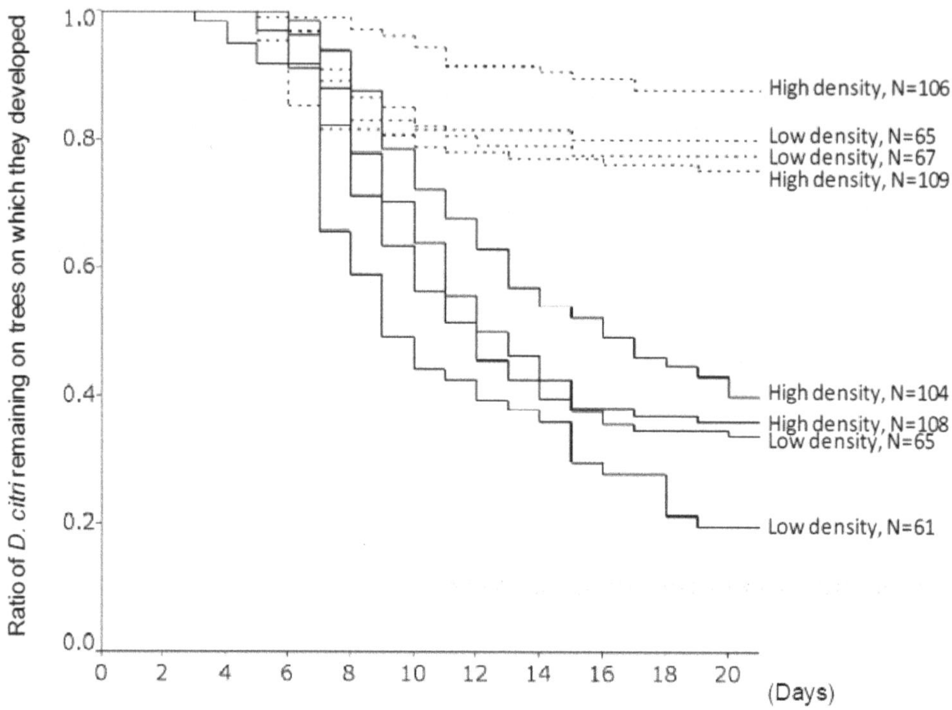

Fig. 2. Relationship between the number of days and the number of migrating adult *D. citri*.

Solid lines: trees with buds; broken lines: trees without buds; N: initial number of adult D. citri on the tree. From Kobori et al. (2011a).

In the 10,000-adult experiment, the mean dispersal distance from the release point was 5-6 m, and in the 1,000-adult experiment it was 6-12 m. In each case, the proportion of *D. citri* found on different days over the experimental periods declined with increasing distance from the release point, and did not change substantially over time (Fig. 4). These results suggested that once the adult *D. citri* arrived on the host plants they hardly moved again. Moreover, the center of distribution would be expected to move with the wind direction, if each *D. citri* continued its random movement, because the same wind direction was observed throughout most of each experimental period; however, the center of distribution in fact stayed close to the release point throughout the experimental periods (Fig. 5). Individuals that moved more than 7.5 m from the release point were found on the lee side. These results suggested that once *D. citri*, here carried by the wind, had arrived on the host plants, they too hardly moved again.

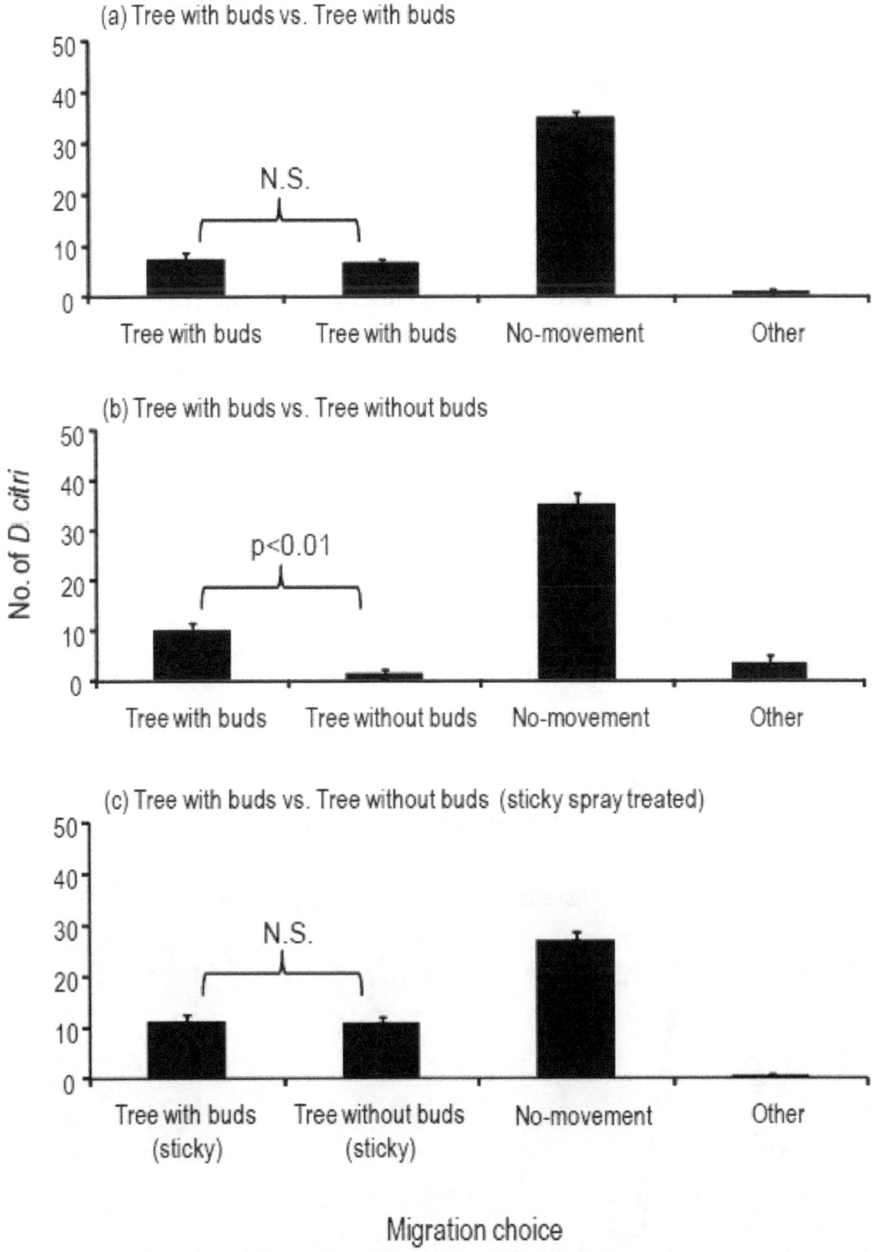

Tree with buds: 8 to 10 buds present on the tree; tree without buds: no buds on the tree; other: individuals found on the side wall of the glasshouse or missing inside the glasshouse. From Kobori et al. (2011a).

Fig. 3. Results of a choice test with adult *D. citri* (mean ± S.E.).

In light of the above results, we determined the dispersal kernel. We did not include the effects of vector insect density in the frequency of movement in the kernel. On the other hand, it is highly possible that the existence of new buds affects movement behavior. However, in the Mekong Delta, Vietnam, our targeted area, there are new buds on the host trees almost the entire year. Thus, we did not include bud effects in our model. The direction of movement was isotropic, and the distance of a given movement, r, follows a probability distribution w(r). For the dispersal kernel, we approximated mobility by means of the formula:

$$w(r) = \gamma\, e^{-\gamma r} \quad (1)$$

γ was estimated at 0.8 from previous experiments (Fig. 4). As a result of a given movement, the insect lands on the nearest-neighbour host tree. If the insect lands on a D tree, the individual moves again without feeding or reproducing, in accordance with the kernel.

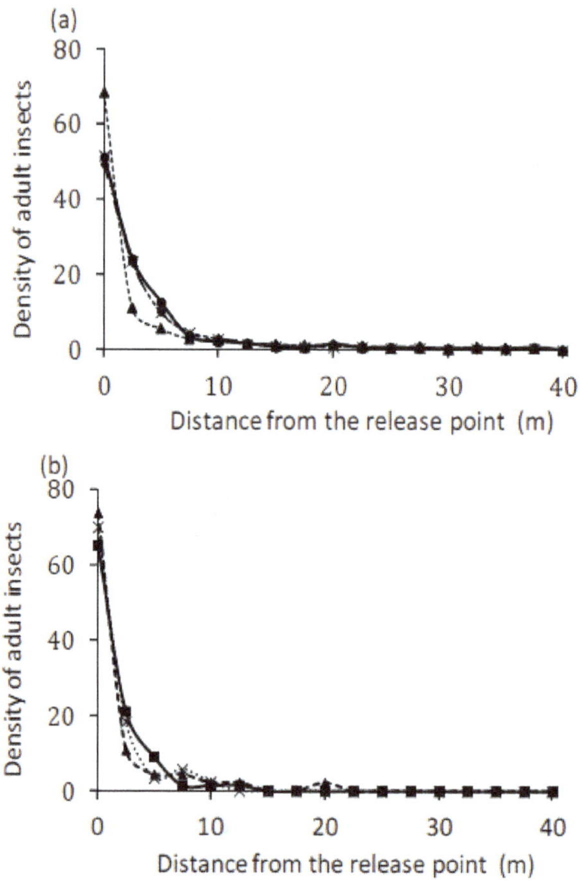

Fig. 4. The density of *D. citri* (*D*) found at different time periods, plotted against distance from the release point; (a) 10,000 adults released, (b) 1,000 adults released.

These values were calculated in a one-dimensional cross-section from south to north in the field. Density of adult insects = (number of individuals / total number of the individuals in field) × 100. ▲: 3days, ■: 7 days, ×: 14 days, ●: (a) 20 days, (b) 21 days after release. From Kobori et al. (2011b).

When the individual lands on an *H*, *LP* or *IP* tree, it does not move again. In the process of moving, individuals have a certain probability (*Dm*) of dying. After reproduction, individuals have a certain probability (*Dl*) of dying; individuals remaining alive (probability of 1-*Dl*) are integrated into the next generation. There is insufficient previous research to provide estimates for these parameters, and we could not estimate their value by certain observation or experiment. Thus, we roughly estimated these parameters at *Dm*: 0.7 and *Dl*: 0.9 by speculative inference from life-table analysis and our own limited observation in this report (e.g. Mead, 1977; Tsai et al., 2002).

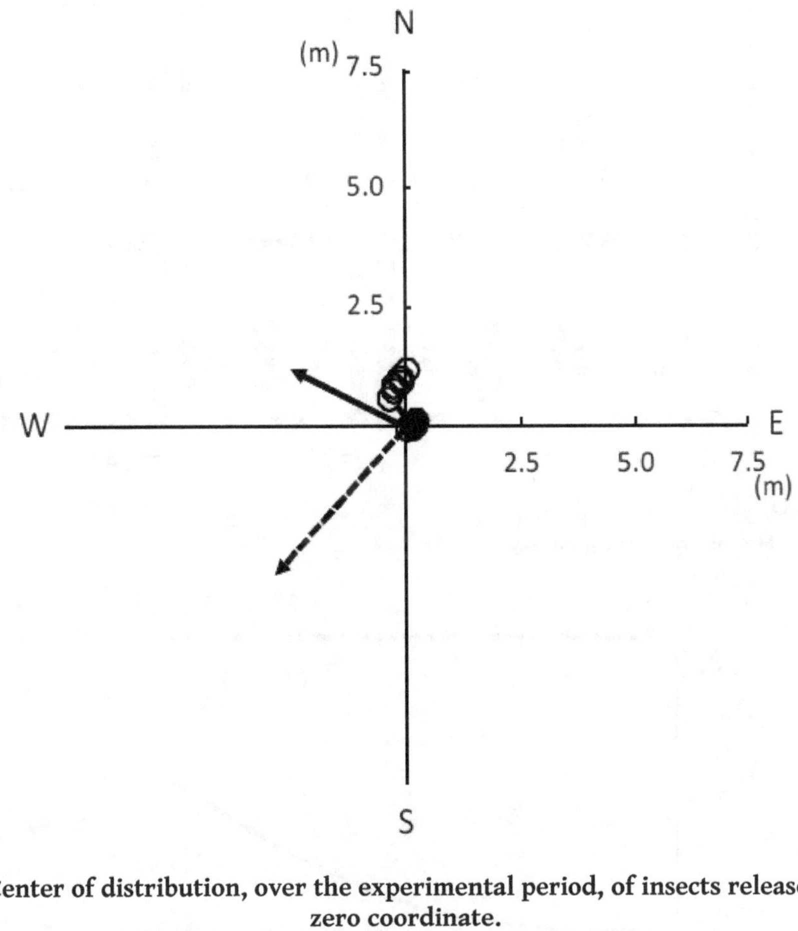

Fig. 5. Center of distribution, over the experimental period, of insects released at the zero coordinate.

Empty and solid circles indicate 10,000 adults and 1,000 adults released, respectively. The two arrows (solid: 10,000 adults; dashed: 1,000 adults) indicate the composite wind vectors over the experimental periods, as calculated from the daily dominant wind directions. The average velocities were 2.42 m/s in the case of 10,000 adults released, and 3.98 m/s in the case of 1,000 adults released. From Kobori et al. (2011b).

3. HLB DISEASE-SPREAD SIMULATION USING THE MODEL

3.1 Spread of HLB in a field under default conditions

We simulated HLB spread under default conditions. Field size was set at $L, M = 100$. Healthy trees (H) were distributed at the center of the field ($L, M = 20$). The remaining trees were D. Then, 100 virulent vector insects were distributed at the center of the H field. We set the model calculation period for 84 months.

Within 12t, more than 10 newly infected trees had appeared in the given field (Fig. 6 (a)). The spread speed increased with time (Fig. 6 (b)). By the end of the calculation period, more than half of the H trees had changed status to *LP, IP* or *D*.

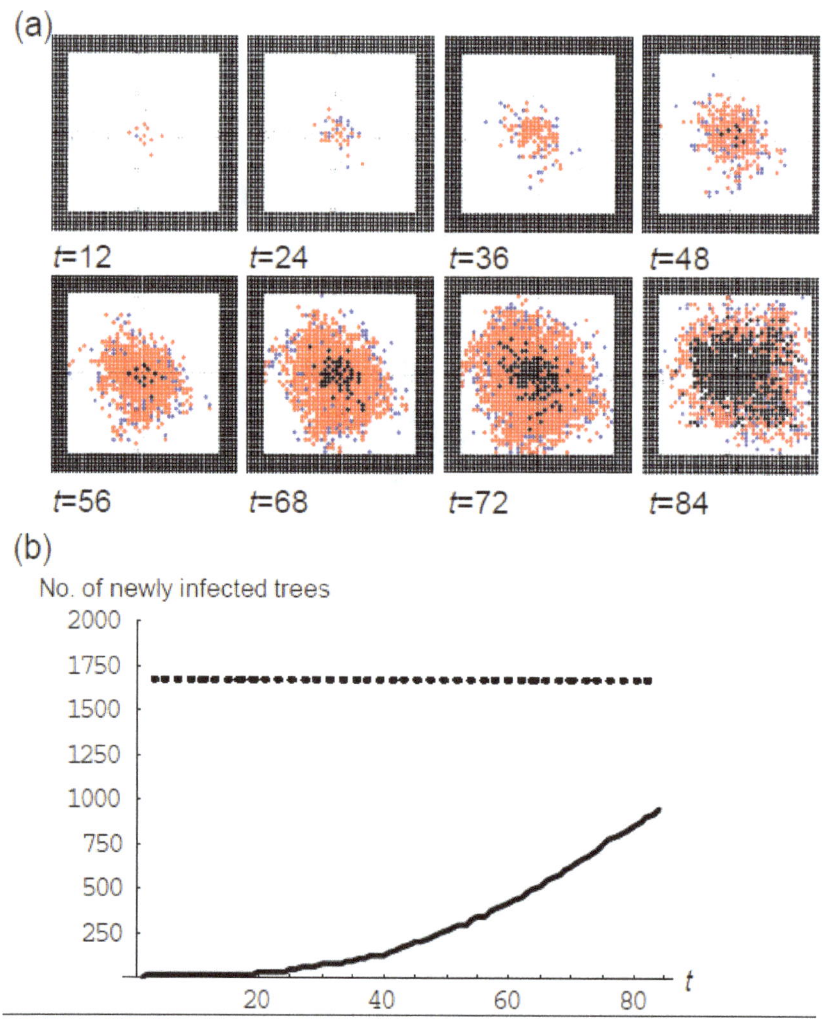

Fig. 6. (a) Snapshots of tree-status distribution for t=12, 24, 36, 48, 56, 68, 72, 84. (b) Number of trees newly infected in the field over time.

Blue dot indicates Latent Period (*LP*); red: Infectious Period (*IP*), black: Dead (*D*). Healthy (*H*) trees are not shown. Solid line: number of trees newly infected; dotted line: number of initial *H* trees in the simulated field (1,681 trees).

3.2 Effectiveness of delaying the latent period (LP), and of removing infectious (IP) trees, in suppressing the spread rate of HLB in an orchard

We estimated the effectiveness of delaying the latent period (*LP*) and removing infectious (*IP*) trees in suppressing the spread rate of HLB in an orchard, compared with the default situation. The field was defined according to the default conditions above. Then, 100 virulent vector insects were distributed at the center of the *H* field. We set the model calculation period for 84 months.

First, we changed the number of time steps required to transition from *LP* to *IP* from the default (3*t*) to 6, 12 and 24*t*, respectively, in order to estimate the effects of delaying the latent period. The model estimation suggested that the spread rate of HLB in the orchard slowed as the time from *LP* to *IP* increased (Fig. 7). The theory that mean generation time has more affect on the intrinsic rate of natural increase than the reproductive rate, would account for this result. Our estimation suggested that cultivars in which the transition from *LP* to *IP* is delayed might show resistance to HLB and thus suppress the rate of disease spread.

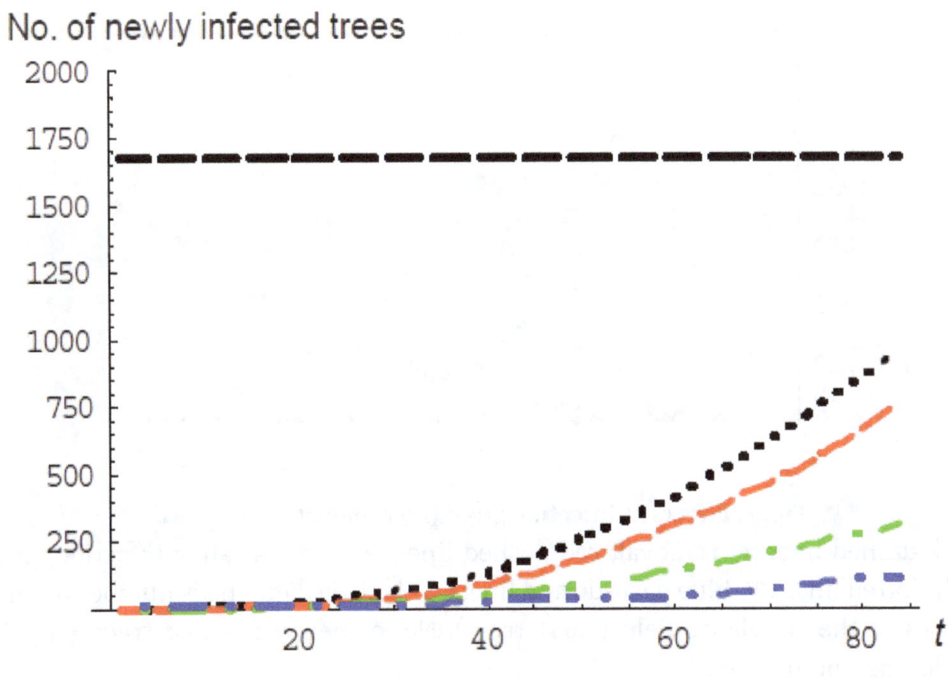

Fig. 7. Effects of delaying the transition from latent period (*LP*) to infectious period (*IP*).

Black dotted line, 3*t* to change from *LP* to *IP* (default parameter); red dashed line: 6*t*; green dash-dotted line: 12*t*; blue dash-dotted line: 24*t*. Black dashed line is number of initial *H* trees in the simulated field (1,681 trees). No. of newly infected trees = *LP*+*IP*+*D* (excluding initial *D* trees).

Second, we estimated the effects of removing infectious (*IP*) trees. We assumed that the *IP* trees were removed at 24, 12 and 6*t*, respectively, after their status had changed from *LP* (default: trees were removed at 45*t*, when their status changed to *D*), and we compared the spread rates of HLB. Under the no-removal scenario (default), the number of newly infected trees was more than 1,000 (Fig. 8). When we removed the *IP* trees 24*t* after their status change, the number of newly infected trees was almost same as under the default condition. However, when we removed the *IP* trees 9*t* after their status change, the number of newly infected trees decreased relative to the default condition. Moreover, when we removed the *IP* trees 6*t* after their status change, the number of newly infected trees barely increased. We may thus predict that the removal of infectious trees will be efficacious in preventing the spread of HLB in a field.

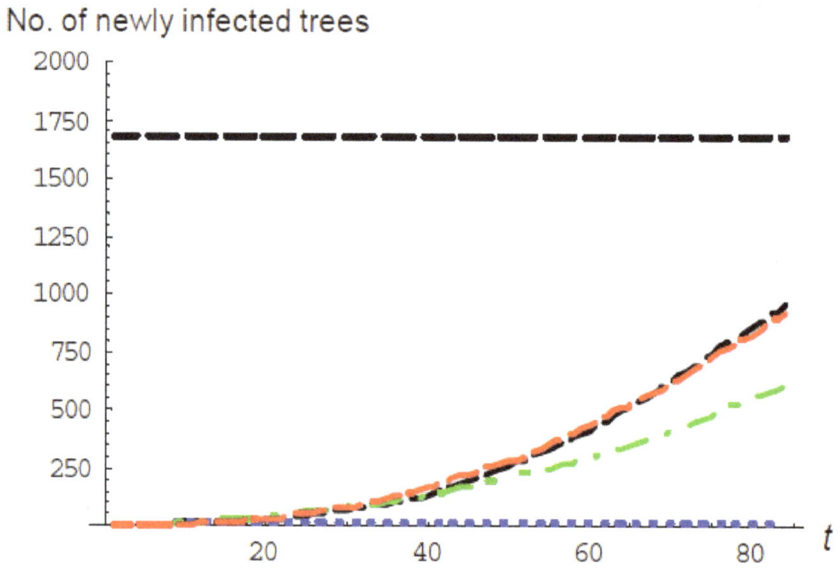

Fig. 8. Effects of infectious tree (*IP*) removal in the field.
Black dashed line: no removal; red dashed line: removal 24*t* after infection; green dash-dotted line: 12*t*; blue dash-dotted line: 6*t*. Black dashed line is number of initial *H* trees in the simulated field (1,681 trees). No. of newly infected trees = *LP*+*IP*+*D* (excluding initial *D* trees).

3.3 Attempts to simulate more realistic scenarios

The vector insect population follows seasonal trends (e.g., Nakahira et al., 2011). Thus, we established high/low reproductive periods and conducted the simulation. The number of newly infected trees increased with time without immigration of virulent vector insects. The model forecast that more than 80% of trees were infected with HLB by 84*t*. Hence, we suggest that the use of disease-free seedlings offers a fundamental technique for preventing HLB spread in an orchard. Additionally, we estimated the suppression of the HLB spread rate through systemic pesticide treatment in a newly planted orchard. Several reports have indicated that systemic pesticide treatment causes high mortality rates in *D. citri* (e.g., Ichinose et al., 2010). We assumed that pesticide applied in the new orchard would be effective for two years. Hence, we assumed that, until 24*t* after planting, 100% of the insects that had migrated from an infected orchard into the newly planted orchard would die after feeding (if an individual was virulent it might transmit HLB, but would not reproduce on the tree). After this period, individuals migrating from the infected orchard would reproduce in the newly planted orchard. The number of newly infected trees was lower in the plantings treated with systemic pesticide than in those with the negative controls. We concluded that treatment with systemic pesticide was effective in suppressing the spread rate of HLB in a newly planted orchard, and the effect was definitive when the virulent *D. citri* migrated from a nearby orchard (Fig. 9). Part of this study have been reported in Kobori et al. (2011c).

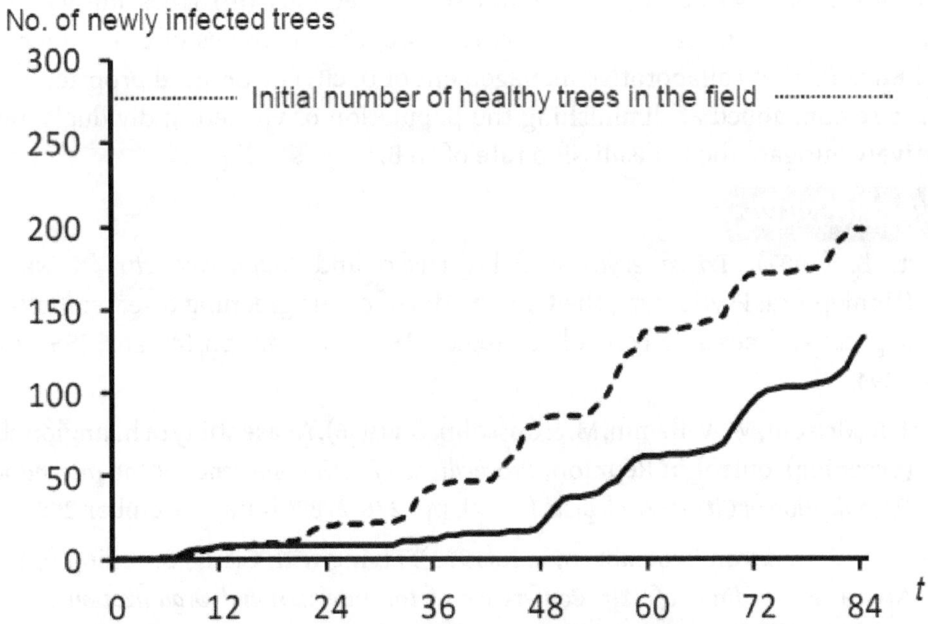

Fig. 9. Effects of systemic pesticide to suppress HLB in the field.

Dashed line, negative control (did not treat systemic pesticide); soiled line, treated with systemic pesticide (effective for 24*t*). Number of initial *H* trees in the simulated field was 256. No. of newly infected trees = *LP+IP+D* (excluding initial *D* trees). Modified from Kobori et al. (2011).

Despite the useful ideas derived from these data, an issue arose within our simulated conditions. In several replications, the vector insect became extinct. In empirical observation, even insects absent in a particular season may be present in subsequent observations. A potential source of this may be the insular nature of our simulation system. Vector insects may migrate into the area. We are presently augmenting the model to include this potentiality. Furthermore, the dispersal kernel was formulated for the computation of single-field dynamics. We must modify the kernel to replicate larger scenarios, including those with many, distinct fields.

4. CONCISE SUMMARY

This chapter detailed the evolution of an individual-based modeling technique to mimic disease transmission patterns. Our model successfully generated parameters for each citrus tree and its corresponding vector insect, enabling the analysis of disease propagation dynamics in the simulation field through the computation of cumulative outcomes based on individual actions. This model is applicable to other insect-vectored diseases, although it was specifically created for HLB. Our simulated findings indicated that both postponing the transition from the latent phase (LP) to the infectious period (IP) and eliminating infectious (IP) trees mitigated the spread rate of HLB in an orchard. Furthermore, the outcomes of our exploratory trial indicated that collaborative management of D. citri by orchard proprietors in a specific region, aimed at diminishing the population of virulent individuals, might effectively mitigate the transmission rate of HLB.

5. REFERENCES

Aubert, B. (1987). *Trioza erytreae* del Guercio and *Diaphorina citri* Kuwayama (Homoptera: Psylloidea), the two vectors of citrus greening disease: Biological aspects and possible control strategies. *Fruits, Vol.* 42, pp.149-162, ISSN 0248-1294

Aubert, B., Grisoni, M., Villemin, M. & Rossolin. G. (1996). A case study of huanglongbing (greening)control in Réunion, *Proceedings of 13th Conference of the International Organization of Citrus Virologists (IOCV)*, pp. 276-278, China, November 2005

Buitendag, C. H. & von Broembsen, L. A.(1993). Living with Citrus Greening in South Africa, *Proceedings of 12th Conference of the International Organization of Citrus Virologists (IOCV)*, pp. 269-273, India, November 1992

Bové, J. M., Garnier, M., Ahlawat, Y. S., Chakraborty, N. K. & Varma, A. (1993). Detection of the Asian Strains of the Greening BLO by DNA-DNA Hybridization in Indian Orchard Trees and Malaysian *Diaphorina citri* Psyllids, *Proceedings of 12th Conference of the International Organization of Citrus Virologists (IOCV)*, pp. 258-263, India, November 1992

Capoor, S. P., Rao, D. G. & Viswanath, S. M. (1974). Greening disease of citrus in the Deccan Trap Country and its relationship with the vector, *Diaphorina citri* Kuwayama, *Proceedings of 6th Conference of the International Organization of Citrus Virologists (IOCV)*, pp. 43-49, Swaziland, August 1972

Chavan, V. M. & Summanwar, A. S. (1993). Population dynamics and aspects of the biology of citruspsylla, *Diaphorina citri* Kuw., in Maharashtra, *Proceedings of 12th Conference of the International Organization of Citrus Virologist (IOCV)*, pp. 286-290, India, November 1992

Garnier, M. & Bové, J. M. (1993). Citrus Greening Disease and the Greening Bacterium, *Proceedings of 12th Conference of the International Organization of Citrus Virologists (IOCV)*, pp. 212-219, India, November 1992

Halbert, S. E. & Manjunath, K. L. (2004). Asian citrus psyllids (Sternorrhyncha: Psyllidae) and greening disease of citrus: A literature review and assessment of risk in Florida. *Florida Entomologist*, Vol. 87, pp. 330-353, ISSN 0015-4040

Huang, C. H., Tsai, M. Y. & Wang, C.L. (1984). Transmission of citrus likubin by a psyllid, *Diaphorina citri*. *Journal of Agricultural Research of China*, Vol.33, pp..15-72.

Ichinose, K., Bang, D. V., Tuan, D. H. & Dien, L. Q. (2010). Effective use of neonicotinoids for protection of citrus seedlings from invasion by *Diaphorina citri* (Hemiptera: Psyllidae). *Journal of Economic Entomology*, Vol.103, pp. 127-35, ISSN 0022-0493

Inoue, H., Ohnishi, J., Ito, T., Tomimura, K., Miyata, S., Iwanami, T. & Ashihara, W. (2009). Enhanced proliferation and efficient transmission of *Candidatus* Liberibacter asiaticus by adult *Diaphorina citri* after acquisition feeding in the nymphal stage. *Annals of Applied Biology*, Vol. 155, pp. 29-36, ISSN 0003-4746

Kobori, Y., Nakata, T. & Ohto Y. (2011a). Estimation of dispersal pastern of adult Asian psyllid, *Diaphorina citri* Kuwayama (Hemiptera: Psyllidae). *Japanese journal of Applied Entomology and Zoology*, Vol. 55, pp. 177-181, ISSN: 0021-4914 (In Japanese with English summary, tables and figures)

Kobori, Y., Nakata, T., Ohto, Y. & Takasu, F. (2011b). Dispersal of adult Asian citrus psyllids, *Diaphorina citri* Kuwayama (Homoptera: Psyllidae), the vector of citrus greening disease, in artificial release experiments. *Applied Entomology and Zoology*, Vol. 46, pp. 27-30, ISSN 0003-6862

Kobori, Y., Ohto, Y., Nakata, T & Ichinose, K. (2011c). Spread risk estimation of citrus greening disease vectored by *Diaphorina citri* (Hemiptera: Psyllidae) in a citrus orchard using an individual-based model. *JIRCAS Working Report No.72*, pp. 39-43, ISSN 1341-710X

Kobori, Y., Ohto, Y., Nakata, T. & Takasu, F. (2010). Development of individual based-model for simulating the spread of Citrus greening disease by the vector insects. In: Research Highlights 2010., Available from http://www.jircas.affrc.go.jp/kankoubutsu/highlight/highlights2010/2010_30.ht ml

Lin, K. H. & Lin, K. H. (1990). The citrus huanglongbing (greening) disease in China, *Rehabilitation of Citrus Industry in the Asia Pacific Region. Proceedings of Asia Pacific International Conference on Citriculture*, pp. 1-26, Thailand, February 1990

Liu, Y. H. & Tsai, J. H. (2000). Effects of temperature on biology and life table parameters of the Asian citrus psyllids, *Diaphorina citri* Kuwayama (Homoptera: Psyllidae). *Annals Applied Biology*, Vol. 137, pp. 201-216, ISSN 0003-4746

Mead, F. W. (1977). The Asiatic citrus psyllid, *Diaphorina citri* Kuwayama (Homoptera: Psyllidae). Entomology Circular 180. Florida Department of Agriculture and Consumer Services, Division of Plant Industry, pp. 4

Nakahira K., Kobori Y., Ohto Y. , Chau N. M. & Dien L. Q. (2011). Population dynamics of the Asian citrus psyllid (*Diaphorina citri* Kuwayama, Hemiptera) in a king mandarin (*Citrus nobilis* Loureiro) orchard in Mekong Delta. *JIRCAS Working Report No.72*, pp. 39-43, ISSN 1341-710X

Nakata, T. (2006). Temperature-dependent development of the citrus, *Diaphorina citri* (Homoptera: Psylloidea), and the predicted limit of its spread based on overwintering in the nymphal stage in temperature regions of Japan. *Applied Entomology and Zoology*, Vol. 41, pp. 383–387, ISSN 0003-6862

Nakata, T. (2008). Effectiveness of micronized fluorescent powder for marking citrus psyllid, *Diaphorina citri*. *Applied Entomology and Zoology*, Vol. 43, pp. 33-36, ISSN 0003-6862

Roistacher, C. N. (1996). The Economics of Living with Citrus Diseases: Huanglongbing (Greening) in Thailand, *Proceedings of 13th Conference of the International Organization of Citrus Virologists (IOCV)*, pp. 279-285, China, November 2005

Su, H.-J., Cheon, J.-U. & Tsai, M.-J. (1986). Citrus greening (Likubin) and some viruses and their control trials, pp. 143-147 In *Plant Virus Diseases of Horticultural Crops in the Tropics and Subtropics. FFTC Book Series No. 33.* (Asian and Pacific Council. Food & Fertilizer Technology Center ed.), Food and Fertilizer Technology Center for the Asian and Pacific Region, Taiwan

Takasu, F. (2009). Individual-based modelling of the spread of pine wilt disease: vector beetle dispersal and the Allee effect. *Population Ecology, Vol.* 51, pp. 399-409 ISSN 1438-3896

Takasu, F. (2008). An introduction to individual-based modelling - a framework to implement IBM. In: Research activities, 10. 4. 2008, Available from http://gi.ics.nara-wu.ac.jp/~takasu/research/IBM/ibm.html

Tian, Y., Ke, S. & Ke, C. (1996). Polymerase chain reaction for detection and quantitation of *Liberobacter asiaticum*, the bacterium associated with Huanglongbing (Greening) of citrus in China, *Proceedings of 13th Conference of the International Organization of Citrus Virologists (IOCV)*, pp. 252-257, China, November 2005

Toorawa, P. (1998). La maladie du huanglongbing (greening) des agrumes a L'Île Maurice. Detectionde "*Candidatus* Liberobacter asiaticum" et"Candidatus Liberobacter africanum" dans lesagrumes et les insects vecteurs, *Doctoral Thesis, L' University de Bordeaux*, pp. 186

Tsai, J. H., Wang, J. J. & Liu, Y. H. (2002). Seasonal abundance of the Asian citrus psyllid, *Diaphorina citri* (Homoptera: Psyllidae) in southern Florida. *Florida Entomologist,* Vol. 85, pp. 446-451, ISSN 0015-4040

Van Vuuren, S. P. (1993). Variable transmission of African greening to sweet orange, *Proceedings of 12th Conference of the International Organization of Citrus Virologists (IOCV)*, pp. 264-268, India, November 1992

Wang, Z., Yin, Y., Hu, H., Yuan, Q., Peng, G. & Xia, Y. (2006) Development and application of molecular-based diagnosis for '*Candidatus* Liberibacter asiaticus', the causal pathogen of citrus huanglongbing. *Plant Pathology*, Vol. 55, pp. 630–638, ISSN 1365- 3059

Xu, C. F., Xia, Y. H., Li, K.-B. & Ke, C. (1988). Further study of the transmission of citrus huanglongbing by a psyllid, *Diaphorina citri* Kuwayama, *Proceedings of 10th Conference of the International Organization of Citrus Virologists (IOCV)*, pp. 243-248, Spain, November 1986

Yang, Y., Huang, M., Beattie, G. A.C., Xia, Y., Ouyang, G. & Xiong, J. (2006). Distribution, biology, ecology and control of the psyllid *Diaphorina citri* Kuwayama, a major pest of citrus: A status report for China. *International Journal of Pest Management*, Vol. 52, pp. 343-352, ISSN 0967-0874

CHAPTER 6

CURRENT STATUS OF ENTOMOPATHOGENIC FUNGI AS MYCOINSECTICIDES AND THEIR COST-EFFECTIVENESS

1. INTRODUCTION: AN OVERVIEW

Synthetic chemical pesticides have been the primary method of pest eradication for over 50 years. Nonetheless, insecticide resistance, pest resurgence, safety hazards for humans and domestic animals, groundwater contamination, biodiversity decline, and other environmental issues have prompted researchers to develop environmentally benign pest control strategies, including the utilization of biological control agents. Naturally occurring biological control agents are significant regulatory elements in insect populations. Numerous species are utilized as biological control agents for insect pests in glasshouses, row crops, orchards, ornamental plants, rangelands, turf and lawns, stored products, and forestry, as well as for the mitigation of pest and vector insects of veterinary and medical significance (Burges, 1981; Lacey & Kaya, 2000; Tanada & Kaya, 1993). The utilization of microbes for the management of insect pests was suggested by prominent early figures in invertebrate pathology, including Agostino Bassi, Louis Pasteur, and Elie Metchnikoff (Steinhaus, 1956, 1975). Biological control agents, including viruses, bacteria, protozoa, nematodes, and fungus, significantly regulate target populations. Entomopathogenic fungi represent the most extensive category of insect pathogens among micro-organisms. Typically, two categories of fungi are identified as pathogens in insects. Entomopathogenic fungi are classified within the orders Entomophthorales and Hypocreales (formerly referred to as Hyphomycetes). Various entomopathogenic fungi from other taxonomic groupings are also recognized. To date, more than 700 species of fungi have been identified as parasites of insects (Wraight et al, 2007). Insecticidal fungi offer significant benefits. Firstly, they serve as significant natural antagonists of arthropods (Chandler et al, 2000), capable of directly infecting them via the integument. Furthermore, the cultivation of these fungi and the manufacture of infective conidia are straightforward and relatively inexpensive (Roberts & Hajek, 1992). Entomogenous fungus can be located in many ecological circumstances (Ferron, 1978). Distinct from other entomopathogens, fungus can penetrate their hosts straight via the exoskeleton or cuticle without requiring ingestion. Consequently, entomopathogenic fungi are capable of infecting non-feeding life stages, including eggs and pupae. The insect cuticle serves as the primary barrier against biological pesticides. The insect cuticle primarily consists of three layers: the epicuticle, procuticle, and epidermis. Each layer possesses distinct chemical structures and characteristics (Juárez & Fernández, 2007). The epicuticle is exceedingly thin (0.1–3 μm) and composed of many layers. The uppermost layer of the epicuticle is the lipid layer, which is largely resistant to enzymatic destruction and possesses features such as serving as a water barrier; unless physically compromised, it can inhibit the penetration of cuticle-degrading fungal enzymes. Insects often invade by the mouthparts, inter-segmental folds, or spiracles. In these locations, elevated humidity facilitates conidial germination, and the cuticle is non-sclerotized and more readily penetrable (Clarkson & Charnley, 1996; Hajek & St. Leger, 1994).

Upon landing on a potential host, conidia launch a sequence of events that may result in a compatible (infection) or incompatible (resistance) response. In a compatible reaction, fungal identification and adhesion lead to germination on the host cuticle. Upon breaching the epicuticle, the advancement of the penetration peg through the cuticle may occur directly via penetrant hyphae, whereas penetrant structures may also extend laterally (Hajek & St. Leger, 1994). Fargues (1984) posited that adhesion transpires in three sequential phases: (1) the adsorption of fungal propagules onto the cuticular surface; (2) the adhesion or consolidation of the interface between pre-germinant propagules and the epicuticle; (3) the germination and development of the fungus on the insect cuticular surface, culminating in the formation of an appressorium to initiate the penetration phase. Zacharuk (1970b) postulated an active adhesion mechanism for *M. anisopliae* upon seeing epicuticle breakdown and the infiltration of mucoid substances into the pore canals. Infection will ensue following successful penetration. In terrestrial environments, fungal conidial germination involves the development of a germ tube (Boucias & Pendland, 1991) or appressorium (Madelin et al., 1967; Zacharuk, 1970a), which produces a slender penetration peg that penetrates the insect cuticle through mechanical (turgor pressure) or enzymatic methods (proteases) (Zacharuk, 1970b). Exocellular mucilage, suggested to improve adhesion to the host cuticle, is also excreted by many entomogenous fungi during the development of infective structures (Boucias and Pendland, 1991). In M. anisopliae, the formation of appressoria, production of hydrophobins, and expression of cuticle-degrading proteases are induced by reduced nutrition availability (St. Leger et al, 1992), indicating that the fungus detects environmental circumstances or host signals at the onset of infection. The synthesis of cuticle-degrading enzymes, including chitinases, lipases, and proteases, has been acknowledged as a crucial factor in the infection process of various fungi, aiding in penetration and supplying sustenance for subsequent development (Charnley, 1984; Dean & Domnas, 1983; Hussain et al., 2010b; Samsináková et al., 1971). The spore-bound Pr1 protease in entomopathogenic fungi has been well described, and its function in cuticle penetration has been confirmed (Hussain et al., 2010b; St. Leger, 1994). Ultra-structural investigations of M. anisopliae penetration sites on Manduca sexta larvae have revealed elevated concentrations of Pr1 corresponding with the hydrolysis of cuticular proteins (Goettel et al, 1989; St. Leger et al, 1989). Pr1 inhibition tests shown a postponement of death in M. sexta larvae, attributed to a delayed penetration of the cuticle (St. Leger et al, 1988). Moreover, the development of an M. anisopliae strain featuring several copies of the Pr1 gene and exhibiting over-expression of the protease led to a 25% decrease in time to mortality in M. sexta compared to those infected by the wild-type strain (St. Leger et al, 1996).

Additionally, it has been documented that repeated *in vivo* passage augmented the fungus's ability to induce infection (Daoust et al, 1982; Hussain et al, 2010b), hence elevating the activity of spore-bound Pr1 (Shah et al, 2007). Following penetration of the cuticle, the conidia infiltrate the hemocoel, resulting in a robust mycelial proliferation (Zimmerman, 1993). In addition to penetration, fungi generate secondary metabolites, which are derivatives of different intermediates, some of which exhibit insecticidal properties (Vey et al, 2001). Experimental evidence indicates that entomopathogens manufacturing these toxins lead to accelerated host mortality (McCauley et al., 1968) in contrast to strains that do not synthesize these metabolites (Kershaw et al., 1999; Samuels et al., 1988). Destruxins, cyclic depsipeptide toxins from Metarhizium spp., have insecticidal characteristics as described by Kodaira (1961) and have been demonstrated to be generated in wax moth and silkworm larvae by Roberts (1966) and Suzuki et al. (1971). Moreover, these poisons have been evaluated against multiple insect species (Roberts, 1981). As of now, more than 28 distinct destruxins have been identified, primarily from Metarhizium spp, exhibiting diverse levels of efficacy against various insects (Vey et al, 2001). Destruxin levels have been associated with virulence (Al-Aïdroos & Roberts, 1978) and host specificity (Amiri-Besheli et al., 2000). Research on destruxin activity has demonstrated regulation of the host cellular immune system, including the reduction of nodule development (Huxham et al., 1989; Vey et al., 2001) and suppression of phagocytosis (Vilcinskas et al., 1977) in infected insects. Destruxins are synthesized during the growth of the mycelium within the insect. Additional representative toxins generated by entomopathogenic fungi encompass oosporein, beauvericin, and bassianolide from Beauveria spp. (Eyal et al, 1994; Gupta et al, 1994; Suzuki et al, 1977), efrapeptins (Dtolypin) from Tolypocladium spp. (Weiser & Matha,

1988), and hirsutellin from Hirsutella thompsonii (Mazet & Vey, 1995). Within the insect haemocoel, the fungus transitions from filamentous hyphal development to yeast-like hyphal entities that circulate in the hemolymph.

The expansion of these hyphal bodies transpires via budding (Boucias & Pendland, 1982). Subsequently, the fungus reverts to a filamentous phase and infiltrates interior tissues and organs (Mohamed et al, 1978; Prasertphon & Tanada, 1968). The fungus subsequently penetrates the cuticle, and an external mycelium envelops the entire host, producing infective spores under suitable climatic circumstances (Boucias & Pendland, 1982; McCauley et al., 1968). Under suboptimal conditions, certain fungi develop resting structures within the cadaver, as exemplified by Nomuraea rileyi in environments of low relative humidity and temperature (Pendland, 1982). The life cycle of the fungus concludes when the hyphal bodies produce spores on the host's carcass. The exterior hyphae generate conidia that mature and are discharged into the environment. This facilitates horizontal transmission of the disease among insects (Khetan, 2001). Of the 85 genera of entomopathogenic fungi, only six species are commercially accessible for field application (Table 1). Nonetheless, relatively few have been examined as prospective mycoinsecticides. Fungal pathogens, specifically B. bassiana, I. fumosorosea, and M. anisopliae, are being assessed for their efficacy against several agricultural and urban insect pests. Numerous species within the order Isoptera (Hussain et al., 2010a; Hussain et al., 2011), Lepidoptera (Hussain et al., 2009), Coleoptera (Ansari et al., 2006), Hemiptera (Leite et al., 2005), and Diptera (St. Leger et al., 1987) exhibit susceptibility to diverse fungal infections. This has resulted in several endeavors to employ entomopathogenic fungi for pest management, yielding variable levels of efficacy.

Table 1. Commercial formulations of entomopathogenic fungal pesticides

Fungus	Product and Company	Formulation
Aeschersorzia aleyrodis	Koppert / Holland	Wettable powder
Beauveria bassiana	Naturalis™, Troy Bio-Science, USA.	Liquid formulation
B. bassiana	Conidia, AgrEvo, germany, Columbia	Suspendible granules
B. bassiana	Brocani™, Laverlam, Columbia	Wettable powder
B. bassiana	Boverol / Czeck Republic	Wettable powder and dry pellets
B. bassiana	Mycontrol-WP / Mycotech. Corp. USA	Wettable powder
B. bassiana	Ostrinil / natural Plant Protection / France	Microgranules of mycelium

Fungus	Product and Company	Formulation
B. brongniarti	Betel / Natural Plant Protection / France	Microgranules of mycelium
B. brongniarti	Engerlingspilz / Andermatt - Biocontrol / Switerzarland	Barley kernels colonized with the fungus
M. anisopliae	Bio -path™ / Eco Science / OSA	Conidia on a medium placed in trap / chamber
M. anisopliae	Biogreen / Biocare Technology Pvt. Ltd / Australia	Conidia produced on grains
M. anisopliae	Biologic Bio 1020 / Bayer AG Germany	Granules of mycelium
Paecilomyces fumosoroseus	Pfr 21 / WR Grace USA	Wettable powder
Verticillium lecanii	Mycotal / Koppert / Netherlands	Wettable powder
V. lecanii	Vertalec / koppart / Netherlands	Wettable powder

(Bhattacharyya et al., 2004)

The majority of fungal production systems consist of two stages system in which fungal inoculum of hyphal bodies is produced in liquid culture and then transferred to a solid substrate for production of aerial spores (Devi, 1994). For practical use of entomopathogenic fungi as bio-insecticides at each stage, it is necessary to develop culture medium and method that produce high concentrations of viable and virulent propagules at low cost (Jackson, 1997). These goals can be achieved by using the most favorable inexpensive components for fungal growth at the lowest concentration that afford high yield. Most common compounds for fungal entomopathogens include agro-industrial by-products such as corn steep liquor (Zhao et al., 2010) and sugarcane molasses (Hussain et al., 2011). Our previous investigations showed that both the by-products stimulate the growth of the propagules of entomopathogenic fungi. Corn steep solid and cotton seed flour with yeast extract and KCl, NaCl etc., optimized blastospore production under water stress conditions (Ypsilos & Magan, 2005).

Entomopathogenic fungi infect insects in an aggressive manner by secreting cuticle degrading enzymes such as esterases, lipases, N-acetylglucosaminidases and chitinases (St. Leger et al., 1986). However, the production of extracellular protease Pr1, a major virulence determinant, plays an important role in the success of entomopathogenic fungi in insect penetration, which leads to the subsequent pathogenicity in the target host (Hussain et al., 2010b; Shah and Butt 2005). Previously, agro-industrial by-products such as corn-steep liquor and molasses have been used as alternative growth substrates to produce exopolysaccharides (Fusconi et al., 2008; Sutherland, 1996). Sugarcane molasses (SM), an industrial by-product rich in fermentable sugars, was proposed as a nutritious medium to produce bacterial cellulose by *Zoogloea* sp. (Paterson-Beedle et al., 2000). While, the dried powder of corn steep liquor was used as an inexpensive substitute for beef extract in the medium, which enhanced the lipase production of the strain of *Serratia marcescens* (Zhao et al., 2010). Historically, there have been no reports regarding the activity of the extracellular protease Pr1 enzyme derived from the spores of entomopathogenic fungi growing on rice, which was previously grown on media with varying compositions. This study aims to assess the effects of three distinct nitrogen sources and two sugar sources in various combinations and concentrations to determine i) production in liquid media and ii) the activity of extracellular protease Pr1 from locally isolated strains of entomopathogenic fungi.

2. MATERIALS AND METHODS

2.1 Culturing of entomopathogenic fungi

The entomopathogenic fungus *M. anisopliae* (EBCL 02049), *B. bassiana* (EBCL 03005), and *I. fumosorosea* (EBCL 03011) were initially isolated from *C. formosanus* in China. The strains were submitted to the European Biological Control Laboratory in France. The strains were sequentially sub-cultured on Potato Dextrose Agar (PDA, Difco Laboratories, Detroit, MI, US) at 26 ± 0.5 ºC, in total darkness. The maintenance of fungal strains was consistent with our prior research, which was thoroughly detailed (Hussain et al., 2009). In summary, 24-day-old spores from the examined strains grown on PDA served as the inoculum for all growth media.

2.2 Influence of liquid media composition on the *in vitro* growth of entomopathogenic fungi

The three sources of nitrogen: peptone (Sigma), yeast extract (Sigma) and corn steep liquor (Shanghai Xiwang Starch Sugar Co., Ltd.), and two sources of sugar: glucose (Sigma) and sugar molasses, were used in different combinations as shown in Table 2. In all the treatments, the following salts were used at the concentration, $CaCl_2 . 2H_2O$ (0.06%), KCl (0.28%), $MgCl_2 . 6H_2O$ (0.16%), $MgSO_4 . 7H_2O$ (0.2%), $NaHCO_3$ (%0.03) and $NaH_2PO_4 . H_2O$ (0.1%). To avoid reactions among the salts, they were prepared in compatible pairs: $CaCl_2. 2H_2O$ with KCl; $MgCl_2. 6H_2O$ with $NaH_2PO_4. H_2O$; $MgSO_4. 7H_2O$ with $NaHCO_3$. The other precautionary measures to avoid precipitation were adopted as described by Leite et al. (2005). Media preparations were finalized by adjusting the pH to 6.2, with filter-sterilized HCl (0.1%) and NaOH (10%).

Table 2. Composition of the media of shake flask cultures in agitated liquid cultures of entomopathogenic fungi

GROWTH MEDIA	SUGAR AND NITROGEN SOURCES				
	Glucose (G) %	Sugar molasses (SM) %	Corn steep liquor (CSL) %	Peptone (PE) %	Yeast extract (YE) %
G + CSL	2.66		1		
G + PE	2.66			1	
G + YE	2.66				1
G + CSL + PE	2.66		0.50	0.50	
G + CSL + YE	2.66		0.50		0.50
G + PE + YE	2.66			0.50	0.50
G + CSL + PE + YE	2.66		0.333	0.333	0.333
SM + CSL		2.66	1		
SM + PE		2.66		1	
SM + YE		2.66			1
SM + CSL + PE		2.66	0.50	0.50	
SM + CSL + YE		2.66	0.50		0.50
SM + PE + YE		2.66		0.50	0.50
SM + CSL + PE + YE		2.66	0.333	0.333	0.333
G+SM+CSL+PE+YE	1.33	1.33	0.333	0.333	0.333

The media were poured into 250 ml Erlenmeyer flasks and autoclaved. After cooling, all the flasks were inoculated with one milliliter of spore suspension (1×10^6 spores/ml) in 0.03 % Tween 80 (Sigma-Aldrich, St Louis, MO, US) from 24-day-old cultures of *B. bassiana*, *M. anisopliae* and *I. fumosorosea* grown on PDA. Four replicates were used for each growth medium. After inoculation, cultures (100 ml) were grown in 250 ml shake flasks at 120 rpm on a rotary shaker, 25 ± 0.5 ºC and 16 h fluorescent light photophase. After 120 h of growth, fungal biomass of each flask was evaluated separately. The fungal biomass was filtered through Whatman No. 1 filter paper. After filtration, the filtrates were dried for 24 h at 70 ºC and weighed.

2.3 Influence of liquid media composition on the growth of blastospores of entomopathogenic fungi

The optical densities of the inocula of all the studied fungi grown on different media after different time intervals such as 20 h, 40 h and 60 h were measured at 600 nm (OD600) by using a spectrophotometer (Shimadzu UV-1800). The media without inoculation were used as control.

2.4 Production on solid substrates

The entomopathogenic fungi were grown on solid substrate (rice grains) as described in our previous study (Hussain et al., 2011). In brief, 50 % diluted 10 ml inoculum obtained from above media was added into a self-aerating bag containing 20 g of sterilized par boiled rice, separately. Inoculated rice granules were mixed thoroughly from the outside of the bag. Inoculated rice bags were then incubated for 18 days at ambient conditions (24 ± 2 ºC, 75- 85% RH). The sporulating rice in bags from each growth medium was then allowed to dry for 10 d at 30 °C. The spores were separated from the rice by sieving through a 300 µm mesh. A collecting vessel, such as a bucket was fitted to the plastic sheeting at the bottom of the sieve to create a funnel into the collecting vessel. The sieve was shaken until all the loose spores had been removed from the rice and had collected in the vessel below. The spores were then further sieved using a 106 µm sieve to separate the larger rice dust particles from the spores. The spores as powder were kept at 4 ºC for subsequent analysis of the activity of extracellular protease Pr1.

2.5 Influence of artificial media on the activity of extracellular protease Pr1

The activity of Pr1 protease bound to 10 mg spores harvested from the rice of above mentioned media were assayed using a modified method of St. Leger et al. (1987). Briefly, 10 mg spores were washed once in 0.3% aq Tween 80 solution and twice in distilled water, were then incubated in 1 ml of 0.1M Tris HCl (pH 7.95), containing 1mM Succinyl-ala-ala- pro-phe-*p*-nitroanilide (Sigma) for 5 min at room temperature. The spores were pelleted by centrifugation for 5 min at 12000 g (Fastwin Bio-Tech Company Limited). A 200 µL of supernatant was transferred to quartz cuvette. The absorbance was measured at 405 nm by using a spectrophotometer (Shimadzu UV-1800). Buffered substrate was used as a reference. The amount of spore bound extracellular protease Pr1 is expressed as micromoles of nitroanilide (NA) released per minute.

2.6 Statistical analyses

All experiments were repeated four times except the activity of spore bound Pr1, repeated three times. Data were analyzed by analysis of variance using the ANOVA procedure of SAS (SAS Institute, 2000) for a completely randomized design. When the effect was significant ($P < 0.05$), means were separated using Duncan's Multiple Range Test.

3. RESULTS

3.1 Effect of different sources of sugar and nitrogen on the dried fungal biomass of entomopathogenic fungi

Significant differences in dried biomass of *M. anisopliae* (F = 10.632; df = 14, 45; P < 0.001), *B. bassiana* (F = 9.286; df = 14, 45; P < 0.001) and *I. fumosorosea* (F = 9.596; df = 14, 45; P < 0.001) were observed when fungi were grown on media containing different sources of nitrogen and sugar in different combinations. The media supplemented with sugarcane molasses (SM) afforded comparatively higher growth of *M. anisopliae* and *I. fumosorosea* than glucose (Fig. 1a, c). The growth medium contained G + CSL + YE, showed the lowest growth (12.95 mg/ml) of *M. anisopliae*. The glucose in combination with CSL, PE and YE showed higher growth; while their combination among them did not afford higher fungal growth (Fig. 1a). The growth medium such as G + CSL + PE, G + CSL + YE and G + PE + YE, afforded the lowest growth of *I. fumosorosea* and non significant differences were observed among them (Fig. 1c). *B. bassiana* grown on media supplemented with SM in combination with CSL, PE and YE showed significantly higher growth compared to the media containing glucose as sugar source (Fig. 1b). In contrast, their combinations (CSL, PE and YE) with one another in the presence of glucose as sugar source produced higher growth compared with SM (Fig. 1b). The medium SM + CSL afforded higher growth ranges from 31.46-35.28 mg/ml in all the studied fungal strains (Fig. 1a-c).

(a)

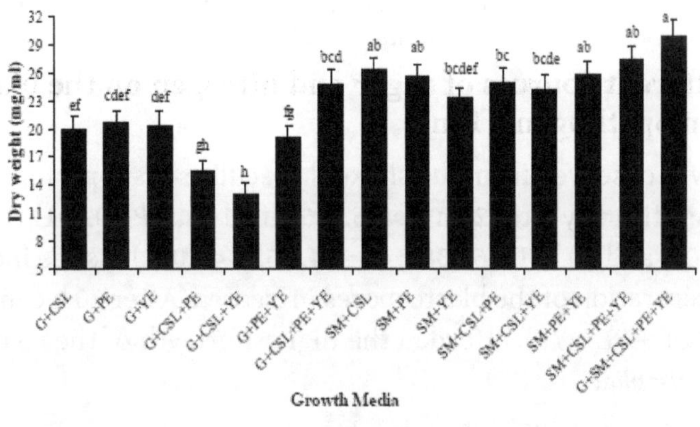

(b)

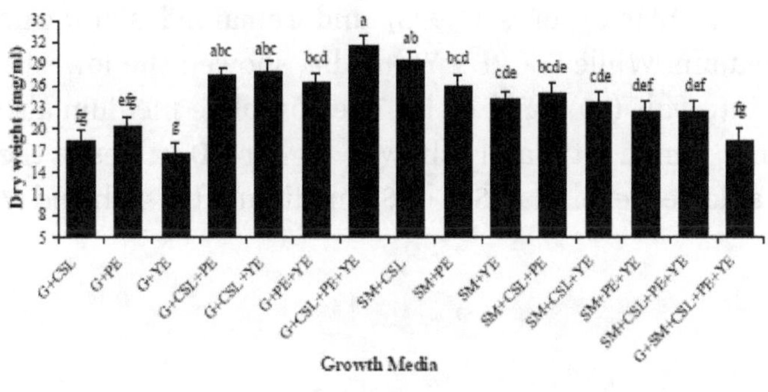

(c)

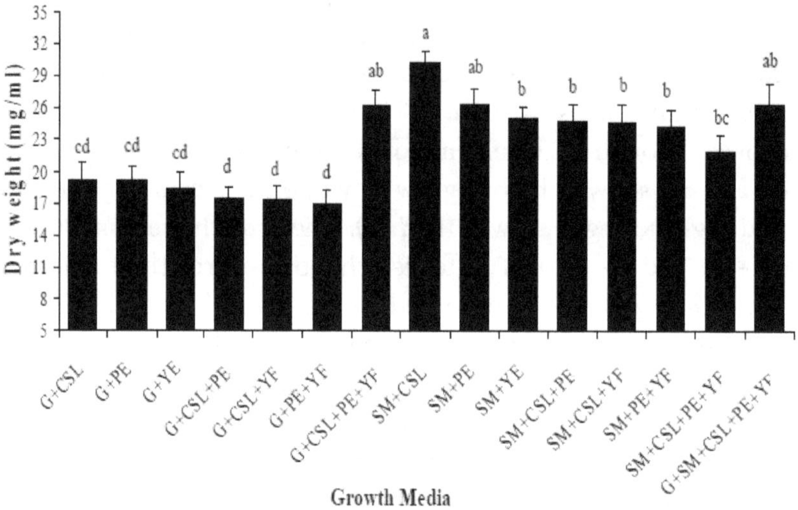

Fig. 1. Influence of media composition on the dried biomass (mg/ml) of (a) *M. anisopliae* (b) *B. bassiana* and (c) *I. fumosorosea*. Mean ± SE values with the same letter(s) along the bars of different growth media are not significantly different ($P < 0.05$). For detail of treatments see table 1.

3.2 Effect of different sources of sugar and nitrogen on the blastospores growth of entomopathogenic fungi

M. anisopliae growth observed from the shake flask cultures supported with different media differed significantly after 20 h ($F = 96.535$; df = 14, 45; $P < 0.001$), 40 h ($F = 77.536$; df = 14, 45; $P < 0.001$) and 60 h ($F = 67.381$; df = 14, 45; $P < 0.001$). As the incubation time elapsed, the concentration of the blastospores increased. After 20 h complete growth media (G + SM + CSL + PE + YE), afforded the highest growth of the yeast like hyphal bodies of the *M. anisopliae*.

While, G + PE, G + YE and G + PE + YE showed the lowest optical density of the growth media. After 40 and 60 h of incubation, the medium SM + CSL produced higher concentration of the blastospores (1.995) and remained significantly at higher level than any medium. While G + PE + YE media, showed the lowest values at both the studied time intervals (40 h and 60 h). The complete medium also afforded the growth of the fungi, which ultimately showed higher OD values of the blastospores but significantly at lower level than SM + CSL medium after 40 h and 60 h (Fig. 2a).

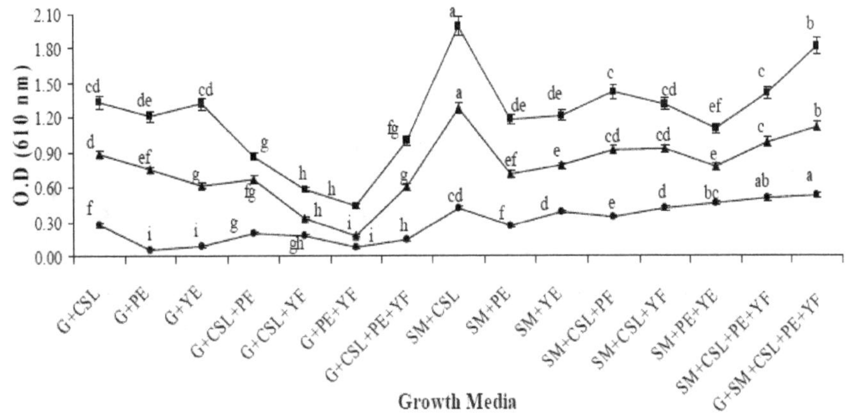

Fig. 2a. Blastospores concentration (OD) of *M. anisopliae* grown on different media supplemented with different sources of nitrogen and sugar after —■— 20 h, —■— 40 h and —■— 60 h incubation. Mean ± SE values with the same letter along the bars of different growth media are not significantly different ($P < 0.05$). For detail of treatments see table 1.

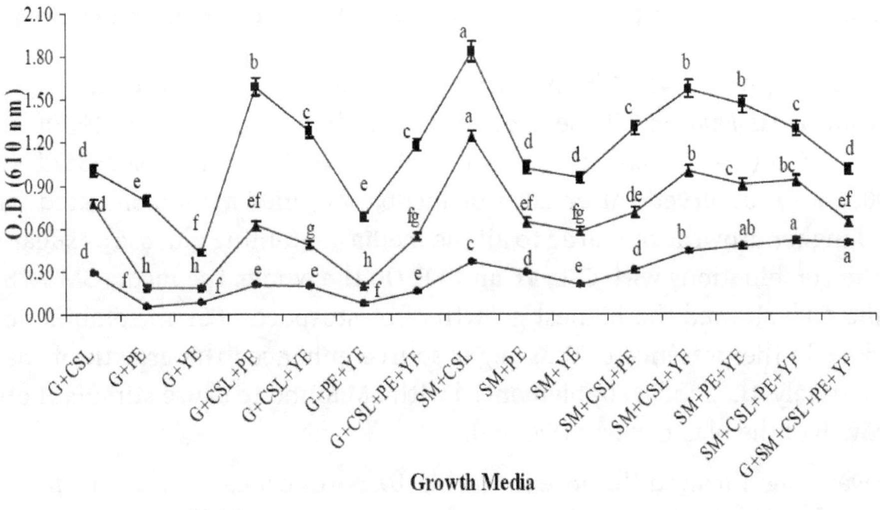

Fig. 2b. Blastospores concentration (OD) of *B. bassiana* grown on different media supplemented with different sources of nitrogen and sugar after —■— 20 h, —■— 40 h and —■— 60 h incubation. Mean ± SE values with the same letter along the bars of different growth media are not significantly different ($P < 0.05$). For detail of treatments see table 1.

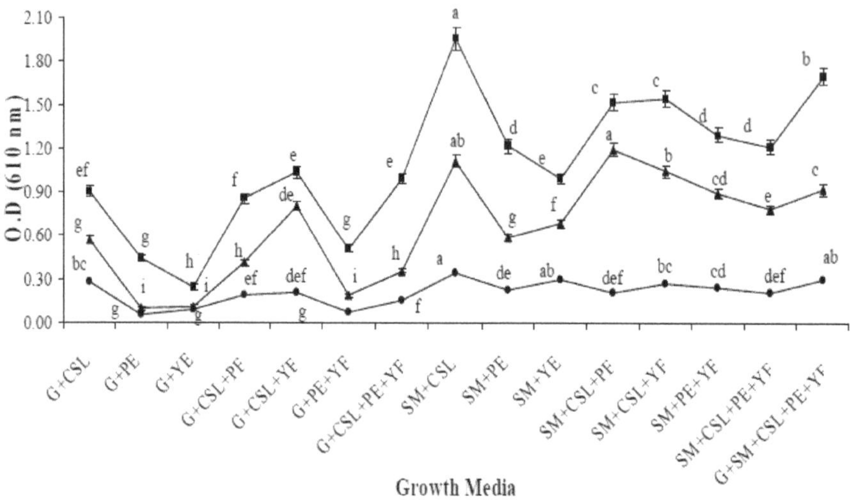

Fig. 2c. Blastospores concentration (OD) of *I. fumosorosea* grown on different media supplemented with different sources of nitrogen and sugar after —■— 20 h, —■— 40 h and —■— 60 h incubation. Mean ± SE values with the same letter along the bars of different growth media are not significantly different ($P < 0.05$). For detail of treatments see table 1.

Blastospores growth was highly variable and significant differences among the growth of *B. bassiana* on all the growth media after 20 h ($F = 172.619$; df = 14, 45; $P < 0.001$), 40 h ($F = 103.584$; df = 14, 45; $P < 0.001$) and 60 h ($F = 62.153$; df = 14, 45; $P < 0.001$) were observed. After 20 h of incubation, media supplemented with SM showed higher growth compared to all the media containing glucose as sugar source in all the combinations with CSL, PE and YE. On the whole, the media SM + CSL after 40 h and 60 h showed the highest growth of blastospores. CSL in combination with PE and YE in the presence of G as sugar source enhanced the growth of the fungi. Comparatively, the media supplemented with SM found to cause stimulant effect on the growth of the blastospores (Fig. 2b).

There was a significant difference in the blastospores concentration of *I. fumosorosea* after 20 h ($F = 36.819$; df = 14, 45; $P < 0.001$), 40 h ($F = 145.915$; df = 14, 45; $P < 0.001$) and 60 h ($F = 110.554$; df = 14, 45; $P < 0.001$) grown on different media supplemented with different sources of nitrogen and sugar in all possible combinations. *I. fumosorosea* grown on SM + CSL exhibited the highest growth not only after 20 h of incubation but also after 60 h (Fig. 2c). While, SM + CSL also promoted the growth resulting higher blastospores but remained significantly lower than SM + CSL + PE, which showed the highest growth after 40 h of incubation. On the whole, media supplemented with SM as sugar source showed higher growth of the blastospores compared with the media containing glucose as sugar source (Fig. 2c).

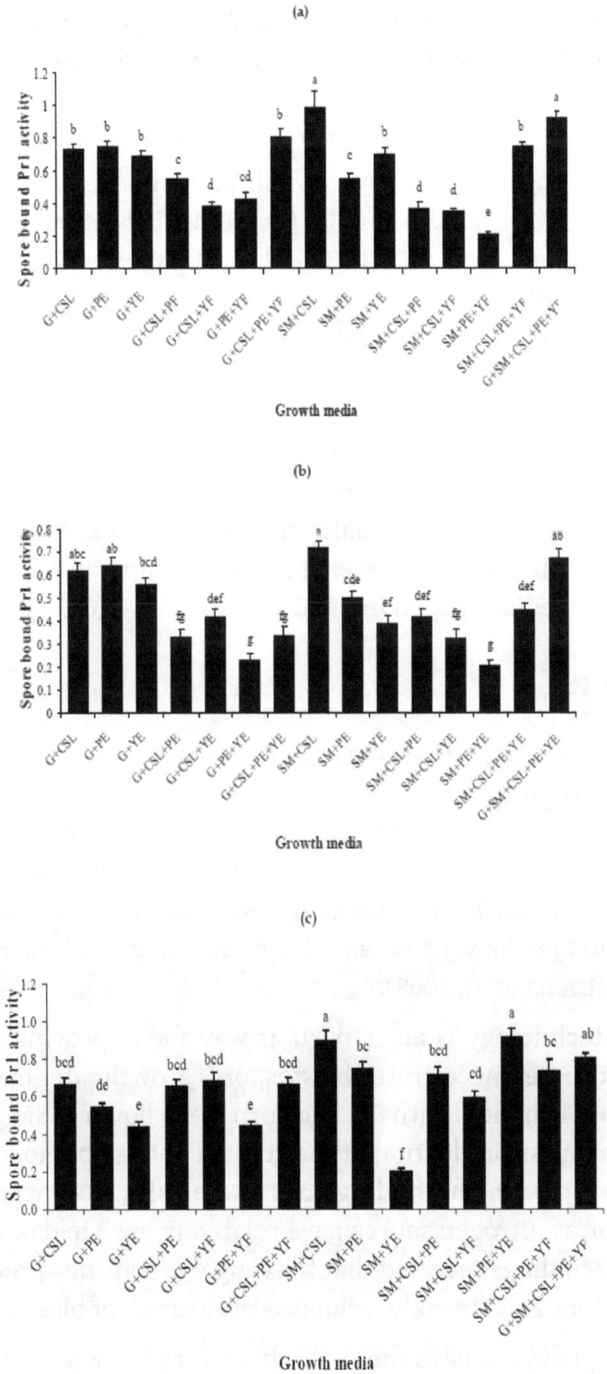

Fig. 3. Extracellular protease Pr1 activity (μmol NA ml^{-1} min^{-1}) of (a) *M. anisopliae* (b) *B. bassiana* and (c) *I. fumosorosea* grown on media supplemented with different sources of nitrogen and sugar. Mean ± SE values with the same letter along the bars of different growth media are not significantly different ($P < 0.05$). For detail of treatments see table 1.

3.3 Effect of different sources of sugar and nitrogen on the enzymatic activity (μmol NA ml^{-1} min^{-1}) of spore bound Pr1 of entomopathogenic fungi

Significant differences in enzymatic activity of spore bound Pr1 were observed when different sources of sugar and nitrogen in all possible combinations were used to grow *M. anisopliae* ($F = 28.945$; df = 14, 30; $P < 0.001$), *B. bassiana* ($F = 20.302$; df = 14, 30; $P < 0.001$) and *I. fumosorosea* ($F = 21.376$; df = 14, 30; $P < 0.001$). The spores of all the studied entomopathogenic fungi cultivated from rice inoculated with medium supplemented with SM + CSL showed the highest activity of Pr1 (Fig. 3a-c). *M. anisopliae* spores grown on complete media also showed higher enzymatic activity (Fig. 3a). The growth medium SM + PE + YE in case of *M. anisopliae,* while SM + PE + YE and G + PE + YE, in case of *B. bassiana* showed the lowest enzymatic activity (Fig. 3a-b). The addition of PE and YE solely in combination with glucose produced spores of *B. bassiana* with relatively higher enzymatic activity compared to SM (Fig. 3b). In case of *I. fumosorosea*, CSL supplemented media showed comparatively higher enzymatic activity except the medium supplemented with SM + YE (0.20 μmol NA ml^{-1} min^{-1}) (Fig. 3c).

4. DISCUSSION

These studies demonstrated that sugar and nitrogen sources significantly effect the growth of blastospores produced by cultures of *B. bassiana, M. anisopliae* and *I. fumosorosea*. Higher blastospores growth and dried fungal biomass was produced in cultures grown on media supplemented with sugarcane molasses (Fig. 1a, c). Previous studies with *Paecilomyces farinosus* (Hotmskiold) and *Paecilomyces lilacinus* (Thom.) showed that media supplemented with SM supported the highest growth of the studied fungi (Leena et al, 2003).

Mass production technology is an important way for improving mycoinsecticides based on the blastospores. Accelerated blastospores growth rates in the current study from the media supplemented with SM and corn steep liquor (CSL) greatly improved the dried fungal biomass production of the studied entomopathogenic fungi. While, our results disagreed with the findings of Leite et al, (2003) who concluded that replacement of CSL as nitrogen source gave relatively low yield of the three studied fungal strains. Our findings revealed that the fungal organisms directly interact with the culture conditions and strongly influence the growth of blastospores.

All the studied fungi did not have the same characteristics when cultivated in media with different complex sources of nitrogen and sugar. Sugarcane molasses, a by-product from the sugar industry, supported higher growth of dried biomass of *M. anisopliae* and *I. fumosorosea*, compared to the media supplemented with glucose. Thus, it may be speculated that SM efficiently enhanced the growth of blastospores, which ultimately led to the production of higher fungal biomass of the fungi. The complete media SM + G + CSL + PE + YE, greatly enhanced the growth of *M. anisopliae* and *I. fumosorosea*. These components when evaluated separately afforded lower growth for *M. anisopliae*. This suggests that these components differ concerning types of nutrients, and therefore provide complete nourishment when offered together.

CSL, peptone and yeast extract afforded higher growth of *B. bassiana* in fungal cultures supplemented with glucose, while these nitrogen sources did not enhance the growth of fungi in the presence of SM. On the other hand, these nitrogen sources when used in combinations (SM + CSL + PE, SM + CSL + YE, SM + PE + YE, SM + CSL + PE + YE), did not increase fungal production of *B. bassiana*. These combinations in the presence of glucose greatly enhanced the dried fungal biomass. Even, the complete medium showed significantly lower growth. On the basis of above findings, we may suggest that nutrition greatly influenced the growth of *B. bassiana*. The result of our study corroborates similar research on the effect of nutrition and propagule production in *Metarhizium* spp. and other entomopathogenic hyphomycetes (Inch et al, 1986; Rombach, 1989; Kleespies & Zimmermann, 1992; Jackson et al, 1997; Vidal et al, 1998).

CSL, a by-product from the corn industry, supported higher growth of the blastospores (Fig. 3a-c), and also ultimately led to the production of spores with higher enzymatic activity of Pr1. Since, peptone and yeast extract are expensive sources of nitrogen; CSL was chosen because of its stimulatory effect and cost economics and could be used to replace the nitrogen sources used previously. Supplementation with 1% CSL with 2.66% SM enhanced not only the growth of all the studied fungi but also enzymatic activity of spore bound Pr1 of only *M. anisopliae*. This is in agreement with the results of McCoy et al, (1988), that nutrition is one of the several factors that may determine the specificity of a fungal pathogen. The results suggest that the growth of blastospores can be efficiently improved from inexpensive CSL and SM, making fermentation an economical and environmental friendly process.

The management of arthropod pests often encompasses preventive strategies and corrective control techniques (Lewis, 1997; Su & Scheffrahn, 1998). Insecticides now registered have undergone extensive field-testing, yielding variable efficacy results. Certain pesticides are costly and exhibit lower persistence, resulting in diminished lifetime and the collapse of the chemical barrier (Su and Scheffrahn, 1998). Moreover, substantial amounts of persistent pesticides are eliciting apprehensions over applicator safety, environmental pollution, and potential harmful effects on non-target species. Considering the aforementioned difficulties, it is imperative to standardize microbe-based products for insect control to ensure the safety of humans, animals, and the environment.

5. CONCISE SUMMARY

In conclusion, substituting nitrogen and sugar sources with CSL and SM, respectively, in the liquid production medium markedly enhanced the development and activity of spore-bound extracellular protease Pr1 of M. anisopliae for the first time. The accelerated growth rate of blastospore production enabled us to identify suitable substrates for the large-scale commercial cultivation of entomopathogenic fungus, facilitating the safe management of insect pests and mitigating the harmful effects of insecticides. Entomopathogenic fungi, as part of an integrated strategy, can offer substantial and targeted pest management. In the foreseeable future, we anticipate synergistic integrations of microbial control agents with various technologies (including semiochemicals, soft chemical pesticides, alternative natural enemies, resistant plant varieties, chemigation, remote sensing, etc.) that will augment the efficacy and sustainability of integrated control strategies.

6. REFERENCES

Al-Aïdroos, K. & Roberts, D.W. (1978). Mutants of *Metarhizium anisopliae* with increased virulence toward mosquito larvae. *Canadian Journal of Genetics and Cytology,* Vol.20, pp. 211-219, ISSN 0008-4093.

Amiri-Besheli, B., Khambay, B. Cameron, M. Deadman, M.L. & Butt, T.M. (2000). Interand intra-specific variation in destruxin production by the insect pathogenic *Metarhizium*, and its significance to pathogenesis. *Mycological Research*, Vol.104, pp. 447-452, ISSN 0953-7562.

Ansari, M.A., Shah, F.A. Tirry, L. & Moens, M. (2006). Field trials against *Hoplia philanthus* (Coleoptera: Scarabaeidae) with a combination of an entomopathogenic nematode and the fungus *Metarhizium anisopliae* CLO 53. *Biological Control,* Vol.39, No.3, pp. 453-459, ISSN 1049-9644.

Bhattacharyya, A., Samal, A.C. & Kar, S. (2004). Entomophagous fungus in pest management. News Letter, Vol.5, pp. 1-4.

Boucias, D.G. & Pendland, J.C. (1991). Attachment of mycopathogens to cuticle: the initial event of mycoses in arthropod hosts. In: *The fungal spore and disease initiation in plants and animals.* G.T. Cole, H.C. Hoch. (Eds.), Plenum Press, New York, pp. 101- 127.

Boucias, D.G. & Pendland, J.C. (1982). Ultrastructural studies on the fungus, *Nomuraea rileyi*, infecting the velvetbean caterpillar, *Anticarsia gemmatalis. Journal of Invertebrate Pathology,* Vol.39, pp. 338-345, ISSN 0022-2011.

Burges, H. D. (1981). Microbial Control of Pests and Plant Diseases 1970–1980. Academic Press, London, ISBN 9780121433604.

Chandler, D., Davidson, G. Pell, J.K. BALL, B.V. Shaw, K. & Sunderland K. D. (2000). Fungal biocontrol of Acari. *Biocontrol Science and Technology,* Vol.10, pp. 357-384, ISSN 0958- 3157.

Charnley, A.K. (1984). Physiological aspects of destructive pathogenesis in insects by fungi: a speculative review. *British Mycological Society Symposium,* Vol.6, pp. 229-270.

Clarkson, J.M. & Charnley, A.K. (1996). New insights into the mechanisms of fungal pathogenesis in insects. *Trends in Microbiology,* Vol.4, pp. 197-203, ISSN 0966-842X.

Daoust, R.A., Ward, M.G. & Roberts, D.W. (1982). Effect of formulation on the virulence of *Metarhizium anisopliae* conidia against mosquito larvae. *Journal of Invertebrate Pathology*, Vol.40, pp. 228-236, ISSN 0022-2011.

Dean, D.D. & Domnas, A. (1983). The extracellular proteolytic enzymes of the mosquito– parasitizing fungus *Lagenidium giganteum. Experimental Mycology,* Vol.7, pp. 31-39, ISSN 0147-5975.

Devi, P.S.V. (1994). Conidia production of the entomopathogenic fungus *Nomuraea rileyi* and its evaluation for control of *Spodoptera litura* (Fab) on *Ricinus communis. Journal of Invertebrate Pathology,* Vol. 63, No. 2, pp. 145-150, ISSN 0022-2011.

Eyal, J., Mabud, M.D.A. Fischbein, K.L. Walter, J.F. Osborne, L.S. & Landa, Z. (1994). Assessment of *Beauveria bassiana* Nov. EO–1 strain, which produces a red pigment for microbial control. *Applied Biochemistry and Biotechnology,* Vol.44, pp. 65-80, ISSN 0273-2289.

Fargues, J. (1984). Adhesion of the fungal spore to the insect cuticle in relation to pathogenicity. In: *Infection processes of fungi.* D.W. Roberts, J.R. Aist (Eds.). The Rockefeller Foundation, New York, pp. 90-110, ISBN-13: 9780813300238

Ferron, P. (1978). Biological control of insect pests by entomogenous fungi. *Annual Review of Entomology,* 23, pp. 409-442, ISSN 0066-4170.

Fusconi, R., Godinho, M.J.L. & Bossolan, N.R.S. (2005). Culture and exopolysaccharide production from sugarcane molasses by *Gordonia polyisoprenivorans* CCT 7137, isolated from contaminated groundwater in Brazil. *World Journal of Microbiology & Biotechnology* Vol.24. No.7, pp. 937-943, ISSN 0959-3993.

Goettel, M.S., St. Leger, R.J. Rizzo, N.W. Staples, R.C. Roberts, D.W. (1989). Ultrastructural localization of a cuticle–degrading protease produced by entomopathogenic fungus *Metarhizium anisopliae* during penetration of the host (*Manduca sexta*) cuticle. *Journal of General Microbiology,* Vol.135, pp.2233-2239, ISSN 0022-1287.

Gupta, S., Montillot, C. & Hwang, Y.S. (1994). Isolation of novel beauvericin analogues from the fungus *Beauveria bassiana*. *Journal of Natural Products,* Vol.58, pp. 733-738, ISSN 0163-3864.

Hadley, N.F. (1981). Cuticular lipids of terrestrial plants and arthropods: a comparison of their structure, composition and waterproofing barrier. *Biology Reviews,* Vol.56, pp. 23-47, ISSN1469-185X.

Hajek, A.E. & St. Leger, R.J. (1994). Interaction between fungal pathogens and insect hosts. *Annual Review of Entomology,* Vol.39, pp. 293-322. ISSN 0066-4170.

Hussain, A., Ahmed, S. & Shahid, M. (2011) Laboratory and field evaluation of *Metarhizium anisopliae var. anisopliae* for controlling subterranean termites. *Neotropical Entomology,* Vol.40, No.2 pp. 244-250, ISSN 1519-566X.

Hussain, A., Tian, M.Y. He, Y.R. Bland, J.M. & Gu, W.X. (2010a). Behavioral and electrophysiological responses of *C. formosanus* towards entomopathogenic fungal volatiles. *Biological Control,* Vol.55, pp. 166-173, ISSN 1049-9644.

Hussain, A., Tian, M.Y. He, Y.R. & Lin, R. (2010b). *In vitro* and *in vivo* culturing impacts on the virulence characteristics of serially passed entomopathogenic fungi. *Journal of Food Agriculture & Environment,* Vol.8, No.3&4, pp. 481-487, ISSN 1459-0255.

Hussain, A., Tian, M.Y., He, Y.R. & Ahmed, S. (2009) Entomopathogenic Fungi disturbed the larval growth and feeding performance of *Ocinara varians* Walker (Lepidoptera: Bombycidae) Larvae. *Insect Science,* Vol.16, No.6, pp. 511–517, ISSN 1672-9609.

Huxham, I.M., Lackie, A.M. & McCorkindale, N.J. (1989). Inhibitory effects of cyclodepsipeptides, destruxins, from the fungus *Metarhizium anisopliae,* on cellular immunity in insects. *Journal of Insect Physiology,* Vol.35, pp. 97-105, ISSN 0022-1910.

Inch, J.M.M., Humphreys, A.M, Trinci, A.P.J. & Gillepsie, A.T. (1986) Growth and blastospore formation by *Paecilomyces flimosoroseus,* a pathogen of brown planthopper (*Nilaparvata lugens*). *Transactions of the British Mycological Society,* Vol.87, pp. 215–222, ISSN 0007-1536.

Jackson, M.A. (1997) Optimizing nutritional conditions for the liquid culture production of effective fungal biological control agents. *Journal of Industrial Microbiology and Biotechnology,* Vol.19, No.3, pp. 180–187, ISSN 1367-5435.

Jackson, M.A., McGuire, M.R, Lacey, L.A. & Wraight S.P. (1997) Liquid culture production of desiccation tolerant blastospores of the bioinsecticidal fungus *Paecilomyces fumosoroseus. Mycological Research,* Vol.101, pp. 35-41, ISSN 0953-7562.

Juárez, M.P. & Fernández, G.C. (2007). Cuticular hydrocarbons of triatomins. Comparative Biochemistry and Physiology – Part A: Molecular and Integrative Physiology. Vol.147, No.3, pp. 711-730, ISSN: 1095-6433.

Kershaw, M.J., Moorhouse, E.R. Bateman, R. Reynolds, S.E. & Charnley, A.K. (1999). The role of destruxins in the pathogenicity of *Metarhizium anisopliae* for three species of insect. *Journal of Invertebrate Pathology,* Vol.74, pp. 213-223, ISSN 0022-2011.

Khetan, S.L. (2001). Microbial pest control. Marcel Dekker, Inc, New York, pp. 211-256, ISBN 9780824704452.

Kleespies, R.G. & Zimmermann, G. (1992) Production of blastospores by three strains of *Metarhizium anisopliae* (Metch.) Sorokin in submerged culture. *Biocontrol Science and Technology,* Vol.2, No.2, pp. 127-135, ISSN 0958-3157.

Kodaira, Y. (1961). Biochemical studies on the muscardine fungi in silkworms, *Bombyx mori.* In: J. Fac. (Ed.), Text. Sci. Tech. 5. Sinshu. University Sericulture. pp. 1-68.

Krasnoff, S.B., Gupta, S. St. Leger, R.J., Renwick, J.A.A. & Roberts, D.W. (1991). Antifungal and insecticidal properties of efrapeptins: metabolites of the fungus *Tolypocladium niveum. Journal of Invertebrate Pathology,* Vol.58, pp. 180-188, ISSN 0022-2011.

Lacey, L.A. & Kaya, H.K. (Eds.). (2000). Field Manual of Techniques in Invertebrate Pathology: Application and Evaluation of Pathogens for Control of Insects and Other Invertebrate Pests. Kluwer Academic, Dordrecht, ISBN 9781402059322.

Leena, M.D., Easwaramoorthy S. & Nirmala R. (2003) *In vitro* production of entomopathogenic fungi *Paecilomyces farinosus* (hotmskiold) and *Paecilomyces lilacinus* (Thom.) Samson using byproducts of sugar industry and other agro-industrial byproducts and wastes. *Sugar Tech,* Vol.5, pp. 231-236, ISSN 0972-1525.

Leite L.G., Alves S.B, Filho A.B. & Roberts D.W. (2005) Simple, inexpensive media for mass production of three entomophthoralean fungi. *Mycological Research,* Vol.109, No.3, pp. 326-334, ISSN 0953-7562.

Leite, L.G., Alves, S.B, Filho, A.B. & Roberts, D.W. (2003) Effect of salts, vitamins, sugars and nitrogen sources on the growth of three genera of Entomophthorales: *Batkoa, Furia,* and *Neozygites. Mycological Research,* Vol.107, No.7, pp. 872-878, ISSN 0953-7562.

Lewis, V.R. (1997). Alternative control strategies for termites. *Journal of Agricultural Entomology,* Vol.14, pp. 291- 307, ISSN 0735-939X.

Madelin, M.F., Robinson, R.F. & Williams, R.S. (1967). Appressorium like structures in insect parasitizing deuteromycetes. *Journal of Invertebrate Pathology*, Vol.9, pp. 404–412, ISSN 0022-2011.

Mazet, I. & Vey, A. (1995). Hirsutellin A, a toxic protein produced *in vitro* by *Hirsutella thompsonii*. *Microbiology*, Vol.141, pp. 1343-1348, ISSN 1350-0872.

McCauley, V.J.E., Zacharuk, R.Y. & Tinline, R.D. (1968). Histopathology of the green muscardine in larvae of four species of Elateridae (Coleoptera). *Journal of Invertebrate Pathology*, Vol.12, pp. 444–459, ISSN 0022-2011.

McCoy, C.W., Samson, R.A. & Boucias, D.G. (1988). Entomogenous fungi. In: *Handbook of Natural Pesticides vol V. Microbial insecticides, part A. Entomogenous protozoa and fungi.* C.M. Ignoffo. (Ed.), 151-236, ISBN CRC Press, Boca Raton, FL, ISBN 9780849336607. Mohamed, A.K.A., Sikorowski, P.P. & Bell, J.V. (1978). Histopathology of *Nomuraea rileyi* in larvae of *Heliothis zea* and *in vitro* enzymatic activities. *Journal of Invertebrate Pathology*, Vol.31, pp. 345-352, ISSN 0022-2011.

Paterson-Beedle, M., Kennedy, J.F, Melo, F.A.D, Lloyd, L.L. & Medeiros, V. (2000) A cellulosic exopolysaccharide produced from sugarcane molasses by a *Zoogloea* sp. Carbohydrate Polymers, Vol.42, No.4, pp. 375-383, ISSN 0144-8617.

Pendland, J.C. (1982). Resistant structures in the entomogenous hypomycete, *Nomuraea rileyi*: an ultrastructural study. *Canadian Journal of Botany*, Vol.60, pp. 1569-1576, ISSN 1480-3305.

Prasertphon, S. & Tanada, Y. (1968). The formation and circulation, in *Galleria*, of hyphal bodies of entomophthoraceous fungi. *Journal of Invertebrate Pathology*, Vol.11, pp. 260-280, ISSN 0022-2011.

Roberts, D.W. & Hajek, A.E. (1992). Entomopathogenic fungi as bioinsecticides. In: *Frontiers in Industrial Mycology.* G. Leathan (Ed.). Chapman and Hall, New York, pp. 144-159.

Roberts, D.W. (1966). Toxins from the entomogenous fungus *Metarhizium anisopliae.* II. Symptoms and detection in moribund hosts. *Journal of Invertebrate Pathology*, Vol.8, pp. 222-227, ISSN 0022-2011.

Roberts, D.W. (1981). Toxins of entomopathogenic fungi. In: Microbial control of pests. H.D. Burges (Ed.). Academic Press, London, pp. 441-464, ISBN 9780121433604.

Rombach, M.C. (1989) Production of *Beauveria bassiana* [*Deuteromycotina, Hyphomycetes*] sympodulo-conidia in submerged culture. *BioControl*, Vol.34, No.1, pp.45-52, ISSN 1386-6141.

Samsináková, A., Misiková, S. & Leopold, J. (1971). Action of enzymatic systems of *Beauveria bassiana* on the cuticle of the greater wax moth larvae (*Galleria mellonella*). *Journal of Invertebrate Pathology*, Vol.18, pp. 322-330, ISSN 0022-2011.

Samuels, R.J., Charnley, A.K. & Reynolds, S.E. (1988). The role of destruxins on the pathogenicity of 3 strains of *Metarhizium anisopliae* for the tobacco hornworm *Manduca sexta*. *Mycopathologia*, Vol.104, pp. 51-58. ISSN 0301-486X.

SAS Institute. 2000. SAS user's guide: statistics, SAS Institute, Cary, North Carolina.

Shah, F.A. Allen, N. Wright, C.J. & Butt, T.M. (2007). Repeated *in vitro* subculturing alters spore surface properties and virulence of *Metarhizium anisopliae*. *FEMS Microbiology Letters*, Vol.276, pp. 60-66, ISSN 0378-1097.

Shah, F.A. & Butt, T.M. (2005) Influence of nutrition on the production and physiology of sectors produced by the insect pathogenic fungus *Metarhizium anisopliae*. *FEMS Microbiology Letters*, Vol.250, No.2, pp. 201-207, ISSN 0378-1097.

St. Leger, R.J., Allee, L.L. May, B. Staples, R.C. & Roberts, D.W. (1992). World-wide distribution of genetic variation among isolates of *Beauveria* spp. *Mycological Research*, Vol.96, pp. 1007-1015, ISSN 0953-7562.

St. Leger, R.J., Butt, T.M. Staples, R.C. & Roberts, D.W. (1989). Synthesis of proteins including a cuticle-degrading protease during differentiation of the entomopathogenic fungus *Metarhizium anisopliae*. *Experimental Mycology*, Vol.13, pp. 253-262, ISSN 0147-5975.

St. Leger, R.J., Durrands, P.K. Charnley, A.K. & Cooper, R.M. (1988). Role of extra-cellular chymoelastase in the virulence of *Metarhizium anisopliae* for *Manduca sexta*. *Journal of Invertebrate Pathology*. Vol.52, pp. 285-293, ISSN 0022-2011.

St. Leger, R.J., Joshi, L. Bidochka. & M. Roberts, D.W. (1996), Construction of an improved mycoinsecticide over expressing a toxic protease. *Proceedings of National Academy of Sciences USA*, Vol.93, pp. 6349–6354, ISSN 0027-8424.

St. Leger, R.J. (1994). The role of cuticle degrading proteases in fungal pathogenesis of insects. *Canadian Journal of Botany*. Vol.73 (Suppl. 1), pp. 1119-1125, ISSN 1480-3305.

St. Leger, R.J., Charnley A.K. & Cooper R.M. (1986) Cuticle degrading enzymes of entomopathogenic fungi: Synthesis in culture on cuticle. *Journal of Invertebrate Pathology*, Vol.48, No.1, pp. 85-95, ISSN 0022-2011.

St. Leger, R.J., Cooper, R.M. & Charnley, A.K. (1987). Production of cuticle-degrading enzymes by the entomopathogen *Meterhizium anisopliae* during infection of cuticles from *Calliphora vomitoria* and *Manduca sexta*. *Microbiology*, Vol.133, No.5, pp. 1371– 1382, ISSN 1350-0872.

Steinhaus, E.A. (1956). Microbial control: The emergence of an idea. *Hilgardia*, Vol.26, pp. 107-160. ISSN 0073-2230.

Steinhaus, E.A. (1975). Disease in a Minor Cord. Ohio State Univ. Press, Columbus, OH, ISBN 9780814202180.

Su, N.Y. & Scheffrahn, R.H. (1998). A review of subterranean termite control practices and prospects for integrated pest management programs. *Integrated Pest Management Reviews,* Vol.3, pp. 1-13. ISSN 1572-9745.

Sutherland, I.W. (1996) Microbial biopolymers from agricultural products: production and potential. *International Biodeterioration & Biodegradation* 38: 249–261, ISSN 0964-8305.

Suzuki, A., Kanaoka, M. Isogai, A. Murakoshi, S. Ichinoe, M. & Tamura, S. (1977). Bassianolide, a new insecticidal cyclodepsipeptide from *Beauveria bassiana* and *Verticillium lecanii. Tetrahedron Letters,* Vol.18, pp. 2167-2170, ISSN 0040-4039.

Suzuki, A., Kawakami, K. & Tamura, S. (1971). Detection of destruxins in silkworm larvae infected with *Metarhizium anisopliae. Agricultural and Biological Chemistry,* Vol.35, pp. 1641-1643. ISSN 0002-1369.

Tanada, Y. & Kaya, H.K. (1993). Insect Pathology. Academic Press, New York, ISBN 0-12- 683255-2.

Vey, A., Hoagland, R.E. & Butt, T.M. (2001). Toxic metabolites of fungal control agents. In: *Fungi as Biocontrol Agents.* T.M. Butt, C. Jackson, N. Magan, (Eds.), CAB International, New York, pp. 311-346, ISBN 0-85199-356-7.

Vidal C., Fargues, J. Lacey, L.A. & Jackson, M.A. (1998) Effect of various liquid culture media on morphology, growth, propagule production, and pathogenic activity to *Bemisia argentifolii* of the entomopathogenic hyphomycete, *Paecilomyces fumosoroseus. Mycopathologia,* Vol. 143, No. 1, pp. 33-46, ISSN 0301-486X.

Vilcinskas, A., Matha, V. & Gotz, P. (1977). Inhibition of phagocytic activity of plasmatocytes isolated from *Galleria mellonella* by entomogenous fungi and their secondary metabolites. *Journal of Insect Physiology,* Vol.43, pp. 475-483, ISSN 0022-1910.

Weiser, J. & Matha, V. (1988). Tolypin: A new insecticidal metabolite of fungi of the genus *Tolypocladium. Journal of Invertebrate Pathology,* Vol.51, pp. 94-96, ISSN 0022-2011.

Wraight, S.P., Inglis, G.D. & Goettel, M.S. (2007). Fungi, In: *Field manual of techniques in invertebrate pathology,* L.A. Lacey & H.K. Kaya, (Eds.), 223-248, 2nd edition, Springer, Dordrecht, ISBN 978-1-4020-5931-5..

Ypsilos, I.K. & Magan, N. (2005). Characterization of optimum cultural environmental conditions for the production of high numbers of *Metarhizium anisopliae* blastospores with enhanced ecological fitness. *Biocontrol Science and Technology,* Vol. 15, No. 7, pp. 683-699, ISSN 0958-3157.

Zacharuk, R.Y. (1970a). Fine structure of the fungus *Metarhizium anisopliae* infecting three species of larval Elateridae (Coleoptera). II. Conidial germ tubes and appressoria. *Journal of Invertebrate Pathology,* Vol.15, pp. 81-91, ISSN 0022-2011.

Zacharuk, R.Y. (1970b). Fine structure of the fungus *Metarhizium anisopliae* infecting three species of larval Elateridae (Coleoptera): III. Penetration of the host integument. *Journal of Invertebrate Pathology,* Vol.15, 372-396, ISSN 0022-2011.

Zhao, L.L., Chen, X.X. & Xu, J.H. (2010). Strain improvement of *Serratia marcescens* ECU1010 and medium cost reduction for economic production of lipase. *World Journal of Microbiology and Biotechnology,* Vol. 26, No. 3, pp. 537-543, ISSN 0959-3993.

Zimmermann, G. (1993). The entomopathogenic fungus *Metarhizium anisopliae* and its potencial as a biocontrol agent. *Pesticide Science*, Vol.37, pp. 375-379, ISSN 1096- 9063.

CHAPTER 7

NEUROPHYSIOLOGICAL RECORDING TECHNIQUES APPLIED TO INSECT CHEMOSENSORY SYSTEMS

1. INTRODUCTION: AN OVERVIEW

This chapter aims to examine contemporary electrophysiological recording techniques employed in the investigation of olfactory and gustatory information processing in insects. We will delineate strategies utilized (a) to ascertain the physiological features of gustatory receptor cells (GRCs) and olfactory receptor cells (ORCs) in the peripheral nervous system and (b) to physiologically characterize identifiable olfactory neurons in the central nervous system. Utilizing animal models, such as insects, can be beneficial for experimenting with the anatomical and functional organization of the nervous system. Insect neural systems serve as exemplary models for studying gustatory and olfactory mechanisms due to their very basic peripheral and central nervous systems relative to vertebrates. They possess readily accessible sensory organs (sensilla) and distinct brain structures that may be individually identified. The quantity of receptor cells facilitating the processes associated with olfaction and gustation is rather limited. Notably, certain brain regions in both invertebrates and vertebrates exhibit significant morphological and physiological similarities, hence the insect nervous system offers valuable insights into the fundamental principles of taste and odor coding present in higher vertebrates. Gustatory and olfactory stimuli are crucial in influencing insect behavior. Insects depend on these senses for food source sampling and selection, avoidance of harmful or poisonous substances, mating, and identifying oviposition sites. GRCs and ORCs in sensilla, mostly situated on the mouthparts, legs, and antennae, detect these chemical signals and transduce them into a neural code of action potentials, which is transmitted to the central nervous system for processing. GRCs and ORCs serve as sensory filters for ambient olfactory and gustatory cues in insects. They constitute the initial layer of the decision-making process and convey information straight to brain areas.

2. INSECT CHEMOSENSORY ORGANS

Insects serve as exemplary models for olfactory and gustatory research. They possess several sensory organs or sensilla that enable continuous monitoring and response to alterations in their internal and external habitats, hence ensuring optimal conditions for survival. Sensilla facilitate sensory perception for olfaction, gustation, auditory stimuli, tactile sensation, vision, proprioception, and georeception, thermoreception, and hygrometry. These unique cuticular structures exhibit variability in size and shape, serving as the initial mechanism for environmental awareness. Receptor cells that innervate these sensilla are specialized to detect environmental conditions and alterations, transmitting information about the nature of these changes to the central nervous system. In insects, behavior is governed by environmental taste (gustatory) and smell (olfactory) inputs. Gustatory and olfactory sensilla are generally innervated by several bipolar sensory neurons. These neurons possess dendrites encased by auxiliary or sheath cells. Sensory information is converted by receptor cells into electrical impulses, leading to the creation of nerve signals made up of action potentials. The absolute frequency and temporal dispersion of action potentials within a spike train provide information regarding the stimulus. Axons of sensory neurons convey this information to modality-specific brain areas. Deciphering the sensory code involves stimulating particular sensilla and electrophysiologically measuring the sequences of action potentials (input), and assessing the resultant behavior (output).

2.1 Peripheral olfactory processing

Olfactory stimuli play an important role in the orientation of many animals in their environment. Moths detect odor cues with their main olfactory organs, paired antennae. The antennae detect diverse mixtures of volatiles by means of ORCs residing in various types of sensilla. These sensilla are the crucial interface between the outer world and the central nervous system of the moth. The olfactory systems of both invertebrates and vertebrates share many similarities (see Hildebrand and Shepherd, 1997) and are capable of detecting and discriminating among a large number of odorants that differ in size, shape, and complexity. The olfactory organs of invertebrates (e.g., paired antennae of insects) and vertebrates (e.g., nose in mammals) are adapted to detect a vast array of odorants by means of receptors that are located on ORCs. These ORCs are associated with various types of sensilla in invertebrates (e.g., insects) or the olfactory epithelium lining a portion of the nasal cavity of vertebrates (e.g., mammals). Clyne et al. (1999) and Vosshall et al. (1999) identified a novel family of seven-transmembrane-domain proteins, which are encoded by 100-200 genes and are likely to function as *Drosophila melanogaster* olfactory receptors. An individual ORC in the antenna of *D. melanogaster* is thought to express one or a few of the candidate olfactory receptor genes and therefore, each ORC is functionally distinct (Vosshall, 2001).

Because of the fundamental morphological and physiological similarities between invertebrate and vertebrate olfactory systems and since current evidence indicates that basic olfactory processing is similar across all phyla (Hildebrand and Shepherd, 1997), it is feasible to use insects, such as the sphinx moth *Manduca sexta*, as a model system and thereby gain insights into the neural mechanisms of odor recognition and discrimination common to insects and to other animals (Hildebrand, 1995; 1996; Hildebrand and Shepherd, 1997). Insects are ideal experimental models because they possess readily accessible olfactory sensilla, have relatively simple peripheral and central nervous systems, possess individually identifiable neural structures, and have a relatively small number of sensory cells mediating olfactory mechanisms. In moths, the olfactory system comprises two parallel subsystems. One system processes information about plant associated volatiles odors and is very similar to the main olfactory pathway in vertebrates. The other system is narrowly specialized to detect and respond to information regarding the sex pheromone emitted by conspecific females. Many researchers have classified ORCs based on their individual response profiles and have used terms such as "specialist" and "generalist" to describe those ORCs. Specialist ORCs have been described as those responding with a high sensitivity and selectivity to a single type of odor molecule, such as a pheromone component or a narrow range of related odorants, whereas generalist ORCs have been classified as those responding with relatively lower sensitivity to a broad range of odorants, such as plant-associated volatiles (Schneider et al., 1964). Interestingly, several researchers have found, that some insect ORCs respond to plant-associated volatiles and exhibit relatively high selectivity and sensitivity for effective stimulus molecules (e.g., Priesner, 1979; Dickens, 1990; Anderson et al., 1995; Heinbockel and Kaissling, 1996; Pophof, 1997; Hansson et al., 1999; Shields and Hildebrand, 2001a).

2.2 Antennal morphology

The antennae of both sexes of *M. sexta* comprise three segments, two small, basal segments (scape and pedicel) and a long distal segment (flagellum) (Fig. 1A, B). In adult *M. sexta*, the antennal flagellum is about 2 cm long, comprises approximately 80 subsegments called annuli (or flagellomeres) (Fig. 1C), and is sexually dimorphic (Sanes and Hildebrand, 1976). Each male and female antenna has approximately 4×10^5 ORCs (Sanes and Hildebrand, 1976) and 3.0×10^5 - 3.4×10^5 ORCs (Oland and Tolbert, 1988), respectively. The antenna of each sex is associated with about 10^5 sensilla (Sanes and Hildebrand, 1976; Keil, 1989; Lee and Strausfeld, 1990; Shields and Hildebrand, 1999a; 1999b) and each annulus may bear approximately 2100-2200 sensilla (Fig. 1C, D) (Lee and Strausfeld, 1990; Shields and Hildebrand, 1999b). Male flagella possess long and hair-like (trichoid) male-specific sensilla that house ORCs specialized to detect components of the conspecific female's sex pheromone (Sanes and Hildebrand, 1976; Kaissling et al., 1989). Females also possess trichoid sensilla (i.e., trichoid type-A) (Fig. 1C, D). They are much shorter than in males and respond to plant volatiles. In addition, both male and female antennae carry several other types of olfactory sensilla, some of which resemble short pegs, thought to contribute to the detection of plant-associated odorants. In total, there are five types present in males (Sanes and Hildebrand, 1976; Lee and Strausfeld, 1990) and six types present in females (Shields and Hildebrand, 1999a; 1999b). In the case of trichoid type-A sensilla, circumferential cuticular ridges are present which form a helical pattern (Fig. 1E). The cuticular shaft of all olfactory sensilla is pierced by a multitude of pores (Fig. 1E). These pores extend through the entire thickness of the shaft. It is generally believed that olfactory molecules gain entry through these pores and, in order to reach the dendritic receptor sites of an ORC, the small volatile lipid-soluble odor molecules must traverse an aqueous phase, perhaps with the aid of odorant-binding proteins.

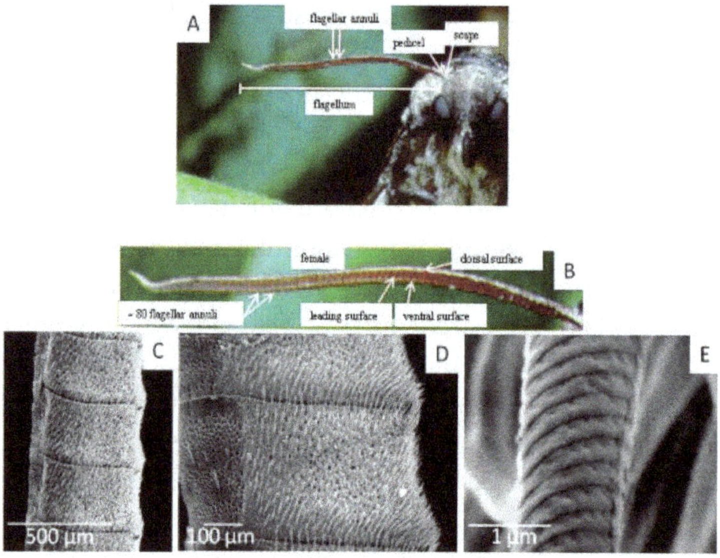

Fig. 1. Female *Manduca sexta* antenna. (A) Light micrographs showing a female *M. sexta* moth and its antenna. The antenna comprises three segments, two small, basal segments (scape and pedicel) and a long distal segment (flagellum). The antennal flagellum is about 2 cm long and comprises approximately 80 subsegments called annuli (or flagellomeres). (B) Higher magnification view of (A) showing the multitude of annuli. (C) Scanning electron micrograph of a portion of an adult female antennal flagellum showing a higher magnification of three annuli. (D) Higher magnification view of a single annulus. Long, hair-like sensory organs (sensilla) called trichoid sensilla are abundant on the surface of an annulus. (E) Higher magnification view of a single trichoid type-A sensillum showing the cuticular shaft and pores that extend through its entire thickness. The shaft bears circumferential cuticular ridges which form a helical pattern. Odorant molecules diffuse through these pores and interact with the underlying dendrites.

The binding of these odorant ligands to odor-specific receptors coupled to G-proteins initiates a cascade of intracellular second messengers (e.g. cyclic AMP, IP3) that ultimately activate cyclic nucleotide-gated cation-permeable channels (e.g. Ca^{2+}, Cl^-, K^+) (reviewed by e.g. Firestein, 1992; Shepherd, 1994). These events result in the generation of action potentials in temporal patterns (in each ORC axon) and spatial patterns (across the array of ORC axons) that represent features of the stimulus and travel along the ORC axons to the primary olfactory center in the brain (i.e., the antennal lobe, AL, of insects or the olfactory bulb of vertebrates). The axons of antennal ORCs project to and terminate in compartments of condensed synaptic neuropil (glomeruli) in the AL (Hildebrand and Shepherd, 1997). Mounting evidence indicates that glomeruli are discrete anatomical and functional units, each dedicated to collecting and processing olfactory information about a subset of odor molecules (see below). Information about the odor stimulus is conveyed to a particular glomerulus by the axons of ORCs that express a particular olfactory receptor protein (Buck, 1996; Hildebrand and Shepherd, 1997; Mombaerts, 1996).

2.3 Peripheral gustatory processing

Insect larvae depend largely on their sense of taste and smell to find food. These larvae possess elaborate sensory organs (i.e., sensilla) located on the antennae and mouthparts that serve to gather olfactory and gustatory information on the chemical composition of the food plant. The feeding response in lepidopterous larvae is controlled by input from gustatory sensilla located on the mouthparts (Schoonhoven and Dethier, 1966; Shields, 1994). Food plant recognition is thought to be primarily mediated by the input from a bilateral pair of styloconic sensilla (Schoonhoven and Dethier, 1966; de Boer et al., 1977; de Boer and Hanson, 1987) located on the mouthparts. Each sensillum houses four GRCs (Figs. 2, 6) that are thought to play a primary role in hostplant discrimination. They have been referred to as the salt- sugar-, inositol-, and deterrent-sensitive cells (Schoonhoven, 1972; Schoonhoven et al., 1992), since they typically respond to salt, sweet, inositol, and bitter compounds, respectively (e.g. Schoonhoven, 1972; Frazier, 1986; Shields and Mitchell, 1995; Bernays et al., 1998; Glendinning et al., 1999). During feeding, the sensilla are in continuous contact with the sap liberated from the plant leaf and are capable of detecting different chemicals (i.e., phytochemicals) present in the plant. This gustatory sensory input is encoded as patterns of nerve impulses by GRCs and this information is then transferred to taste centers in the brain of the insect. Therefore, GRCs form the first layer of the decision-making process that ultimately determines whether food is acceptable or should be rejected. Thus, the insect faces the task to decipher individual tastants in a complex multimolecular mixture and to make appropriate feeding choices.

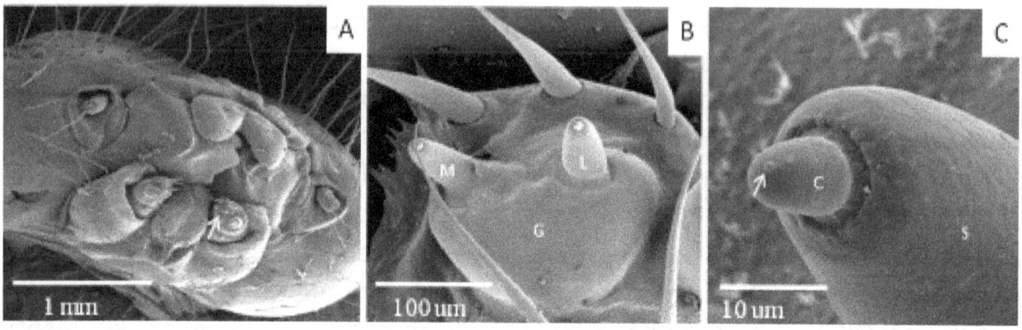

Fig. 2. Scanning electron micrographs showing the (A) whole head of a gypsy moth, *Lymantria dispar*, larva. The arrow denotes the location of the styloconic sensilla. (B) Higher magnification view of the lateral (L) and medial (M) styloconic sensilla located on the galea (G). (C) Higher magnification view of a lateral styloconic sensillum. The sensillum is comprised of a cone (C) or peg inserted into a style (S) or column. A terminal pore (arrow) is visible at the apex of the cone.

Insect GRCs transduce the quality and quantity of the complex plant chemistry into a neural code of action potentials. Complex stimuli resulting from e.g. plant saps often evoke spike trains in several receptor cells innervating one or more sensilla. The frequency of action potentials and the temporal distribution of action potentials in a spike train contain information about the stimulus. The axons of GRCs travel to, and converge in, the first relay station, the subesophageal ganglion (SOG), without intermittent synapses. Unraveling the sensory code occurs by analyzing "input-output" relationships (Schoonhoven and van Loon, 2002) and can be achieved by stimulating specific sensilla and quantifying electrophysiologically the trains of action potentials (input), as well as quantifying the behavior (output) on the basis of how much food is consumed (Bernays and Chapman, 1994).

To better understand the neural communication between the chemosensory organs and the central nervous system that results in acceptance or rejection behavior, sensory responses have been categorized as (1) labeled line system, (2) across-fiber patterning, and (3) temporal patterning. The first theory suggests that the more important a single compound is for controlling or modifying behavior, the more likely its detection will be coded by a single cell (Stadler, 1984). This "labeled line" (i.e., line or axon along which information is transferred to the brain) to the central nervous system would only carry information from cells with a narrow and well defined sensitivity spectrum of a specific chemical (or family of chemicals) and would be directly linked to a specific behavioral response (Schoonhoven and Blom, 1988). Such chemosensory cells seem to be quite unique for specialized herbivorous insects and have not been documented for other animal groups, such as vertebrates. The second theory suggests that the nervous system bases its decision for behavioral output by evaluating the responses from many individual sensory cells with different but overlapping response spectra and the central nervous system extracts meaningful information by reading and processing the simultaneous inputs across all afferent sensory fibers (axons) (across- fiber patterning) (Dethier and Crnjar, 1982), also known to occur in vertebrates (Dethier, 1982). The third theory suggests that temporal patterning may be superimposed on across- fiber patterning, suggesting that the ratios of firing across different cells changes with time and can modify a particular message (Schoonhoven, 1982). Most importantly, it should be noted that all three theories (code types) are not mutually exclusive and can be amalgamated into one model (Schoonhoven et al., 1992).

Sensory codes mediating acceptance can: (i) stimulate specific sugar cells coding for acceptance profile; (ii) stimulate broad spectrum sugar cells that the CNS recognizes as an acceptance profile (Schoonhoven, 1982; 1987), and (iii) inhibit specific phagodeterrent receptors; this contributes to the neural coding of acceptance (Schoonhoven et al., 1998). Feeding deterrents may alter sensory input by: (i) stimulating specific deterrent receptors; (ii) stimulating broad spectrum receptors; (iii) stimulating some cells and inhibiting others, thereby changing complex and subtle codes; (iv) inhibiting specific phagostimulant receptors; this contributes to the neural coding of deterrence, and (v) evoking highly unnatural impulse patterns, often at high frequency (Schoonhoven et al., 1998). The ability of a deterrent neuron to respond to a wide range of chemicals is due to it having a diverse range of receptor sites, each with its own structure-function specificity, or due to the active chemicals having common features making them able to interact with a single receptor site (Blaney et al., 1988).

Deterrent cells possess a number of unique characteristics: (i) they generally adapt more slowly than cells which respond to phagostimulatory compounds; (ii) the tonic activity of the deterrent receptor stabilizes at a higher level than in other cell types; (iii) there may be a relatively long latency period prior to the tonic response; (iv) there may be a slow increase in spike frequency following stimulus application, and (v) there may be an increase in spike amplitude with stimulus concentration (Schoonhoven, 1982); Hanson and Peterson, 1990). Schoonhoven (1982) used differential adaptation rates to explain that the sensory code changes with time, with the result that the deterrent receptor activity gradually becomes more pronounced in the sensory message sent to the brain. Food, which at the beginning of a meal may be acceptable, soon becomes unacceptable because of the more prominent share of the deterrent in the total sensory impression.

Recently, work by Wanner and Robertson (2008) revealed a family of 65 gustatory receptor (Gr) genes from the silkworm moth, *Bombyx mori*, genome. These authors revealed Gr genes for sugar, as well as those for cuticular hydrocarbons and carbon dioxide. Interestingly, they also found 55 Gr genes that are predominantly bitter receptors involved in the detection of a large variety of secondary plant chemicals and suggested that these Gr genes mediate food choice and avoidance, as well as oviposition site preference. This finding may provide new tools for controlling pest damage and lead to better understanding of the peripheral taste system and is noteworthy since about 99% of the 150,000 described species of lepidopterous insects are phytophagous feeders (Grimaldi and Engel, 2005). Functional characterization of GRCs may also provide a better understanding of the molecular and cellular basis of taste coding. Interestingly, Clyne et al. (2000) found a large Gr gene family and characterized the GRCs expressing divergent GR genes in the fruit fly, *D. melanogaster*. In the sequenced *D. melanogaster* genome, 68 receptors, encoded by 60 genes, were identified and were predicted to encode G protein-coupled receptors (Robertson et al., 2003). Expression and behavioral studies of two *D. melanogaster* Gr genes, *Gr5a* and *Gr66a*, revealed that inactivation of *Gr5a*- positive neurons resulted in a diminished behavioral responses to sugars and low concentrations of salt, whereas inactivation of Gr66a-positive neurons lowered behavioral responses to some bitter compounds (Thorne et al., 2004; Wang et al., 2004). Interestingly, molecular studies revealed that these "sugar" and "bitter" neurons also project to distinct and non-overlapping regions within the SOG (Dunipace et al., 2001; Scott et al., 2001; Thorne et al., 2004; Wang et al., 2004; Dahanukar et al., 2007).

3. EXTRACELLULAR RECORDING TECHNIQUES FOR OLFACTORY STUDIES

A potent instrument for investigating olfactory ORCs is physically recording from individual sensilla through a method known as single-unit or single-sensillum recording (Figs. 3, 4). This approach observes the electrical responses generated by ORCs in reaction to various olfactory stimuli. Although numerous GRCs exist within a single sensillum, the unique electrophysiological responses of each ORC can be differentiated by specialized computer-aided software for spike sorting based on varying spike amplitudes. Understanding the quantity of ORCs located within a certain sensillum, as shown using transmission electron microscopy, can significantly enhance the interpretation and validation of electrophysiological findings.

To record from olfactory ORCs to test the effect of volatile compounds, female *M. sexta* moths reared on artificial diet (17h light: 7h dark; ca., 60% relative humidity), 1-2 days post- eclosion were used for these studies (Bell and Joachim, 1976; Sanes and Hildebrand, 1976). Recordings were carried out at ambient temperatures (24-26ºC). To record extracellulary from ORCs, the cut-tip recording technique was used (Van der Pers and Den Otter, 1978; Kaissling, 1995) (Fig. 4B). This technique involved restraining each moth in a plastic tube (5.2 x 1.2 cm i.d.) so that the head protruded from one end (Fig. 3). One of the moth's antennae was stabilized with a minimal amount of low melting point paraffin wax to allow for easier manipulation of the sensilla. Alternatively, double-sided tape can be used. The tip of a trichoid type-A sensillum, positioned on the distal or proximal margins of the dorsal, ventral, and leading surfaces of an annulus, was cut between two sharpened glass knives (2-mm diameter, Corning Inc., Horseheads, NY) positioned at 90 degrees with respect to each other (micro knife-beveling device; type EG-03, Syntech, Hilversum, The Netherlands) (Van der Pers and Den Otter, 1978; Kaissling, 1995). This method is typically feasible for sensilla less than 50 µm in length. For longer sensilla, an alternative method can be used, whereby the sensillum is cut by pinching off the tip using sharpened forceps (Fig. 4A), thus preventing the opening and subsequent damage of the ORC dendrites due their depolarization. The flat side of the forceps is coated with a thin film of Vaseline to prevent desiccation of the sensillum (i.e., loss of the sensillum lymph surrounding the dendrites) (Kaissling, 1995).

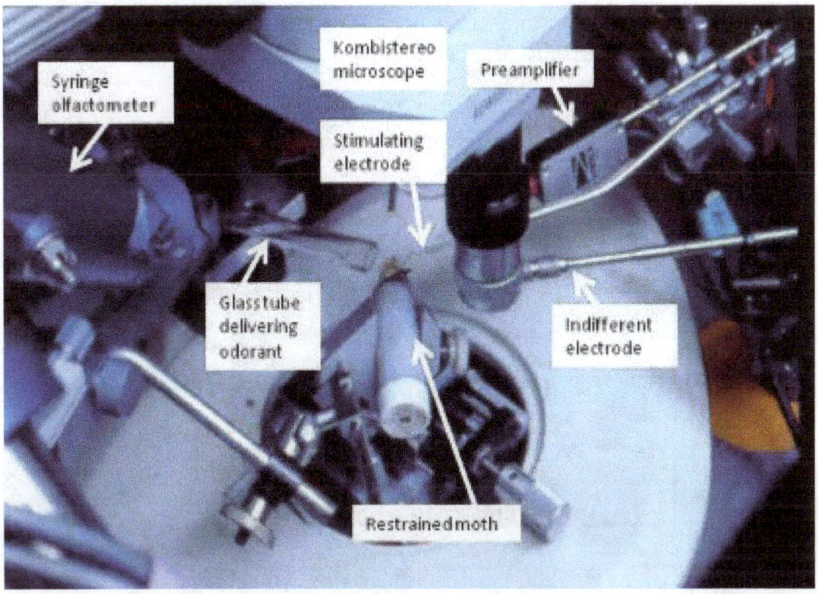

Fig. 3. Typical olfactory extracellular recording set-up.

Recording and ground electrodes (tip diameters: 2.5-3.5 µm and 4-5 µm, respectively) were made from borosilicate glass tubing (1.5 mm o.d., 1.10 mm i.d., Sutter Instrument Co., Novato, CA) on a Flaming-Brown micropipette puller (model P-97, Sutter, Novato, CA). The recording electrode was filled with sensillum-lymph saline solution (Kaissling and Thorson, 1980) and positioned over the tip of a trichoid type-A sensillum with the aid of a micromanipulator (Leitz) and a Wild M3Z Kombistereo microscope (Leica, Germany; total magnification, 640X) (Figs. 3, 4). The ground electrode was filled with hemolymph saline solution (Kaissling and Thorson, 1980) and inserted into the moth's eye or body. Both ground and recording electrodes were held in place by mounting them in electrode holders (1.5 mm i.d., Syntech) containing 0.25 mm silver wire (>99.99 %, Aldrich Chemical Co., Milwaukee, WI) (Fig. 3). Difficulties in cutting the tip of olfactory sensilla can arise if they are very short and small. In this case, a modification of the tip-cutting technique can be used called sidewall recording. In this method, a sharpened tungsten electrode is plunged into the base of the sensillum to establish contact with the underlying olfactory ORCs. In either case, these electrophysiological recording techniques record the contribution of one or all of the ORCs present within a sensillum and preparations can remain stable for several hours. The electrophysiological components of the set-up were placed on a vibration-free air table (Figs. 3).

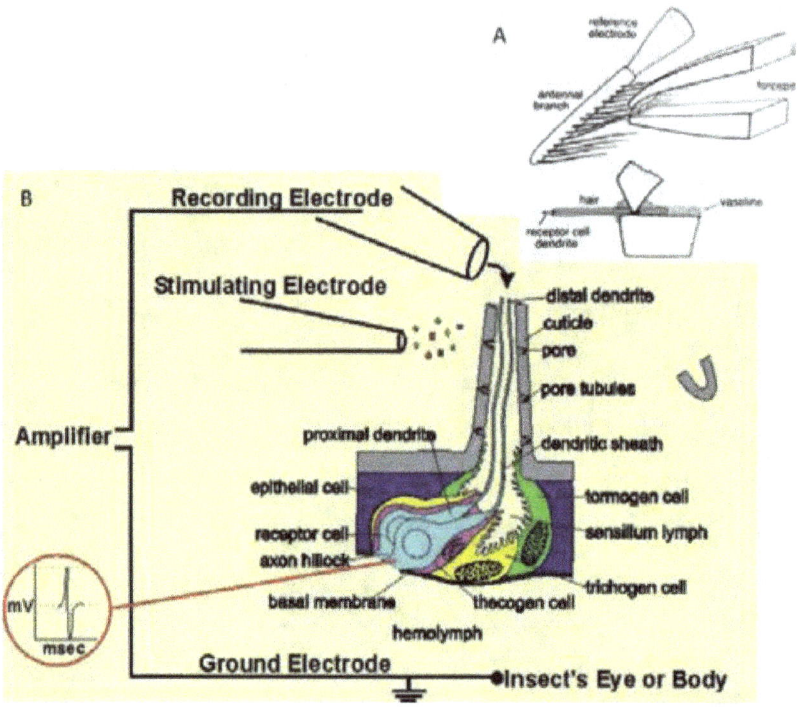

Fig. 4. Extracellular recording technique. (A) After the sensillum is cut, shown here using sharpened forceps, the recording electrode (B) is placed over the tip of the sensillum. The stimulating electrode, containing the odorant, is used to deliver the odorant to the sensillum. Odor stimuli are presented randomly to the sensillum. Action potentials are amplified, recorded, and relayed to a computer for further analyses. Fig. (A) was taken from Kaissling, 1995.

4. GUSTATORY EXTRACELLULAR RECORDING METHODS

Randomly selected, fifth instar, 12-24 hours post-molt, *Lymantria dispar* larvae, reared on artificial diet (12h light: 12h dark; ca., 60% relative humidity) (Shields et al., 2006) were used for all experiments. The larvae were food deprived at least 12 hours prior to the experiments and were naive to the test compounds. Recordings were carried out at ambient temperatures (24-26ºC). Each larva was transected just behind the head and a blunt-tipped saline-filled glass electrode was inserted into the head with sufficient force to cause the eversion of the lateral and medial styloconic sensilla. This pipette was filled with an electrically conductive solution (typically KCl or NaCl) and served as the ground or indifferent electrode (Figs. 5, 6). The cut end of the head was sealed with a minimal amount of melted bee's wax (Fig. 5A). The stimulating or recording electrode was filled with a test solution and was positioned over the tip of a single styloconic sensillum (Figs. 5A, 6) with the aid of a micromanipulator (Leitz) under visual control on a vibration-free air table (Fig. 5B). The preparation lasted, on average, one to two hours.

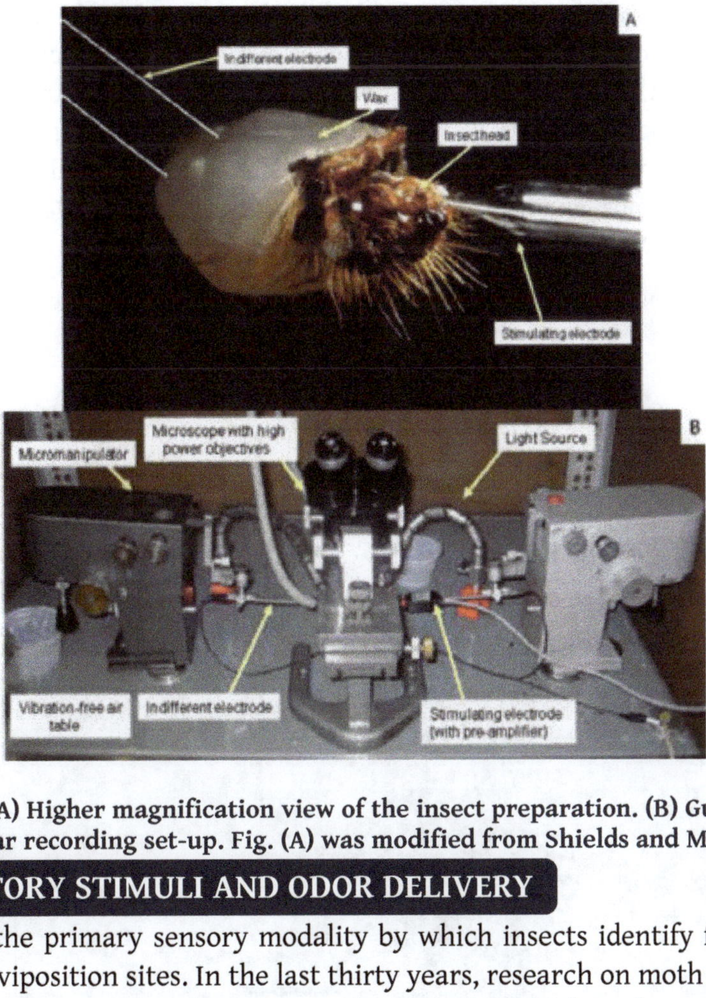

Fig. 5. (A) Higher magnification view of the insect preparation. (B) Gustatory extracellular recording set-up. Fig. (A) was modified from Shields and Martin, 2012.

5. OLFACTORY STIMULI AND ODOR DELIVERY

Olfaction is the primary sensory modality by which insects identify food sources, mates, and oviposition sites. In the last thirty years, research on moth olfaction has predominantly concentrated on the mechanisms by which male moths perceive, interpret sensory information regarding, and behaviorally respond to the sex pheromones released by female conspecifics. Conversely, limited knowledge exists regarding analogous mechanisms pertaining to non-pheromonal odors, including plant volatiles, and the processes through which female moths identify and differentiate plant-associated volatiles (odorants) for foraging and oviposition, as well as how this information is processed by the olfactory system.Previous studies have indicated that scent emitted by sphinx moth-pollinated flowers (Knudsen and

Tollsten, 1993; Raguso and Willis, 1997; Raguso and Light, 1998), as well as floral and vegetative volatiles of tobacco, tomato, and other hostplants (Andersen et al., 1988; Buttery et al., 1987a; 1987b) attract adult *M. sexta* for feeding (Morgan and Lyon, 1928; Yamamoto et al., 1969; Raguso et al., 1996; Raguso and Light, 1998). Flowers having a "white floral" scent, such as that associated with many night-blooming moth-pollinated flowers, possess acyclic terpene alcohols (e.g. linalool, nerolidol, and farnesol), as well as the corresponding hydrocarbons, aromatic alcohols, and esters derived from them, in addition to esters of salicylic acid (Knudsen and Tollsten, 1993). Also thought to play a significant role in the attraction of insects to, and their recognition of, their hostplants are various plant- associated odorants, such as green-leaf volatiles (mainly saturated and unsaturated C6 alcohols and aldehydes), terpenoids, and benzenoid compounds (e.g. Boeckh, 1974; Visser and Avé, 1978; Renwick, 1989; Heath et al., 1992; Knudsen and Tollsten, 1993; Raguso and Willis, 1997).

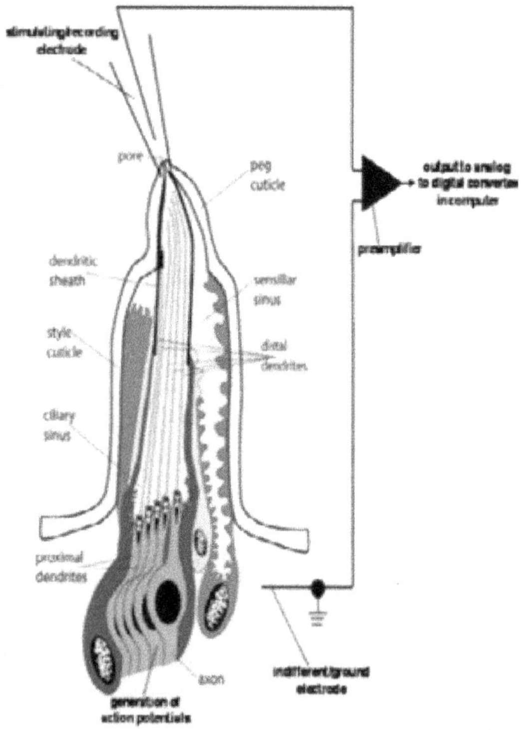

Fig. 6. Schematic diagram showing extracellular recording from a single gustatory styloconic sensillum showing the stimulating electrode containing the stimulus positioned over the tip of the sensillum. This figure was taken from Shields and Martin, 2010.

The majority of the selected, purified, and synthetic odorants tested (chemical purity of 95- 99.9%) (Shields and Hildebrand, 2001a) represented floral headspace volatiles of native, night-blooming flowers, such as *Datura wrightii* (jimson weed), *Hymenocallis sonorensis* (spider lily), and *Oenothera caespitosa* (evening primrose), to which *M. sexta* and other sphinx moths are attracted for nectar-feeding and were available from Sigma (St. Louis, MO) or Aldrich (Milwaukee, WI). Complex blends of volatiles emitted by the headspace inflorescence from *D. wrightii* (i.e., jimsonweed) and the foliage of two solanaceous plants (i.e., *Lycopersicon* (tomato) and *Nicotiana* (tobacco) species were also tested. This was carried out by placing the entire inflorescence or two to three undamaged leaves in a 20 ml capped syringe, which was allowed to equilibrate at room temperature for 24 h prior to use. Tomato and tobacco foliage are preferred by female *M. sexta* for oviposition (Yamamoto et al., 1969), while jimsonweed provides a nectar feeding source.

For testing individual compounds, each odorant was dissolved in odorless mineral oil (light white oil, Sigma, St. Louis, MO) at a dilution of 1:10 (v:v). A 30-μl aliquot was pipetted onto a 1.5-cm^2 piece of grade-1 filter paper (Whatman, Kent, ME) and inserted into a disposable 20 ml syringe. Control syringes, loaded with 30 μl of mineral oil, were also prepared. All syringes were capped and allowed to equilibrate, similar to that described above. After each experimental session, the capped odorant syringes were stored at 4°C. Some odorants eliciting the strongest responses were tested in dose-response experiments. For these experiments, dilutions were prepared in decadic steps (v:v) over at least four log units from stock solutions. Stimulus loads ranged from 3×10^{-4} to 3 μl of the test compound (approximately 0.3 μg to 3 mg). Odorants were tested in order of increasing concentration to prevent adaptation of the ORCs. Also included among the odorants, were two key components of the female moth's sex pheromone, E10,E12-hexadecadienal (bombykal) and E11,Z13-pentadecadienal ("C15"), a relatively stable mimic of E10,E12,Z14-hexadecatrienal (see below) (Kaissling et al., 1989). Each compound was dissolved in *n*-hexane and applied directly (without mineral oil) at 1 ng·μl^{-1} and 10 ng·μl^{-1} to the filter paper.

The odor was delivered to the female antenna using a stimulating device (type CS-01, Syntech) (Shields and Hildebrand, 2001a). The tip of the syringe containing the odorant was introduced into a 20 cm long, L-shaped glass tube (1 cm i.d.) whose open end was positioned 5-7 mm from the test antenna (Fig. 3). The end of the tube was flattened to 0.5 cm to correspond with the length of the antenna. The antenna was continuously flushed with a 5 cm·s^{-1} stream of charcoal-filtered, humidified air. To deliver a stimulus, 2 ml of odor-laden air was injected at a constant rate over 200 ms from the syringe into the airflow using a motor-driven syringe olfactometer, resulting in increased airflow of 20 cm s^{-1} (measured by a thermo-anemometer) (Fig. 3). A glass funnel (3.5 cm i.d.) attached to an air evacuation line was positioned near the preparation to draw away any odor-bearing air following stimulus delivery. Odor stimuli were presented randomly to the sensilla to be tested. Blank controls were tested repeatedly, but never elicited a response from the ORCs. An interstimulus interval of at least 60 s, or longer, if necessary for spontaneous activity of the ORC to return to their initial value, was allowed between stimulations.

6. OLFACTORY RESPONSES FROM TRICHOID TYPE-A SENSILLA

Alternating current (AC) signals were recorded for 6 s, starting 1 s before stimulation (olfactory experiments) and for 10 s, beginning 0 s before stimulation. In either case, these signals were pre-amplified 10X using a Syntech Probe run through a 16-bit analog-to-digital interface (IDAC-02, olfactory, or IDAC-4, gustatory) (Syntech), and then analyzed off-line with Autospike software (Syntech). This software allowed the user to more easily interpret how many cells were responding to specific compounds, as determined by their size, shape, and firing frequency of the biphasic action potentials. The net number of spikes within the first 500 ms after stimulus delivery was considered to be a measure of the strength of excitation and was calculated as the mean number of spikes within 500 ms after the termination of stimulus delivery minus the mean number of spikes within 500 ms before stimulus onset. '

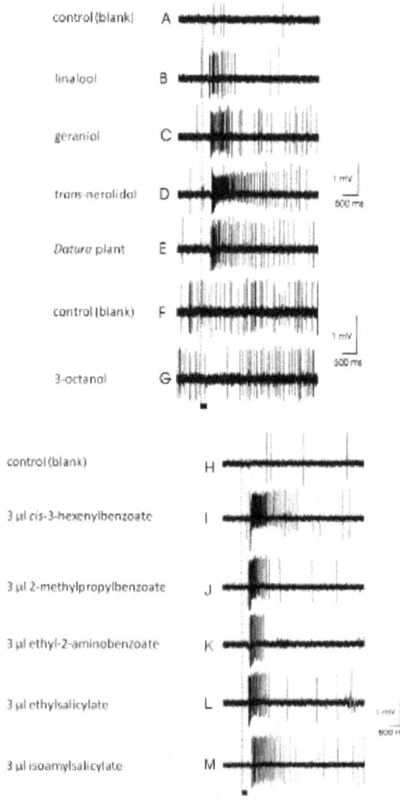

Fig. 7. Representative extracellular olfactory electrophysiological recordings of from three different olfactory receptor cells (ORCs) from female *Manduca sexta* trichoid type-A sensilla showing excitatory (A-F, H-M) and inhibitory (G) responses. The responses in (F) and (G) are from a different ORC than those in (A-E) and (H-M). The stimulus bar is represented by a filled rectangle and represents 200 ms and represents the onset of odor delivery. This figure from modified from Shields and Hildebrand, 2001b.

The responses of ORCs were classified as excitatory (Figs. 7A-F, 7H-M, 8A-N, 8P, 8Q) or inhibitory (Figs. 7G, 8O) if there was cessation in spiking activity for more than 500 ms after stimulation. Dose- response experiments were conducted after the general screening of the odorants was completed and only if the response of an ORC remained constant throughout the experimental session. In order to increase the accuracy of the data analyses, vapor pressure of a particular odorant was taken into consideration when analyzing the sensitivities of selected ORCs.

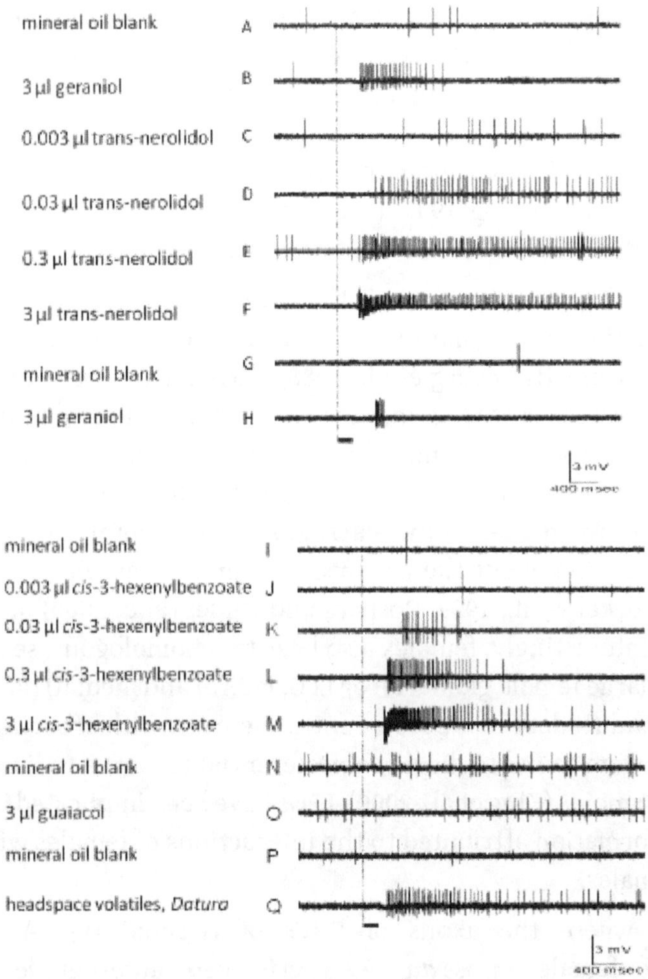

Fig. 8. Representative extracellular olfactory electrophysiological recordings from five different olfactory receptor cells (ORCs) from female *Manduca sexta* trichoid type-A sensilla showing excitatory responses (A-N, P, Q) and inhibitory (O) responses. Note the increasing excitatory activity of the ORC to increasing concentrations of trans-nerolidol (C-F) and *cis*-3- hexenylbenzoate (J-M). Note the two different types of excitatory responses to geraniol (B and H). The stimulus bar is represented by a filled rectangle and represents 200 ms and represents the onset of odor delivery. This figure from modified from Shields and Hildebrand, 2001a.

7. STAINING AND MAPPING OF ORC AXONAL PROJECTIONS

The axons of ORCs project to and terminate in compartments of condensed synaptic neuropil (i.e., glomeruli) in the primary olfactory centers in the CNS of invertebrates and vertebrates (Hildebrand and Shepherd, 1997). Each olfactory glomerulus is a discrete anatomical and functional unit and represents a specific anatomical "address" dedicated to collecting and processing specific molecular features about the olfactory environment, conveyed to it by ORC axons expressing specific olfactory receptor proteins (Buck, 1996; Buonviso and Chaput, 1990; Christensen et al., 1996; Hildebrand and Shepherd, 1997; Mombaerts, 1996). Over approximately the past two decades, there has been mounting evidence that the arrays of glomeruli in the ALs of insects and the olfactory bulbs of vertebrates are organized chemotopically (e.g., Sharp et al., 1975; Rodrigues and Buchner, 1984; Hansson et al., 1992; Mombaerts, 1996; Friedrich and Korsching, 1997; 1998; Galizia et al., 1999), analogous to visuotopy, in visual systems, and tonotopy, of auditory systems.

The axons of antennal ORCs project via the antennal nerve to the ipsilateral AL in *M. sexta*. In the ALs, they form synapses with processes of a subset of the approximately 1,200 central neurons (Homberg et al., 1988; Christensen et al., 1995; Rössler et al., 1998; 1999). In both sexes of *M. sexta*, each AL bears 60 ordinary, sexually isomorphic glomeruli and three sexually dimorphic glomeruli (Rospars and Hildebrand, 2000). Male *M. sexta* moths bear three prominent glomeruli (cumulus, toroid 1, toroid 2), which constitutes the male-specific macroglomerular complex (MGC). This complex processes information about the conspecific female's sex pheromone (Hansson et al., 1991; Heinbockel et al., 1999; Rospars and Hildebrand, 2000) (discussed in more detail, below). Interestingly, females also bear two homologous, sexually dimorphic glomeruli (i.e., large female glomeruli or LFG, lateral and medial) (Rössler et al., 1998; 1999; Rospars and Hildebrand, 2000). Central neurons with arborizations leading to the lateral LFG were found to display a preferential response to linalool and certain other monoterpenoids (King et al., 2000). LFGs have been implicated in being involved in olfactory information attributed to the interactions of females with hostplants or with courting males.

To determine where the axons of ORCs of trichoid type-A sensilla project in the AL of female *M. sexta*, we performed anterograde labeling using dextrantetramethylrhodamine of ORCs in groups of 5-10 sensilla on various surfaces of a single annulus in the middle of an antennal flagellum (Fig. 1C, D). This was carried out by restraining an adult female moth in a plastic tube (Fig. 4) and cutting the tips of 5-10 trichoid type-A sensilla (Fig 4); (see Shields and Hildebrand, 1999a; Shields and Hildebrand, 2001) using the method outlined, above. In order to stain the ORCs associated with these cut sensilla, we created a small well from melted paraffin wax around the selected annulus and filled it with a small amount of 1% solution of dextran-tetramethylrhodamine (3000 MW, anionic, lysine-fixable, D-3308, Molecular Probes, Eugene, OR) until the annulus became completely submerged. The area was then covered with a small amount of petroleum jelly. We transferred the preparation to a humid chamber and kept it in darkness for 2-3 days. Following this period, we excised the brain and fixed it for 24-48 h in 2.5% formaldehyde solution in 0.1 M sodium phosphate buffer (pH 7.4) containing 3% sucrose. Following this, the tissue was dehydrated in a graded ethanol series, cleared in methyl salicylate, and viewed as a temporary whole mount in a laser-scanning confocal microscope.

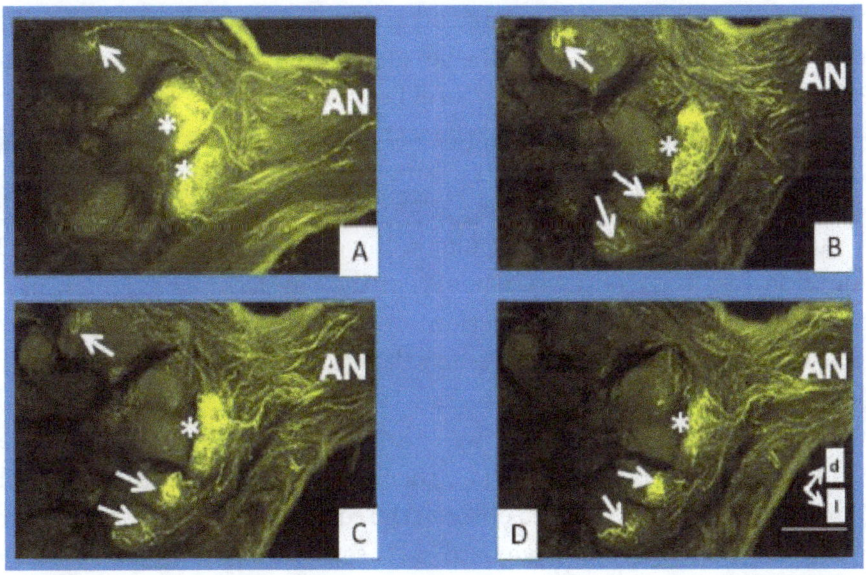

Fig. 9. Confocal microscopic images (A-D) showing serial optical sections of female *Manduca sexta* specimens embedded in plastic and sectioned to improve resolution taken at different depths through the antennal lobe. Images show the central projections of axons from olfactory receptor cells from trichoid type-A sensilla stained with the fluorescent dye dextran- tetramethylrhodamine. Askterisks indicate the sexually dimorphic large female glomeruli located in the dorsolateral region of the antennal lobe, near the site of entry of the antennal nerve (AN). The arrows indicate ordinary sexually isomorphic glomeruli. d, dorsal; l, lateral. Scale bars = 100 µm. Figs. (A) and (B) were modified from Shields and Hildebrand, 2001a.

We found that a majority of axonal projections from these ORCs terminated in the two LFG located in the dorsolateral region of the ipsilateral (AL) (Fig. 9A), more specifically, near the entrance of the antennal nerve into the AL. We also found that in addition to the LFGs, a subset of the other 60 spheroidal, ordinary, sexually isomorphic glomeruli received sparse projections of a subset of ORC afferent axons (Fig. 9). The results of these anatomical studies and our electrophysiological results that some trichoid type-A sensilla are tuned mainly to terpenoids and aromatic esters, provide a basis to hypothesize that information about odorants belonging to those chemical classes is processed in the LFGs (King et al., 2000; Shields and

Hildebrand, 2001a; b). To improve visualization (i.e., resolution) of successfully labeled preparations, samples were embedded in Spurr's low-viscosity embedding medium (Electron Microscopy Sciences, Fort Washington, PA) and sectioned at 48 μm. The preparations were then viewed in a laser scanning confocal microscope (Bio-Rad MRC-600; Cambridge, MA) equipped with a Nikon Optiphot-2 microscope and both 15-mW krypton-argon and 100-mW argon laser light sources and YHS filter cube (excitation wavelength 568 nm). Serial optical sections were collected at 3-μm intervals through the whole mount or 2 μm intervals from the embedded sectioned preparations (Fig. 9). Image processing and analysis were performed using Confocal Assistant 4.02 (copyrighted by Todd Brelje, distributed by Bio Rad, Cambridge, MA), Corel Photopaint 8, and Corel Draw 8 (Corel Corporation, Ottawa, Ontario, Canada).

8. GUSTATORY STIMULI AND RESPONSES FROM THE MEDIAL STYLOCONIC SENSILLA

For all gustatory experiments, stimulus compounds were dissolved in 30 mM potassium chloride (KCl) (control) (Fisher Scientific, Fair Lawn, New Jersey) in distilled water to enhance the electrical conduction of the recording electrode and to improve the signal-to- noise ratio. This inorganic salt was chosen since the hemolymph of plant-feeding (i.e., phytophagous) feeders typically shows high K^+ and low Na^+ concentrations (Kaissling, 1995). KCl was also used to fill the indifferent electrode (Figs. 5, 6). Selected carbohydrates (i.e., sucrose and inositol), as well as alkaloids (Fig. 10) were tested in this study to observe the responses of the deterrent-sensitive, inositol-sensitive, sucrose-sensitive, and salt- sensitive cells (see also Shields and Martin, 2010; 2012). The alkaloids were diluted in 10% ethanol solution, which was added to the electrolyte solution. Ethanol, at this concentration, was found to have no discernible effect on the activity of cell(s) (referred to as KCl-sensitive cell(s)) (see below) responding to the control (i.e., KCl) alone. Each stimulus to be tested was applied to the tip of the styloconic sensillum by way of the stimulating electrode (Figs. 5, 6).

Electrical activity from individual styloconic sensilla was amplified and conditioned (bandpass filter set at 100-1200 Hz) prior to being digitized by a 16-bit analog-to-digital interface (IDAC-4, Syntech). For each electrophysiological recording, we stimulated a sensillum for total of 10 sec. Action potentials generated 50 ms after contact with the sensillum were analyzed off-line using a computer equipped with Autospike (Syntech).A pause of at least 2 minutes was allowed between successive stimulations to minimize any carry-over effects. For dose-response experiments, only those phytochemicals that elicited a robust response were tested. Test stimuli were presented to the sensillum from lowest to highest concentration. To ensure a reproducible response to a particular concentration, 1-2 replications of each were made. The number of action potentials generated 0.05 s after the contact artifact was quantified in 100 ms increments.

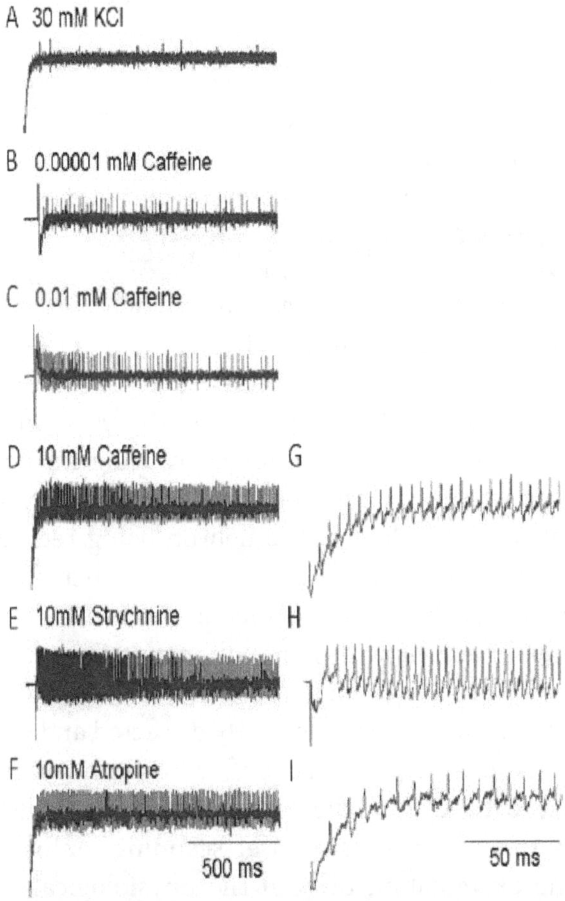

Fig. 10. Representative extracellular gustatory electrophysiological recordings from the medial styloconic sensillum of *Lymantria dispar*. The deterrent-sensitive cell responds to (A) potassium chloride (control) and to the alkaloids, (B-D) caffeine, (E) strychnine, and (F) atropine. The recordings in G, H, and I are higher magnifications of the action potentials from the deterrent-sensitive cell to (D) caffeine, (E) strychnine, and (F), atropine, respectively. The deterrent-sensitive cell displays a typical phasic-tonic response (i.e., a transient pattern of firing changing to that of a more sustained pattern of firing) to these alkaloids. This feature is more apparent at higher concentrations (compare (C) and (D) with (B)) and is shown for caffeine. A weaker concentration of caffeine (B) elicits fewer spikes from the deterrent-sensitive cell. This figure was modified from Shields and Martin, 2010.

We found two cells (i.e., one large spike amplitude and one small spike amplitude) in the medial styloconic sensillum that responded to KCl (Fig. 10A). Another cell in the same sensillum responded robustly to various alkaloids (i.e., deterrent-sensitive cell) (Fig. 10B-F). This cell displayed a typical phasic-tonic response (i.e., the initial high firing frequency of the cell changed rapidly to a more sustained pattern of firing).

The ability of the deterrent- sensitive cell to respond to more than one alkaloid may be due to either the presence of multiple receptor sites, each with their own structure-function specificity, or a broadly tuned single receptor site (Blaney et al., 1988). These results correlated well with feeding behavioral studies using gypsy moth larvae (Shields et al., 2006), where it was shown that an increase in feeding deterrency (decrease in consumption) occurred with increasing alkaloid concentration.

9. INTRACELLULAR RECORDING TECHNIQUES

Olfactory receptor cells (ORCs) in the antenna transmit olfactory information as trains of action potentials to the ipsilateral antennal lobes (ALs) of the insect brain, where the axon terminals of ORCs synapse onto central neurons within neuronal structures termed olfactory glomeruli. Central neurons in the ALs, including projection neurons and local interneurons, can be delineated using intracellular recordings utilizing sharp electrodes. These glass electrodes contain a physiological solution that simulates the intracellular fluid of the neuron being recorded. Furthermore, the electrodes may incorporate intracellular indicators, such fluorescent dyes. The advancement of novel intracellular markers facilitates the fast and comprehensive restoration of individual neurons with minimal or no damage to the neurons. For instance, central neurons are subjected to iontophoretic injection of Lucifer yellow, neurobiotin, or biocytin. Brains are subsequently dissected and preserved overnight in formaldehyde with sucrose in a phosphate buffer. To see biocytin-injected neurons, brains are incubated with Cy3-conjugated streptavidin. Following histological preparation, neurons are examined using laser-scanning confocal microscopy. This method facilitates the examination of both the physiological and morphological characteristics of the recorded neuron in a minimally disturbed in vivo setting, as detailed below.

9.1 Olfactory neurobiology

Female moths release sex pheromones that attract conspecific males over long distances (Kaissling, 1987; Hildebrand, 1996). An outstanding challenge in olfactory neurobiology is to understand how a male moth is able to locate a mate, namely, a conspecific female releasing sex-pheromone. The olfactory brain of a male moth must integrate information about qualitative, quantitative, and spatiotemporal features of an attractive blend of volatile compounds, the sex pheromone, released by a conspecific female (Hildebrand, 1995; 1996; Christensen et al., 1996). Central processing of this information occurs in a sexually dimorphic cluster of olfactory glomeruli, the MGC, in the male moth's AL. Since information about the sex-pheromone is primarily processed in the MGC, it serves as a model for studies of the functional architecture of glomeruli, as well as the physiological relationships between glomeruli in the olfactory system (Hansson and Christensen, 1999;

Christensen and White, 2000; Christensen and Hildebrand, 2002; Reisenman et al., 2008; Lei et al., 2010).

Several species of moths use the same chemical compounds in their sex-pheromone blend, but the attractant signal produced by each species is nevertheless unique because each blend has characteristic proportions of the components (Arn et al., 1992; Kaissling, 1996). Chemical studies revealed that in the sphinx moth *M. sexta*, the sex-pheromone blend comprises eight components. Behavioral data shows that two of the eight components are required to evoke olfaction-modulated flight in males (Tumlinson et al., 1989). The antennae of male *M. sexta* are covered by different types of sensilla (see above). The long trichoid sensilla on the male's antennae house highly selective and sensitive ORCs that detect one or the other of the two key pheromone components (Kaissling et al., 1989). Each of these ORC populations sends their axons to a different glomerulus in the MGC (Christensen et al., 1995). Projection neurons (PNs) that arborize in one of these glomeruli, the toroid 1 [T1] (Strausfeld, 1989; Homberg et al., 1995) respond to antennal stimulation with *E,Z*-10,12-hexadecadienal (bombykal or BAL, one of the two key components of the female's sex pheromone). PNs that arborize in a neighboring glomerulus (the cumulus) respond selectively to *E,E,Z*-10,12,14- hexadecatrienal (the second key component) or its more stable mimic, *E,Z*-11,13- pentadecadienal (C15; Kaissling et al., 1989; Hansson et al., 1991). An MGC-PN with arborizations in both glomeruli is activated by both BAL and C15 (Hansson et al., 1991; Heinbockel et al., 2004).

For a number of moth species it has been demonstrated in behavioral studies in the field and in laboratory wind tunnels that pheromone-modulated upwind flight is regulated in a dose- dependent manner (Baker, 1989). Similarly, electrophysiological studies of AL neurons in male *M. sexta* have shown how the responses of MGC-PNs to pheromone components and extracts of the female's pheromone gland are influenced by stimulus intensity (Christensen and Hildebrand, 1990). In a technological advance, neural-ensemble recordings have revealed concentration-dependent patterns of synchronous firing between MGC-PNs responding to the two key pheromone components (Christensen et al., 2000). A growing body of evidence suggests that glomeruli can receive converging excitatory and inhibitory input. Subpopulations of MGC-PNs receive excitatory input driven by BAL and inhibitory input driven by C15 (or vice versa) (Christensen and Hildebrand, 1987; Hansson et al., 1991; Heinbockel et al., 1999; Lei et al., 2002). Input from both excitatory and inhibitory pathways enhances the ability of MGC-PNs to resolve multiple pulses of pheromone such as intermittent odor stimuli found in natural stimulus situations (Christensen and Hildebrand, 1997; Christensen et al., 1998; Heinbockel et al., 1999). Here, we describe how this processing function is optimized at particular stimulus concentrations or ratios in different PNs (Heinbockel et al., 2004). By using intracellular recording and staining methods, we examined the effect of changing these quantitative attributes of the pheromonal stimulus on the responses of identified PNs innervating different glomeruli in the MGC.

9.2 Experimental preparation

Male *M. sexta* were reared on an artificial diet (modified from that of Bell and Joachim, 1976) at 25°C and 50-60% relative humidity under a long-day photoperiod regimen (17h light : 7h dark) as described previously (Sanes and Hildebrand, 1976; Prescott et al., 1977). Adult moths (1-3 days post-eclosion) were immobilized and prepared by standard methods (Christensen and Hildebrand, 1987). With the antennae intact, the head was separated from the rest of the body and pinned in a Sylgard-coated (Dow Corning, Midland, Michigan) recording chamber (volume <0.5 ml) (Heinbockel and Hildebrand, 1998; Heinbockel et al., 1998). Isolation of the head from the rest of the thorax had no detectable adverse effect on neural responses (Christensen and Hildebrand, 1987). Part of the AL was then desheathed with fine forceps to facilitate insertion of the recording electrode. The brain was superfused constantly with physiological saline solution (modified from that of Pichon et al., 1972; ca. 2 ml/min) containing 150 mM NaCl, 3 mM KCl, 3 mM CaCl2, 10 mM TES buffer (pH 6.9), and 25 mM sucrose to balance osmolarity with that of the extracellular fluid

9.3 Electrophysiological recordings and data analysis

Sharp glass microelectrodes for intracellular recording were produced from borosilicate tubing (o.d. 1.0 mm, i.d. 0.5 mm, World Precision Instruments, Sarasota, Florida) with a Flaming-Brown Puller (P-2000, Sutter Instrument Co., Novato, California). The tip of each electrode was filled with a solution of an intracellular stain (see below), and the shaft was filled with filtered (0.2 μm pore size) 2.5 M KCl. Electrode resistance ranged from 60-100 MΩ, measured in the tissue. Movements of microelectrodes were controlled with a Burleigh Inchworm (Model 6000/ULN; Burleigh Instruments Inc., Fishers, New York) attached to a Leitz micromanipulator (Leitz, Wetzlar, Germany). Recordings were made from neurites in the synaptic neuropil of the MGC. Because the site of the electrode impalement in a neuron can affect the amplitude of postsynaptic potentials, impalements targeted the same area of neuropil in all preparations. Typically, one neuron was recorded per animal, except when a second recorded neuron had different pheromone response patterns (excited by BAL vs. excited by C15). Intracellular recording and current injection were carried out in bridge mode with an Axoclamp-2A amplifier (Axon Instruments – Molecular Devices, Sunnyvale, California), and data were initially stored on magnetic tape (Hewlett Packard Instrumentation Tape Recorder 3968A, Palo Alto, California) and subsequently transferred to a computer and analyzed with Experimenter's Workbench (Datawave Technologies Co., Longmont, Colorado) or Autospike (Syntech, Kirchzarten, Germany).

The data presented below were analyzed statistically for differences using one-way analysis of variance (Kruskal-Wallis One-Way Analysis of Variance by Ranks, ANOVA) (Zar, 1984). A multiple-comparison procedure was used to isolate groups from each other (All Pairwise Multiple-Comparison Procedures, Dunn's Method, or Student-Newman-Keuls Method; $P<0.05$) (Sigma Stat, Jandel Scientific Software, Richmond, California, Version 1.0, Statistical Software).

9.4 Intracellular staining and confocal microscopy

Neurons were injected iontophoretically with either Neurobiotin (Vector Laboratories, Burlingame, California; 3-5% in 2 M KCl with 0.05 M Tris buffer, pH 7.4) or biocytin (Sigma, St. Louis, Missouri; 3-5% in 2M KCl with 0.05 M Tris buffer, pH 7.4). Alternating hyperpolarizing and depolarizing current pulses (30 sec, 1 nA) for about 10 min were used to inject either tracer. Brains were then dissected and fixed overnight in 2.5% formaldehyde with 3% sucrose in 0.1 M sodium phosphate buffer. To visualize injected neurons, brains were incubated with Cy3-conjugated streptavidin (Jackson Immuno Research Laboratories, West Grove, Pennsylvania; diluted 1:100 with 0.2 M sodium phosphate buffer containing 0.3% Triton X-100) for 3 days on a shaker at 4°C. The brains were then dehydrated with increasing concentrations of ethanol and cleared in methyl salicylate. Neurons were further investigated by laser-scanning confocal microscopy (Fig. 11) using a BioRad MRC 600 (Bio- Rad, Cambridge, Massachusetts) with a Nikon Optiphot-2 microscope, a Krypton/Argon (15 mW) laser light source, and appropriate dichromatic filter cubes (Sun et al., 1993). Serial 2-µm optical sections were imaged through whole mounts and saved as series of images on disks.

9.5 Olfactory stimulation

We used synthetic pheromone compounds kindly provided by Dr. J.H. Tumlinson (USDA, Gainesville, FL). The following olfactory stimuli were applied: (1) *E,Z*-10,12-hexadecadienal (bombykal or BAL); (2) *E,Z*-11,13-pentadecadienal (C15), a synthetic mimic of the second pheromone component, *E,E,Z*-10,12,14-hexadecatrienal, and hereinafter referred to as the second component; and (3) mixtures of BAL and C15 with various blend ratios. These compounds were dissolved in *n*-hexane and applied to a piece of filter paper (1 x 2 cm), which was inserted into a glass cartridge (acid-cleaned glass syringe barrel), as described previously (Christensen and Hildebrand, 1987). The stimulus load on the filter paper was reported as the number of ng of compound applied in hexane solution. A pulse of charcoal- filtered, humidified air (1,000 ml/min) moving through the cartridge carried the stimulus, roughly proportional to the loading on the filter paper, to the proximal half of the antenna ipsilateral to the impaled AL. The antenna typically was stimulated with five 50-ms pulses from a cartridge at 5 pulses sec^{-1}.

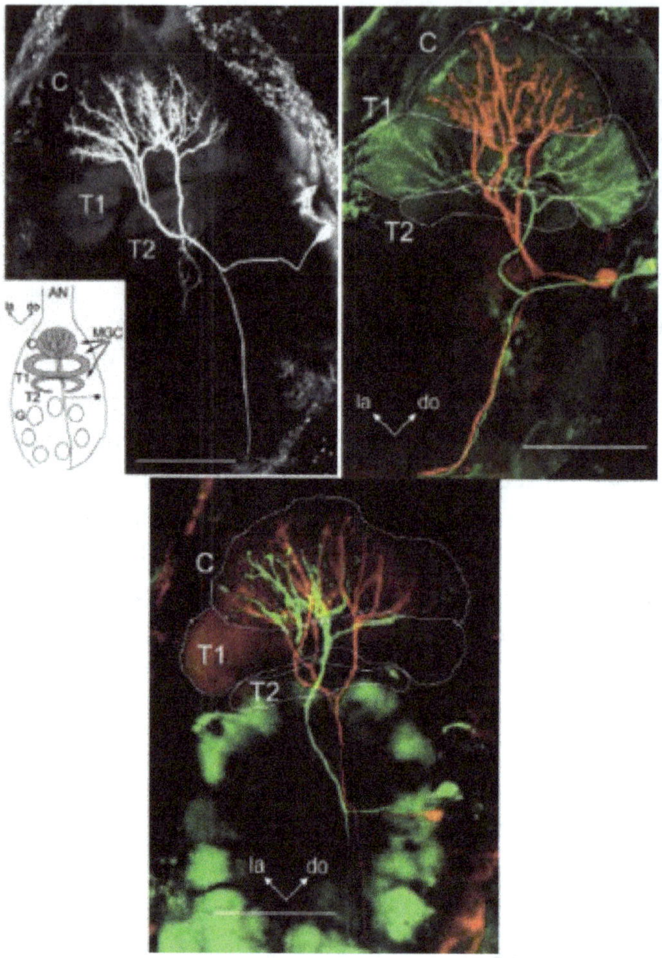

Fig. 11. Laser-scanning confocal micrographs of antennal lobe projection neurons in the moth antennal lobe of male *Manduca sexta*. Left panel: Image of a C15-specialist MGC-PN with arborizations confined to the cumulus. The inset illustrates the organization of the antennal lobe with the macroglomerular complex (MGC) and other glomeruli (G). Right panel: Two specialist MGC-PNs, one neuron, stained with Lucifer Yellow (colored red here), had arborizations confined to the cumulus (C), and the other neuron, stained with biocytin (colored green here) had arborizations confined to the toroid (T1). Areas of apparent overlap between the 2 neurons are shown in yellow and are possible sites of synaptic contact. Bottom panel: Morphological diversity in cumulus neurons. Image of two C15-specialist MGC-PNs with arborizations confined to the cumulus. While the branches of the two neurons apparently overlapped in certain parts of the cumulus (indicated in yellow), other parts were innervated by just one of the two neurons. The green neuron was stained with Lucifer Yellow and the red neuron with biocytin. C - cumulus, T1 - toroid 1, T2 - toroid 2; do, dorsal; la, lateral. Scale bar = 100 µm. Modified from Heinbockel and Hildebrand, 1998 (left panel); Heinbockel et al., 1999 (right and bottom panels).

The two key pheromone components used for sexual communication in *M. sexta* are found in a 2:1 ratio in solvent rinses of intact pheromone glands of calling virgin females (Tumlinson et al., 1989). Because the mimic (C15) of the second key component is less potent than the natural component itself, a 1:1 BAL-to-C15 ratio was used to elicit physiological responses comparable to those recorded in response to the pheromone-gland extract, as found in earlier electrophysiological studies (Christensen and Hildebrand, 1987; Kaissling et al., 1989). In our experiments, different blend ratios above and below 1:1 were tested in sequence at intervals of 1 min. For one series of blend ratios, the BAL stimulus load was held constant (1 ng) while the C15 stimulus load was varied (0.01, 0.1, 1.0, and 10 ng). For the second series of blend ratios, the C15 stimulus load was held constant while the BAL stimulus load was varied in a similar fashion. In addition to these stimulus series, an elevated stimulus level (10 ng) of either BAL or C15 and a blend of BAL and C15 (10 ng of each) were also tested.

9.6 Calculating stimulus "tracking" in MGC-PNs

We applied several criteria to assess the ability of PNs to register each odor pulse of a series of pulses with a discrete burst of action potentials. In most cases, a stimulus pulse was scored as "tracked" if the evoked burst comprised at least five action potentials and was separated from the next burst by a period of inactivity of at least 50 ms, which corresponded to the inter- stimulus interval. Alternatively, in cases where PNs showed a higher level of spontaneous activity, the instantaneous spike frequency (ISF) was calculated for each spike train. A given stimulus pulse was scored as "tracked" if (a) the ISF showed a two-fold increase in response to the odor pulse and (b) there was a clear decrease in ISF before the next odor pulse (<50% of the maximum ISF observed during the previous pulse). The "tracking index" thus ranged from 1 to 5 and was defined as the number of pulses tracked in response to the 5-pulse stimulus train. Essentially identical results were obtained with both methods.

9.7 Olfactory processing

Changes in stimulus intensity can modulate the patterns of glomerular activity and output from the olfactory bulb and AL as has been shown for olfactory systems of vertebrates and invertebrates alike (Anton et al., 1997; Hildebrand and Shepherd, 1997; Christensen and White, 2000; Keller and Vosshall, 2003; Leon and Johnson, 2003; Sachse and Galizia, 2003; Stopfer et al., 2003). Until recently, little was known about how the responses of uniglomerular PNs can be influenced by specific stimulus blends (Wu et al., 1996). Below, we present direct evidence for PNs in the AL of *M. sexta* that the temporal responses of some PNs are optimized for a particular ratio of stimulus compounds in a blend and that altering this ratio dramatically changes the central representation of the blend at the first stage of processing in the brain.

9.7.1 MGC-PNs have different dynamic ranges and response thresholds

When we changed the intensity of the pheromone stimulus blend, we observed pronounced effects on the strength of the responses (IPSPs, EPSPs, number of impulses) in two major types of MGC-PNs: the BAL-selective (*BAL⁺ PNs*) and C15-selective cells (*C15⁺ PNs*) (Fig. 12). However, not all of these neurons exhibited classical concentration-dependent responses in the range of concentrations used in this study. A subset of *BAL⁺ PNs* innervating the T1 glomerulus and *C15⁺ PNs* innervating the cumulus were extremely sensitive at all concentrations tested (Fig. 12, left panel), while others with similar morphology failed to respond except at the highest stimulus concentration (Fig. 12, right panel). Such functional diversity is not an exclusive property of the male MGC as shown in a study in female *M. sexta* (Reisenman et al., 2004). Similar diversity exists in female *M. sexta* among PNs that innervate a single glomerulus and respond selectively to one enantiomer of linalool, a common plant volatile. The results present us with evidence from both male and female moths to clearly demonstrate functional heterogeneity with respect to threshold sensitivity and concentration- response characteristics among PNs innervating a single glomerulus.

The observed diversity in the PN population of a single glomerulus could reflect distinct functional roles for different PNs under diverse environmental conditions, as previously proposed (Christensen et al., 2000). In this scenario, PNs that are little affected by a change in pheromone concentration might signal the presence of the pheromone without regard to its intensity, whereas the PNs with higher response thresholds, recruited into the coding ensemble only at relatively high pheromone concentrations, might therefore function in source location (Murlis, 1997).

Concentration changes may be encoded in two distinct ways by PNs: (1) by individual neurons that give incremental responses to increasing concentrations and/or (2) by recruitment of different populations of PNs at different concentrations. Our data presented evidence for such a dual coding strategy in MGC-PNs (*BAL⁺ PNs* and *C15⁺ PNs*) in that some displayed a monotonic response across all concentrations, whereas others showed concentration-dependent responses over several orders of magnitude of stimulus concentration. The representation of sex-pheromonal information in the AL of *M. sexta* is sparse at low stimulus intensities because only a subset of MGC-PNs is active under these conditions, whereas the representation becomes increasingly combinatorial and complex as pheromone concentration increases because it involves a greater number and functional variety of types of MGC-PNs. This is similar to the neural coding of general olfactory stimuli in AL glomeruli in fruit flies and honey bees (Galizia et al., 1999, Ng et al., 2002; Wang et al., 2003).

A large number of pheromone-responsive afferent neurons converges onto many fewer central neurons. Therefore, threshold values are expected to be lower for

MGC-PNs than for ORCs. Kaissling et al. (1989) determined a 1-ng threshold stimulus load in the stimulus- delivery cartridge for BAL-specific ORCs and a 100-ng load for C15-specific ORCs for *M. sexta*. The threshold stimulus loads found in our experiments are 0.01 ng for *BAL⁺ PNs* and 0.1 ng for *C15⁺ PNs*. These >100-fold lower corresponding thresholds of MGC-PNs are thought to overcome the critical signal-to-noise ratio, when pheromone molecules activate only a few ORCs. Similar response threshold relationships between ORCs and AL neurons have been observed in other insect species (Boeckh, 1984).

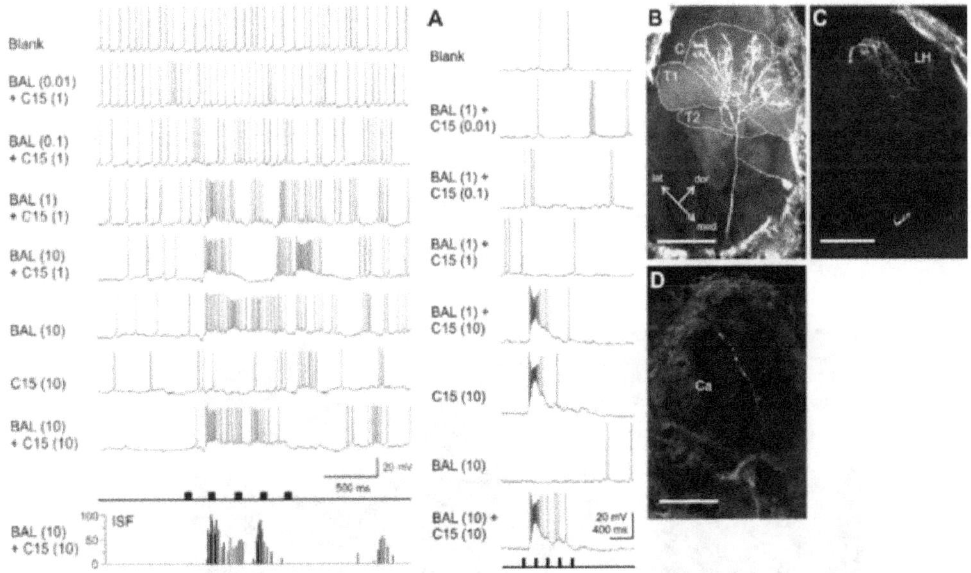

Fig. 12. Panel on previous page: Intracellular recordings from male *Manduca sexta* showing one class of *BAL⁺ PN* that was exclusively responsive to BAL and thus classified as a *BAL⁺- specialist PN*. This type of PN responded to BAL in a dose-dependent manner but is unresponsive to C15, and thus did not show a blend effect. Small BAL stimulus loads evoked only a membrane depolarization (EPSP) accompanied by spiking, but elevated dosages also triggered a brief IPSP preceding the EPSP. In each case, five identical stimulus pulses were delivered to the ipsilateral antenna at a frequency of 5 sec^{-1} (stimulus markers for the 50-ms pulses are shown beneath the records). A plot of instantaneous spike frequency (ISF) calculated from the last record illustrates that the firing dynamics of this type of MGC-PN could not accurately track all pulses of a stimulus train, thus leading to a low tracking index as defined in the text. Panels on this page: (A-D): Olfactory responses and morphology (frontal view) from one class of *C15⁺ PN* of male *M. sexta* that was solely responsive to C15 (*C15⁺-specialist PN*), and therefore did not exhibit a blend effect. (A) This neuron was strongly depolarized by the first pulse of any stimulus that contained 10 ng of C15, but it did not respond to lower stimulus intensities. The neuron showed no clear response to BAL even at the elevated 10-ng load. In each trace, the ipsilateral antenna received five 50-ms stimulus pulses at 5 sec^{-1} (stimulus markers are shown beneath the records). (B) The branches of this PN were confined to the cumulus (*C*) of the MGC. *dor*, dorsal; *lat*, lateral; *med*, medial; *T1*, toroid 1; *T2*, toroid 2. (C) Frontal view of the axon's ramifications in the lateral horn (*LH*) of the protocerebrum. (D) Collateral branches innervating the calyces (*Ca*) of the mushroom body in the posterior protocerebrum. Scale bar = 100 µm. From Heinbockel et al., 2004.

9.7.2 Effects of blend ratio on temporal response dynamics in PNs

In subpopulations of *BAL⁺ PNs* and *C15⁺ PNs*, a change in the ratio of BAL and C15 clearly affected pulse tracking of these PNs (Figs. 13, 14).

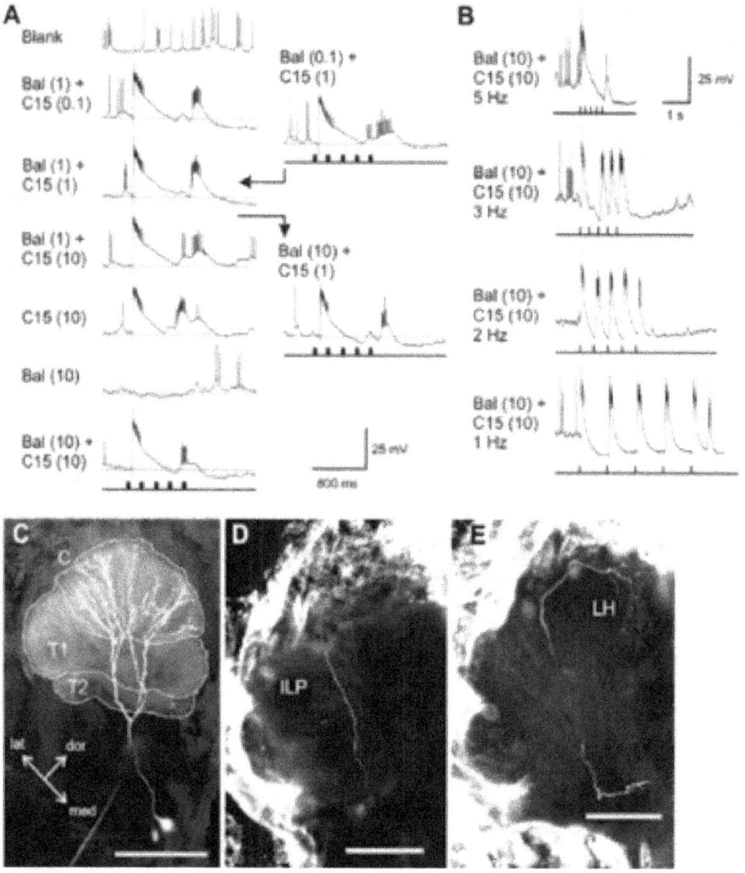

Fig. 13. Responses and anatomy of a *C15⁺ PN* from male *Manduca sexta* that gave depolarizing responses to C15 and inhibitory responses to BAL (*C15⁺ / BAL⁻ PN*). (A) Responses from left to right are to blends containing 1 ng C15 plus increasing amounts of BAL. The ipsilateral antenna received five 50-ms stimulus pulses at 5 pulses sec^{-1} (stimulus markers are shown beneath the records). This neuron had a low threshold for excitation and gave strong responses to stimuli containing only 0.1 ng of C15. The horizontal lines in the records indicate the membrane potential observed prior to stimulation. Notice that BAL alone hyperpolarized the PN and suppressed spiking activity (*C15⁺ / BAL⁻ PN*). (B) Responses of the same PN to different pulse rates of antennal stimulation (BAL, C15, 10 ng each) ranging from 1 to 5 sec^{-1}. At higher frequencies, the PN was unable to repolarize sufficiently to track each of the five stimulus pulses. The tracking index improved dramatically at lower pulse frequencies. (C) The neuron in (A) branched in the cumulus and not in the toroid 1 of the MGC (frontal view). *C*, cumulus; *do*, dorsal; *la*, lateral; *me*, medial; *T1*, toroid; *T2*, toroid 2. (D) More anterior view of the ramifications of the axon in the protocerebrum. *ILP*, inferior lateral protocerebrum. (E) More posterior aspect of protocerebral ramifications of the axon. *LH*, lateral horn. Scale bar = 100 μm. From Heinbockel et al., 2004.

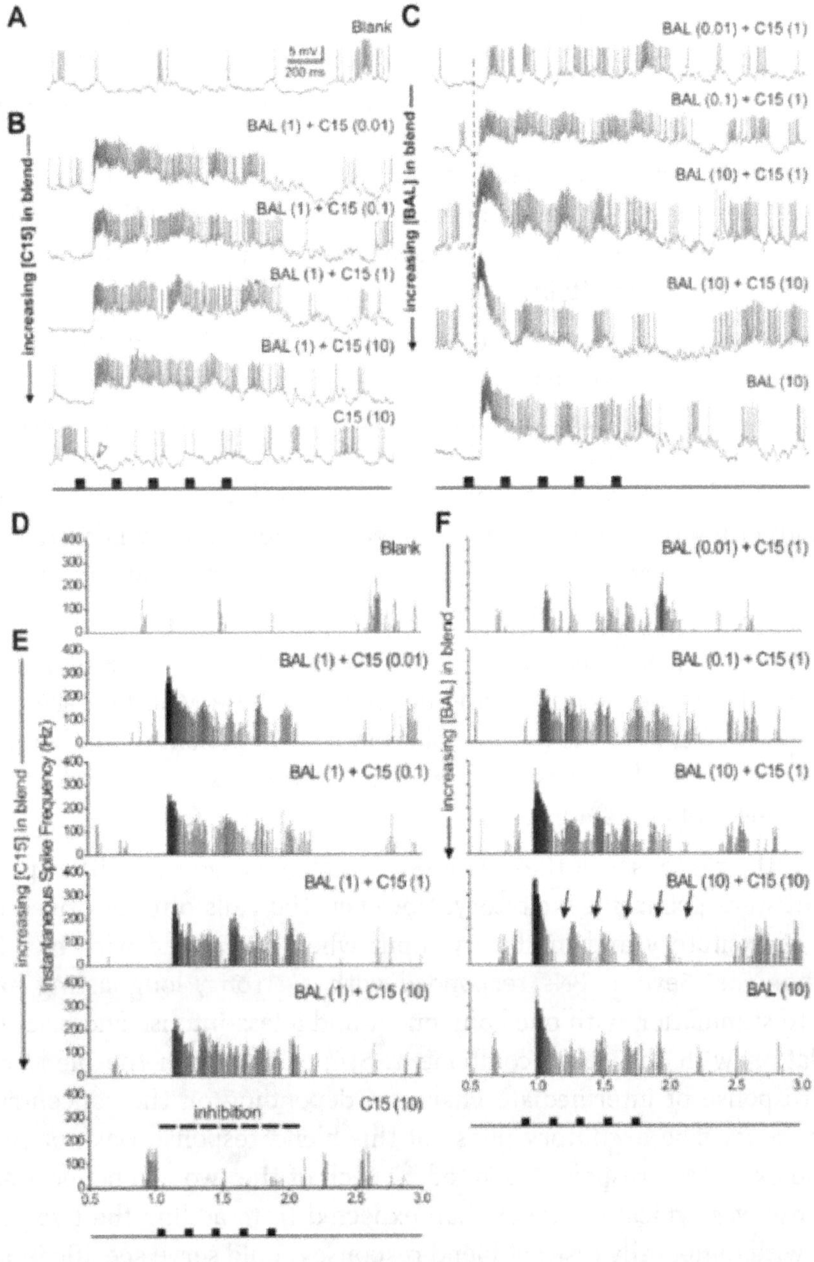

Fig. 14. Intracellular recordings and ISF plots from one *BAL⁺ PN* from male *Manduca sexta* that gave depolarizing responses to BAL and an inhibitory response to C15 (*BAL⁺ / C15⁻ PN*).

Responses to blanks are shown in (A) and (D). (B), (E) Intracellularly recorded responses and corresponding ISF plots to blends containing 1 ng BAL plus increasing amounts of C15. Repetitive stimulus pulses were delivered to the ipsilateral antenna at a frequency of 5 pulses sec^{-1} (stimulus markers are shown beneath the records). Note the modulations of PN firing with each stimulus pulse. C15 at 10 ng led to a hyperpolarization and suppression of firing in the PN (*dashed line*). (C, F) Responses (intracellular records and ISF plots) of the same neuron to blends containing 1 ng C15 plus increasing amounts of BAL. The tracking index of this PN improved at a ratio of 10 ng BAL to 1 ng C15, but was optimized at a 1:1 ratio with 10 ng of each component. Note the distinct periods of inactivity (*arrows*) between spike bursts evoked by consecutive stimulus pulses. The concentration- dependency of this neuron was also indicated by a marked reduction in response latency as the amount of BAL in the blend was increased. From Heinbockel et al., 2004.

These MGC-PNs integrated convergent excitatory and inhibitory afferent input and better resolved intermittent olfactory signals. This processing function was optimized at particular blend ratios for different PNs. *C15⁺ / BAL⁻ PNs* and, particularly, *BAL⁺ / C15⁻ PNs* were best suited to convey the temporal structure of the stimulus to higher brain centers, whereas responses of *BAL⁺*- and *C15-specialist PNs* often lacked the temporal precision to represent rapid changes in pheromone concentration. These results show that information about individual components and stimulus timing is transmitted from the MGC in different output channels.

We also observed blend effects in *BAL⁺ / C15⁺ PNs* and the *C15⁺ / BAL⁺ PNs* (Fig. 15). These blend effects were clearly ratio-dependent. These MGC-PNs are excited by stimulation with either of the two key pheromone components and, therefore, they are considered to be pheromone "generalists" (Christensen and Hildebrand, 1987; 1990; Christensen et al., 1996).

Principally, the responses of these MGC-PNs to stimulation with either pheromone component were primarily excitatory. However, the cells often received different amounts of excitatory and inhibitory input when stimulated with blends of the two components. Several PNs responded with a strong, long-lasting excitatory response to stimulation with one component and a less-intense and brief response to stimulation with the other component. Stimulation with the blend evoked a "mixed" response of intermediate character depending on the concentrations of the components. The excitatory phase of this blend response was not simply the sum of the excitatory responses elicited by each of the two components alone but the response was typically weaker than expected from adding the two responses. MGC-PNs with temporally distinct blend responses could serve specific functions in olfactory information processing. They could function as the neural substrate in the AL for rapid behavioral changes in response to the appropriate component ratio of a pheromone blend (Heinbockel et al., 2004).

9.7.3 Phasis vs. tonic patterning of postsynaptic response

Both key pheromone components (represented in this study by BAL and C15) must be present to elicit upwind olfaction-modulated flight in a wind tunnel in *M. sexta* (Tumlinson et al., 1989). Individual MGC-PNs gave multiphasic responses (inhibition-excitation-afterhyperpolarization) (Figs. 12-15), which are typical of PNs in the AL of *M. sexta* (Christensen and Hildebrand, 1987; Christensen et al., 1998). However, *BAL⁺ PNs* had larger IPSPs, smaller EPSPs, weaker responses during 200 ms after stimulus onset, and stronger responses over 1000 ms after stimulus onset than did *C15⁺ PNs*. This suggests that BAL evokes sustained responses and C15 evokes shorter, more phasic responses. Corresponding response patterns were observed among ORCs in the antenna of *M. sexta* (Kaissling et al., 1989) such that responses of BAL-specific ORCs to antennal stimulation with BAL are typically more phasic-tonic whereas the phasic part of the response is more pronounced in C15-specific receptor cells. Different ORCs with selective responses to the same stimulus compound could nevertheless exhibit different response dynamics as has been shown for pheromone-specific ORCs of moths (Almaas et al., 1991) and ORCs in *D. melanogaster* (de Bruyne et al., 2000). It is tempting to speculate that MGC- PNs with phasic olfactory responses could be synaptic targets of phasic ORCs, whereas PNs giving tonic responses could receive input largely from ORCs that exhibit tonic firing patterns. In this scenario, at least some aspects of temporal coding and coding of stimulus concentration and mixture ratio are regulated through the activation of different populations of ORCs.

If MGC-PNs respond to pheromonal stimulation with sustained firing outlasting the stimulus duration they are likely to report the onset of stimulation rather than its termination and could serve as a substrate for more temporally complex spike codes (Heinbockel and Kloppenburg, 1999; Friedrich and Stopfer, 2001; Laurent et al., 2001; Christensen and Hildebrand, 2002). If MGC-PNs respond with a more phasic spike pattern and show better pulse-following, they could encode both the beginning and the end of each stimulus pulse. These MGC-PNs are able to convey information about rapidly changing signals and by doing so transmit information about the physical dynamics of the stimulus to higher olfactory centers in addition to cues about stimulus identity (Heinbockel et al., 1999; Vickers et al., 2001).

Given that each AL is innervated by less than 50 MGC-PNs (Homberg et al., 1989), each MGC-PN plays a crucial role in the transmission of information from the MGC to higher-order olfactory centers. Various characteristics of the olfactory stimulus are presumably represented by distinct outputs from a single MGC glomerulus, indicating the presence of functional heterogeneity within the population of olfactory output channels from the MGC. In the AL of male M. sexta, both component-specific and blend-specific MGC-PNs are influenced by variations in concentration and/or blend ratio, however the efficacy of distinct PNs is maximized under varying ambient conditions. Improvements in intracellular recording, labeling, confocal imaging, and data analysis enabled us to delineate the responses of individual MGC-PNs and their capacity to encode stimulus properties, such as tracking intermittent olfactory inputs. PNs that incorporated data regarding the two-component pheromone blend (i.e., receiving excitatory input from one component and inhibitory input from the other) were very adept at monitoring a sequence of stimulus pulses. Stimulus-pulse tracking was additionally refined at a synthetic mix ratio that replicates the physiological response to an extract from the female pheromone gland. Our findings indicate that the optimal response of a PN to repeating stimulus pulses is contingent not only upon stimulus intensity but also on the relative strength of the two opposing synaptic inputs absorbed by MGC-PNs.

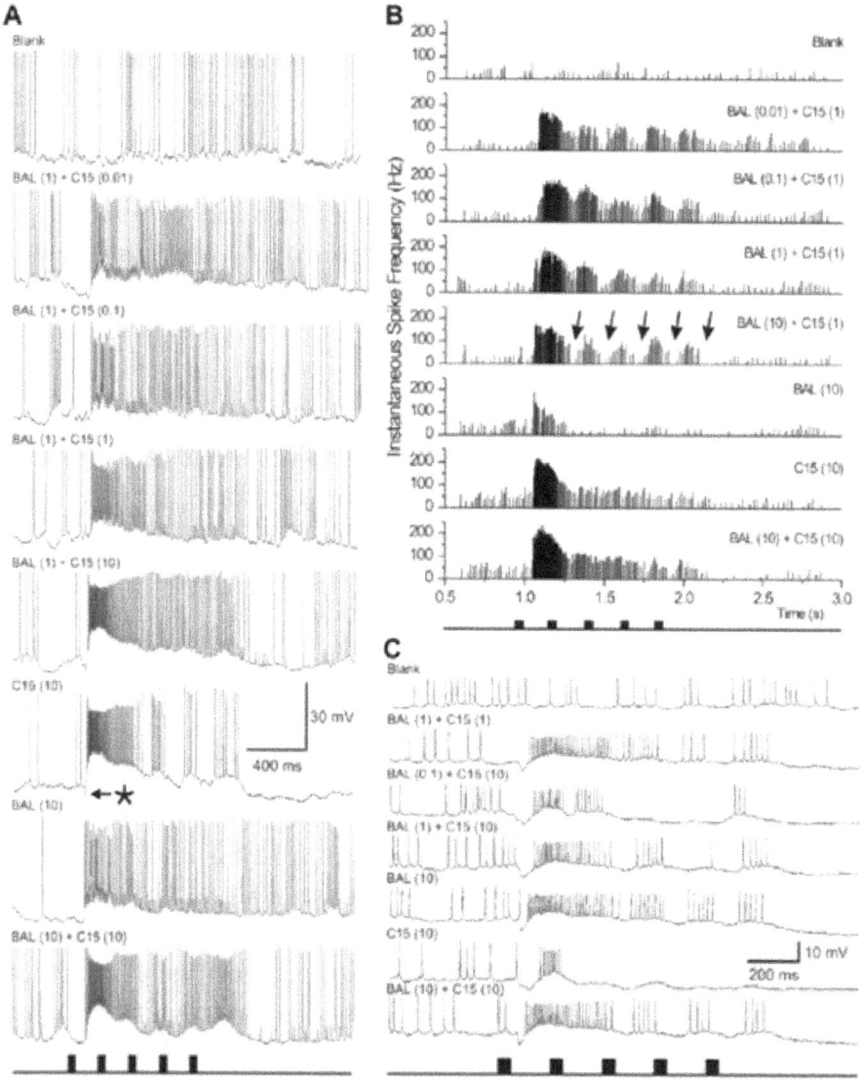

Fig. 15. Responses of *BAL⁺ / C15⁺ PNs* and *C15⁺ / BAL⁺ PNs* from male *Manduca sexta*. These MGC-PNs were depolarized by antennal stimulation with either BAL or C15. (A) This neuron responded with a strong response to C15 and a more sustained response to BAL (*C15⁺ / BAL⁺ PNs*). Stimulation with C15 also evoked a distinct IPSP (marked by asterisk) resulting in a mixed response (inhibition / excitation / inhibition). Varying the blend ratio changed the response character to more phasic or more tonic, depending on the

load in the stimulus blend. (B) ISF response plots of a *C15⁺ / BAL⁺ PN* to various BAL and C15 stimulus loads and ratios revealed a sustained phasic-tonic response to C15 and a brief excitatory response to BAL followed by inhibition. Addition of increasing amounts of BAL to the blend resulted in improved stimulus-pulse tracking. Note the distinct periods of inactivity (*arrows*) between spike bursts evoked by consecutive stimulus pulses. (C) In this neuron, stimulation with C15 evoked a brief excitatory response preceded by an IPSP and followed by a strong and prolonged inhibitory response phase. Stimulation with BAL resulted in a mixed response (inhibition / excitation / inhibition) comprising a strong IPSP and a strong excitatory response phase (*BAL⁺ / C15⁺ PN*). From Heinbockel et al., 2004.

10. CONCISE SUMMARY

This chapter elucidates the extracellular and intracellular recording techniques employed in insect preparations within the domains of insect olfaction and/or gustation. Employing one or both olfactory recording methodologies enhances comprehension of (i) the processing of olfactory information within the insect brain, (ii) the chemical identification of significant plant volatiles pertinent to insect-plant interactions, and (iii) the specific components of the female sex pheromone and plant-associated volatiles that are crucial for the orientation of male and female moths, respectively. The integration of these recording techniques with fluorescent markers facilitates the tracing of axonal projections to the brain and the individual labeling of central neurons. Gustatory extracellular recording techniques can ultimately (i) enhance comprehension of the significance of gustatory cues in larval host-plant interactions, (ii) improve insight into how taste stimuli encode various behavioral responses, and (iii) offer concepts and strategies for safeguarding crops against insect predation. Both extracellular and intracellular methods can enhance our comprehension of the processing of chemosensory information in the insect brain, elucidating how nervous systems recognize, analyze, and respond to intricate sensory stimuli.

11. REFERENCES

Andersen, R.A., Hamilton-Kemp, T.R., Loughrin, J.H., Hughes, C.G., Hildebrand, D.F., & Sutton, T.G. (1988). Green leaf headspace volatiles from *Nicotiana tabacum* lines of different trichome morphology. *Journal of Agricultural and Food Chemistry*, Vol. 36, No. 1, (January 1988), pp. 295-299. ISSN 0021-8561

Anderson, P., Hansson, B.S., & Löfqvist, J. (1995). Plant-odour-specific receptor neurones on the antennae of female and male *Spodoptera littoralis*. *Physiological Entomology*, Vol. 20, No. 3, (September 1995), pp. 189-198

Almaas, T.J., Christensen, T.A., & Mustaparta, H. (1991). Chemical communication in heliothine moths. I. Antennal receptor neurons encode several features of intra- and interspecific odorants in the male corn earworm moth *Helicoverpa zea*. *Journal of Comparative Physiology A*, Vol. 169, No. 3, (September 1991), pp. 259-274. ISSN 0340-7594

Anton, S., Löfstedt, C., & Hansson, B.S. (1997). Central nervous system processing of sex pheromones in two strains of the European Corn Borer *Ostrinia nubilalis* (Lepidoptera: Pyralidae). *Journal of Experimental Biology,* Vol. 200, No. 1, (April 1997), pp. 1073-1087. ISSN 0022- 0949

Arn H., Tóth, M., & Priesner, E. (1992). List of sex pheromones of Lepidoptera and related attractants, 2nd edition, International Organization for Biological Control, West Palearctic Regional Section, ISSN 92-9067-002-9, Montfavet, France

Baker, T.C. (1989). Sex pheromone communication in the Lepidoptera: New research progress. *Experientia,* Vol. 45, No. 3, (March 1989), pp. 248-262. ISSN 0014-4754

Bell, R.A. & Joachim, F.A. (1976). Techniques for rearing laboratory colonies of tobacco hornworms and pink bollworms. *Annals of the Entomological Society of America*, Vol. 69, No. 2, (March 1976), pp. 365-373. ISSN 0013-8746

Bernays, E.A. & Chapman, R.F. (1994). *Host-plant selection by phytophagous insects.* ISSN 0- 412-03131-0, Chapman Hall, New York

Blaney, W.M., Simmonds, M.S.J., Ley, S.V., & Jones, P.S. (1988). Insect antifeedants: a behavioural and electrophysiological investigation of natural and synthetically derived clerodane ditepenoids. *Entomologia Experimentalis et Applicata,* Vol 46, No. 3, (March 1988), pp. 267-274. ISSN 0013-8703

Boeckh, J. (1974). Die Reaktionen olfaktorischer Neurone im Deutocerebrum von Insekten im Vergleich zu den Antwortmustern der Geruchssinneszellen. *Journal of Comparative Physiology*, Vol. 90, pp. 183-205. ISSN 0340-7594

Boeckh, J. (1984). Neurophysiological aspects of insect olfaction. In: *Insect Communication*, Lewis T (ed), pp 83-104, Academic Press, ISSN 0-12-447175-7.

Buck, L.B. (1996). Information coding in the vertebrate olfactory system. *Annual Review of Neuroscience*, Vol., 19, (March 1996), pp. 517–544. ISSN 0147-006X

Buonviso, N. & Chaput, M.A. (1990) Response similarity of odors in olfactory bulb output cells presumed to be connected to the same glomerulus: electrophysiological study using simultaneous single-unit recordings. *Journal of Neurophysiology*, Vol., 63, No. 3, (March, 1990), pp. 447-454. ISSN 0022-3077

Buttery, R.G., Teranishi, R., & Ling, L.C. (1987a) Fresh tomato aroma volatiles: a quantitative study. *Journal of Agricultural and Food Chemistry*, Vol. 35, No. 4, (July 1987), pp. 540- 544. ISSN 0021-8561

Buttery, R.G., Ling, L.C., & Light, D.M. (1987b) Tomato leaf volatile aroma odorants. *Journal of Agricultural and Food Chemistry,* Vol. 35, No. 6, (November 1987), pp. 1039-1042. ISSN 0021-8561

Christensen, T.A. & Hildebrand, J.G. (1987). Male-specific, sex-pheromone-selective projection neurons in the antennal lobes of the moth *Manduca sexta*. *Journal of Comparative* Physiology, *A* Vol. 160, No. 5, (September 1987), pp. 553-569. ISSN 0340-7594

Christensen, T.A. & Hildebrand, J.G. (1990). Representation of sex-pheromonal information in the insect brain. In: *Proceedings of the Tenth International Symposium on Olfaction and Taste, ISOT X*, Døving, K.B. (ed), pp (142-150), University of Oslo, Oslo, Norway

Christensen, T.A. & Hildebrand, J.G. (1997). Coincident stimulation with pheromone components improves temporal pattern resolution in central olfactory neurons. *Journal of Neurophysiology*, Vol. 77, No. 2, (February 1997), pp. 775-781. ISSN 0022-3077

Christensen, T.A. & White, J. (2000). Representation of olfactory information in the brain. In: *The neurobiology of taste and smell*, Vol 2, Finger, T.E., Silver, W.L., Restrepo, D. (eds), pp (201-232), Wiley-Liss, ISBN 0-47125-721-5, New York

Christensen, T.A. & Hildebrand, J.G. (2002). Pheromonal and host-odor processing in the insect antennal lobe: how different? *Current Opinions in Neurobiology*, Vol. 12, No. 4, (August 2001), pp. 393-399. ISSN 0959-4388

Christensen T.A., Mustaparta, H., & Hildebrand, J.G. (1989). Discrimination of sex pheromone blends in the olfactory system of the moth. *Chemical Senses*, Vol. 14, No. 3, (June 1989), pp. 463-477. ISSN 0379-864X

Christensen T.A., Harrow, I.D., Cuzzocrea, C., Randolph, P.W., & Hildebrand, J.G. (1995). Distinct projections of two populations of olfactory receptor neurons in the antennal lobe of the sphinx moth *Manduca sexta*. *Chemical Senses*, Vol. 20, No. 3, (June 1995), 313-323. ISSN 0379-864X

Christensen, T.A., Heinbockel, T., & Hildebrand, J.G. (1996). Olfactory information processing in the brain: encoding chemical and temporal features of odors. *Journal of Neurobiology*, Vol. 30, No. 1, (May 1996), pp. 82-91. ISSN 0022-3034

Christensen, T.A., Waldrop, B.R., & Hildebrand, J.G. (1998). Multitasking in the olfactory system: context-dependent responses to odors reveal dual GABA-regulated coding mechanisms in single olfactory projection neurons. *Journal of Neuroscience*, Vol. 18, No. 15, (August 1998), pp. 5999-6008. ISSN 0270-6474

Christensen, T.A., Pawlowski, V.M., Lei, H., & Hildebrand, J.G. (2000). Multi-unit recordings reveal context-dependent modulation of synchrony in odor-specific neural ensembles. *Nature Neuroscience*, Vol. 3, No. 9, (September 2000), pp. 927-931. ISSN 1097-6256

Clyne, P.J., Warr, C.G., Freeman, M.R., Lessing, D., Kim, J., & Carlson, J.R. (1999). A novel family of divergent seven-transmembrane proteins: candidate odorant receptors in *Drosophila*. *Neuron*, Vol. 22, No. 2 (February 1999), pp. 327-338. ISSN 0896-6273

Clyne, P. J., Warr, C. G., & Carlson, J. R. (2000). Candidate taste receptors in *Drosophila*. *Science*, 287, No. 5497, (December 2000), pp. 1830–1834. ISSN 0036-8075

Dahanukar, A., Lei, Y. T., Kwon, J. Y., & Carlson, J. R. (2007). Two Gr genes underlie sugar reception in *Drosophila*. *Neuron* Vol. 56, No. 3, (November 2007), pp. 503–516. ISSN 0896-6273

de Boer, G., Dethier, V.G., & Schoonhoven, L.M. (1977). Chemoreceptors in the preoral cavity of the tobacco hornworm, *Manduca sexta*, and their possible function in feeding behaviour. *Entomologia Experimatalis et Applicata*, Vol. 21, No. 3, (May 1977), pp. 287-298. ISSN 0013-8703

de Bruyne, M., Foster, K., & Carlson, J.R. (2000). Odor coding in the *Drosophila* antenna. *Neuron*, Vol. 30, No. 2, (November 2000), pp. 537-552. ISSN 0896-6273

Friedrich, R.W. & Stopfer, M. (2001). Recent dynamics in olfactory population coding. *Current Opinions in Neurobiology*, Vol. 11, No. 4, (August 2001), pp. 468-474. ISSN 0959-4388

Dethier, V.G. (1982). Mechanisms of host plant recognition. *Entomologia Experimentalis et Applicata* Vol. 31, No. 1, (January 1982), pp. 49-56. ISSN 0013-8703

Dethier, V.G. & Crnjar, R.M. (1982). Candidate codes in the gustatory system of caterpillars. *The Journal of General Physiology*, Vol. 79, No. 4 (April 1982), pp. 549-569. ISSN 0022-1295

Dickens, J.C. (1990). Specialized receptor neurons for pheromones and host plant odors in the boll weevil, *Anthonomus grandis* Boh. (Coleoptera: Curculionidae). *Chemical Senses*, Vol. 15, No. 3, pp. 311-331. ISSN 0379-864X

Dunipace, L., Meister, S., McNealy, C., & Amrein, H. (2001). Spatially restricted expression of candidate taste receptors in the *Drosophila* gustatory system. *Current Biology*, Vol. 11, No. 11, (June 2001), pp. 822–835. ISSN 0960-9822

Firestein, S. (1992). Electrical signals in olfactory transduction. *Current Opinions in Neurobiology*, Vol. 2, No. 4, (August 1992), pp. 444-448. ISSN 0959-4388

Frazier, J.L. (1986) The perception of plant allelochemicals that inhibit feeding. In: *Molecular Aspects of Insect-Plant Associations*, Brattsten, L.B. and Ahmad, S. (eds), pp. (1-42), Plenum Press, ISSN 03-064-25475, New York

Friedrich, R.W. & Korsching, S.I. (1997). Combinatorial and chemotopic odorant coding in the zebrafish olfactory bulb visualized by optical imaging. *Neuron*, Vol. 18, No. 5, (May 1997), pp. 737-752. ISSN 0896-6273

Galizia, C.G., Sachse, S., Rappert, A., & Menzel, R. (1999). The glomerular code for odor representation is species specific in the honeybee *Apis mellifera. Nature Neuroscience*, Vol. 2, No. 5 (May 1999), pp. 473-478. ISSN 1097-6256

Glendinning, J.I., Tarre, M., & Asaoka, K. (1999). Contribution of different bitter-sensitive taste cells to feeding inhibition in a caterpillar (*Manduca sexta*). *Behavioral Neuroscience,* Vol. 113, No. 4 (August 1999), pp. 840-854. ISSN 0735-7044

Grimaldi, D. & Engel, M.S. (2005). *Evolution of the Insects.* Cambridge University Press, New York, New York

Hansson, B.S. (1995). Olfaction in Lepidoptera. *Experientia* Vol. 51, No. 11, (November 1995), pp. 1003-1027. ISSN 0014-4754

Hansson, B.S. & Christensen, T.A. (1999). Functional characteristics of the antennal lobe. In: *Insect Olfaction*, Hansson, B.S. (ed), pp. (126-161), Springer, ISSN 3-540-65034-2, Berlin

Hansson B.S., Larsson M.C., & Leal, W.S. (1999). Green leaf volatile detecting olfactory receptor neurones display very high sensitivity and specificity in a scarab beetle. *Physiological Entomology,* Vol. 24, No. 2, (June 1999), pp. 121-126. ISSN 0307-6962

Hanson, F.E. & Peterson, S.C. (1990). Sensory coding in *Manduca sexta* for deterrence by a non-host plant, *Canna generalis. Insects-Plants, '89, Symposia Biologica Hungarica,* Vol. 39, pp. 29-37. ISSN 0082-0695

Hansson, B.S., Christensen, T.A., & Hildebrand, J.G. (1991). Functionally distinct subdivisons of the macroglomerular complex in the antennal lobe of the male sphinx moth *Manduca sexta. Journal of Comparative Neurology*, Vol. 312, No. 2, (October 1991), pp. 264-278. ISSN 1096-9861

Heath, R.R., Landolt, P.J., Dueben, B., & Lenczewski, B. (1992). Identification of floral compounds of night-blooming jessamine attractive to cabbage-looper moths. *Environmental Entomology*, Vol. 21, No. 4, (August 1992), pp. 854-859. ISSN 0046-225X

Heinbockel, T. & Kaissling, K.-E. (1996). Variability of olfactory receptor neuron responses of female silkmoths (*Bombyx mori* L.) to benzoic acid and (+)-linalool. *Journal of Insect Physiology*, Vol. 42, No. 6 (June 1996), pp. 565-578. ISSN 0022-1910

Heinbockel T. & Hildebrand, J.G. (1998). Antennal receptive fields of pheromone-responsive neurons in the antennal lobes of the sphinx moth *Manduca sexta. Journal of Comparative Physiology A*, Vol. 183, No. 2, (July 1998), pp. 121-133. ISSN 0340-7594

Heinbockel, T., Kloppenburg, P., & Hildebrand, J.G. (1998). Pheromone-evoked potentials and oscillations in the antennal lobes of the sphinx moth, *Manduca sexta. Journal of Comparative Physiology A,* Vol. 182, No. 6, (May 1998), pp. 703-714. ISSN 0340-7594

Heinbockel, T., Christensen, T.A., & Hildebrand, J.G. (1999). Temporal tuning of odor responses in pheromone-responsive projection neurons in the brain of the sphinx moth *Manduca sexta. Journal of Comparative Neurology*, Vol. 409, No. 1, (June 1999), pp. 1-12. ISSN 1096-9861

Heinbockel, T., Christensen, T.A., & Hildebrand, J.G. (2004). Representation of binary pheromone blends by glomerulus-specific olfactory projection neurons. *Journal of Comparative Physiology A*, Vol. 190, No. 12, (December 2004), pp. 1023-1037. ISSN 0340-7594

Hildebrand, J.G. (1995) Analysis of chemical signals by nervous systems. *Proceedings of the National Academy of Sciences* USA, Vol. 92, No. 1, (January 1995), pp. 67-74. ISSN 0027-8424

Hildebrand, J.G. (1996) Olfactory control of behavior in moths: central processing of odor information and the functional significance of olfactory glomeruli. *Journal of Comparative Physiology A*, Vol. 178, No. 1, (January 1996), pp. 5-19. ISSN 0340-7594

Hildebrand, J.G. & Shepherd, G. (1997). Mechanisms of olfactory discrimination: converging evidence for common principles across phyla. *Annual Review of Neuroscience*, Vol. 20, (March 1997), pp. 595-631. ISSN 0147-006X

Homberg, U., Montague, R.A., & Hildebrand, J.G. (1988). Anatomy of antenno-cerebral pathways in the brain of the sphinx moth *Manduca sexta. Cell and Tissue Research*, Vol. 254, No. 2, (November 1988), pp. 225-281. ISSN 0302-766X

Homberg, U., Christensen, T.A., & Hildebrand, T.A. (1989). Structure and function of the deutocerebrum in insects. *Annual Review of Entomology*, Vol. 34, pp. 477-501. ISSN 0066-4170

Homberg U., Hoskins S.G., & Hildebrand J.G. (1995) Distribution of acetylcholinesterase activity in the deutocerebrum of the sphinx moth *Manduca sexta. Cell and Tissue Research*, Vol. 279, No. 2, (February), pp. 249-259. ISSN 0302-766X

Kaissling K.-E. (1995) Single unit and electroantennogram recordings in insect olfactory organs. In: *Experimental Cell Biology of Taste and Olfaction: Current Techniques and Protocols,* Spielman A.I., Brand, J.G. (eds), pp. 361-377, CRC Press, ISSN 0-8493- 7645-9, New York

Kaissling, K.-E. (1996). Peripheral mechanisms of pheromone reception in moths. *Chemical Senses* Vol. 21, No. 2, (April 1996), pp. 257-268. ISSN 0379-864X

Kaissling K.-E. & Thorson J (1980) Insect olfactory sensilla: structural, chemical and electrical components of the functional organization. In: *Receptors for neurotransmitters, hormones and pheromones in insects*, Sattelle, D.B., Hall, L.M., Hildebrand, J.G. (eds), pp (261-282), Elsevier, ISSN 0444802312, North-Holland, Amsterdam

Kaissling, K.-E., Hildebrand J.G., & Tumlinson, J.H. (1989). Pheromone receptor cells in the male moth *Manduca sexta. Archives of Insect Biochemistry and Physiology*, Vol. 10, No. 4, (April 1989), pp. 273-279. ISSN 0739-4462

Keil, T.A. (1989) Fine structure of the pheromone-sensitive sensilla on the antenna of the hawkmoth, *Manduca sexta. Tissue and Cell,* Vol. 21, No. 1, (February 1989), pp. 139– 151. ISSN 0040-8166

Keller, A. & Vosshall, L.B. (2003). Decoding olfaction in *Drosophila. Current Opinions in Neurobiology*, Vol. 13, No. 1, (February 2003), pp. 103-110. ISSN 0959-4388

King, J. Roche, Christensen, T.A., & Hildebrand, J.G. (2000). Response characteristics of an identified, sexually dimorphic olfactory glomerulus. *Journal of Neuroscience*, Vol. 20, No. 6, (March 2000), pp. 2391-2399. ISSN 0270-6474

Kloppenburg, P. & Heinbockel, T. (2000). 5-hydroxytryptamine modulates pheromone- evoked local field potentials in the macroglomerular complex of the sphinx moth *Manduca sexta. Journal of Experimental Biology*, Vol. 203, No. 11, (June 2000), pp. 1701-1709. ISSN 1477-9145

Knudsen, J.T. & Tollsten, L. (1993). Trends in floral scent chemistry in pollination syndromes: floral scent composition in moth-pollinated taxa. *Botanical Journal of the Linnean Society*, Vol. 113, No. 3, (November 1993), pp. 263-284. ISSN 0024-4074

Laurent, G., Stopfer, M., Friedrich, R.W., Rabinovich, M.I., Volkovskii, A., & Abarbanel,H.D. (2001). Odor encoding as an active, dynamical process: experiments, computation, and theory. *Annual Review of Neuroscience*, Vol. 24, (March 2001), pp. 263-297. ISSN 0147-006X

Lee, J.K. & Strausfeld, N.J. (1990). Structure, distribution and number of surface sensilla and their receptor cells on the olfactory appendage of the male moth, *Manduca sexta. Journal of Neurocytology*, Vol. 19, No. 4 (August 1990), pp. 519–538. ISSN 0300-4864

Lei, H., Christensen, T.A., & Hildebrand, J.G. (2002). Local inhibition modulates odor- evoked synchronization of glomerulus-specific output neurons. *Nature Neuroscience*, Vol. 5, No., 6, (June 2002), pp. 557-565. ISSN 1097-6256

Lei, H., Oland, L.A., Riffell, J.A., Beyerlein, A., & Hildebrand, J.G. (2010). Microcircuits for olfactory information processing in the antennal lobe of *Manduca sexta.* In:

Handbook of brain microcircuits, Shepherd, G.M., Grillner, S. (eds), pp. (417-426), Oxford University Press, ISBN 978-0-19-538988-3, New York

Leon M & Johnson BA (2003) Olfactory coding in the mammalian olfactory bulb. *Brain Research Reviews* Vol. 42, No. 1, (April, 2003), pp. 23-32. ISSN 0165-0173

Mombaerts, P. (1996). Targeting olfaction. *Current Opinions in Neurobiology*, Vol. 6, No. 4, (August 1996), pp. 481-486. ISSN 0959-4388

Morgan, A.C. & Lyon, S.C. (1928). Notes on amyl salicylate as an attractant to the tobacco hornworm moth. *Journal of Economic Entomology*, Vol. 21, pp. 189-191. ISSN 0022- 0493

Murlis, J. (1997). Odor plumes and the signal they provide. In: *Insect pheromone research - new directions.* Cardé R.T., Minks, A.K. (eds), pp (221-231), Chapman & Hall, ISSN 0- 412-99611-1, New York

Ng, M., Roorda, R.D., Lima, S.Q., Zemelman, B.V., Morcillo, P., & Miesenböck, G. (2002). Transmission of olfactory information between three populations of neurons in the antennal lobe of the fly. *Neuron*, Vol. 36, No. 3 (October 2002), pp. 463-474. ISSN 0896-6273

Oland, L.A. & Tolbert, L.P. (1988). Effects of hydroxyurea parallel the effects of radiation in developing olfactory glomeruli. *Journal of Comparative Neurology*, Vol. 278, No. 3, (December 1988), pp. 377–387. ISSN 1096-9861

Pichon, Y., Sattelle, D.B., & Lane, N.J. (1972). Conduction processes in the nerve cord of the moth *Manduca sexta* in relation to its ultrastructure and haemolymph ionic composition. *Journal of Experimental Biology*, Vol. 56, No. 3 (June 1972), pp. 717-734. ISSN 1477-9145

Pophof, B. (1997). Olfactory responses recorded from sensilla coeloconica of the silkmoth *Bombyx mori. Physiological Entomology*, Vol. 22, No. 3, (September 1997), pp. 239-248. ISSN 0307-6962

Priesner, E. (1979). Progress in the analysis of pheromone receptor systems. *Annals of Zoology, Ecology and Animals*, Vol. 11, pp. 533-546.

Prescott, D.J., Hildebrand, J.G., Sanes, J.R., & Jewett, S. (1977). Biochemical and developmental studies of acetylcholine metabolism in the central nervous system of the moth *Manduca sexta. Comparative Biochemistry and Physiology Part C: Comparative Pharmacology*, Vol. 56, No. 2, pp. 77-84. ISSN 1532-0456

Raguso, R.A. & Willis, M.A. (1997). Floral scent and its role(s) in hawkmoth attraction. *Chemical Senses*, Vol. 22, No. 6, (December 1997), pp. 774-775. ISSN 0379-864X Raguso, R.A. & Light, D.M. (1998). Electroantennogram responses of male *Sphinx perelegans* hawkmoths to floral and "green-leaf volatiles." *Entomologia*

Experimentalis et Applicata, Vol. 86, No. 3, (March 1998), pp. 287-293. ISSN 0013-8703

Reisenman, C.E., Christensen, T.A., & Hildebrand, J.G. (2004). Enantioselectivity of projection neurons innervating identified olfactory glomeruli. *Journal of Neuroscience*, Vol. 24, No. 11, (March 2004), pp. 2602-2611. ISSN 0270-6474

Reisenman, C.E., Heinbockel,T., & Hildebrand, J.G. (2008). Inhibitory interactions among olfactory glomeruli do not necessarily reflect spatial proximity. *Journal of Neurophysiology*, Vol. 100, No. 2, (August 2008), pp. 554-564. ISSN 0022-3077

Renwick, J. A. A. (1989). Chemical ecology of oviposition in phytophagous insects. *Experientia*, Vol. 45, No. 3, (March 1989), pp. 223-228. ISSN 1420-682X

Robertson, H. M., Warr, C. G., & Carlson, J. R. (2003). Molecular evolution of the insect chemoreceptor gene superfamily in *Drosophila melanogaster*. *Proceedings of the National Academy of Sciences U.S.A.*, Vol. 100, Suppl. 2, (November 2003), pp. 14537– 14542. ISSN 0027-8424

Rodrigues, V. & Buchner, E. (1984). [^{3}H]2-deoxyglucose mapping of odor-induced neuronal activity in the antennal lobes of *Drosophila melanogaster*. *Brain Research*, Vol. 324, pp. 374-378. ISSN 0006-8993

Rospars, J.P. & Hildebrand, J.G. (2000). Sexually dimorphic and isomorphic glomeruli in the antennal lobes of the sphinx moth *Manduca sexta*. *Chemical Senses*, Vol. 25, No. 2, (April 2000), pp. 119-129. ISSN 0379-864X

Rössler, W., Tolbert, L.P., & Hildebrand, J.G. (1998). Early formation of sexually dimorphic glomeruli in the developing olfactory lobe of the brain of the moth *Manduca sexta*. *Journal of Comparative Neurology*, Vol. 396, No. 4, (July 1998), pp. 415-428. ISSN 1096-9861

Rössler, W., Randolph, P.W., Tolbert, L.P., & Hildebrand, J.G. (1999). Axons of olfactory receptor cells of transsexually grafted antennae induce development of sexually dimorphic glomeruli in *Manduca sexta*. *Journal of Neurobiology*, Vol. 38, No. 1, (January 1999), pp. 1-541. ISSN 0022-3034

Sachse, S. & Galizia, C.G. (2003). The coding of odour-intensity in the honeybee antennal lobe: local computation optimizes odour representation. *European Journal of Neuroscience*, Vol. 18, No. 8, (October 2003), pp. 2119-2132. ISSN 0953-816X

Sanes, J.R. & Hildebrand, J.G. (1976). Structure and development of antennae in a moth, *Manduca sexta*. *Developmental Biology* Vol. 51, No. 2, (July 1976), pp. 282-299. 0012-1606

Schneider, D., Lacher, V., & Kaissling, K.-E. (1964). Die Reaktionsweise und Reaktionsspektrum von Riechzellen bei *Antheraea pernyi* (Lepidoptera,

Saturniidae). *Zeitschrift fuer Vergleichende Physiologie,* Vol. 48, No. 6, (November 1964), pp. 632-662. ISSN 0340-7594

Schoonhoven, L.M. (1973). Plant recognition by lepidopterous larvae. *Symposia of the Royal Entomological Society of London,* Vol. 6, pp. 87-99. ISSN 0080-4363

Schoonhoven, L.M. (1982). Biological aspects of antifeedants. *Entomologia Experimentalis et Applicata,* Vol. 31, No. 1, (January 1982), pp. 57-69. ISSN 0013-8703

Schoonhoven, L.M. (1987). What makes a caterpillar eat? The sensory coding underlying feeding behavior. In: *Perspectives in Chemoreception and Behavior,* Chapman, R.F., Bernays, E.A., and Stoffolano, J.G., (eds), pp. (69-97), Springer-Verlag, ISSN 9780387963747, New York

Schoonhoven, L.M. & Dethier, V.G. (1966). Sensory aspects of host-plant discrimination by lepidopterous larvae. *Archives Néerlandaises de Zoologie,* Vol. 16, No. 4, pp. 497-530. ISSN 0365-5164

Schoonhoven, L.M. & Blom, F. (1988). Chemoreception and feeding behaviour in a caterpillar: towards a model of brain functioning in insects. *Entomologia Experimentalis et Applicata,* Vol. 49, No. 1-2, (November 1988), pp. 123-129. ISSN 0013-8703

Schoonhoven L.M. & van Loon, J.J.A. (2002) An inventory of taste in caterpillars: each species its own key. *Acta Zoologica Academiae Scientiarum Hungaricae, Vol.* 48, No. 3, pp. 215-263. ISSN 1217-8837

Schoonhoven, L.M., Blaney, W.M., & Simmonds, M.S.J. (1992). Sensory coding of feeding deterrents in phytophagous insects. In: *Insect-Plant Interactions,* Vol. 4, Bernays, E.A. (ed), pp. (59-79), CRC Press, ISSN 0-8493-4124-8, Boca Raton, FL

Schoonhoven, L.M., Jermy, T., & van Loon, J.J.A. (1998) *Insect-Plant Biology. From Physiology to Evolution.* ISSN, 9780412587009, Chapman and Hall, London.

Scott, K., Brady, J. R., Cravchik, A., Morozov, P., Rzhetsky, A., Zuker, C., & Axel, R. (2001). A chemosensory gene family encoding candidate gustatory and olfactory receptors in *Drosophila. Cell,* Vol. 104, No. 5 (March 2001), pp. 661–673. ISSN 0092-8674

Sharp, F.R., Kauer, J.S., & Shepherd, G.M. (1975). Local sites of activity-related glucose metabolism in rat olfactory bulb during olfactory stimulation. *Brain Research,* Vol. 98, No. 3, (November 1975), pp. 596-600. ISSN 0006-8993

Shepherd, G.M. (1994). Discrimination of molecular signals by the olfactory receptor neuron. *Neuron* Vol. 13, No. 4 (October 1994), pp. 771-790. ISSN 0896-6273

Shields, V.D.C. (1994). Ultrastructure of the uniporous sensilla on the galea of larval *Mamestra configurata* (Walker) (Lepidoptera: Noctuidae). *Canadian Journal of Zoology,* Vol. 72, No. 11, (November 1994), pp. 2016-31. ISSN 0008-4301

Shields, V.D.C. & Mitchell, B.K. (1995). Responses of maxillary styloconic receptors to stimulation by sinigrin, sucrose and inositol in two crucifer-feeding, polyphagous lepidopterous species. *Philosphical Transactions of the Royal Society of London B.*, Vol. 347, No. 1322, (March 1995), pp. 447-457. ISSN 0962-8436

Shields V.D.C. & Hildebrand J.G. (1999a). Fine structure of antennal sensilla of the female sphinx moth, *Manduca sexta* (Lepidoptera: Sphngidae). I. trichoid and basiconic sensilla. *Canadian Journal of Zoology,* Vol. 77, No. 2, (August 1999), pp. 290–301. ISSN 0008-4301

Shields V.D.C. & Hildebrand J.G. (1999b). Fine structure of antennal sensilla of the female sphinx moth, *Manduca sexta* (Lepidoptera: Sphingidae). II. auriculate, coeloconic, and styliform complex sensilla. *Canadian Journal of Zoology,* Vol. 77, No. 2, (August 1999), pp. 302–313. ISSN 0008-4301

Shields V.D.C. & Hildebrand J.G. (2001a). Responses of a population of antennal olfactory receptor cells in the female moth *Manduca sexta* to plant-associated volatile organic compounds. *Journal of Comparative Physiology A*, Vol. 186, No. 12, (February 2001), pp. 1135–1151. ISSN 0340-7594

Shields V.D.C. & Hildebrand J.G. (2001b). Recent advances in insect olfaction, specifically regarding the morphology and sensory physiology of antennal sensilla of the female sphinx moth *Manduca sexta. Microscopy Research and Technique*, Vol. 55, No. 5, (December 2001), pp. 307–329. ISSN 1059-910X

Shields, V.D.C. & Martin, T.L. (2010). The effect of alkaloids on the feeding of lepidopteran larvae. In: *Alkaloids: Properties, Applications and Pharmacological Effects,* Chapter 6, Cassiano, Nicole M. (ed), pp. (109-138), Nova Science Publishers, Inc., ISSN 9781617611308, Hauppauge, New York

Shields, V.D.C. & Martin, T.L. (2012). The structure and function of taste organs in caterpillars. In: *The Sense of Taste,* Chapter 11, Lynch, E.J., Petrov, A.P. (eds), pp. (147-166), Nova Science Publishers, Inc., ISSN 978-1-61209-748-0, Hauppauge, New York

Shields, V.D.C., Rodgers, E.J., Arnold, N.S., & Williams, D. (2006). Feeding responses to selected alkaloids by gypsy moth larvae, *Lymantria dispar* (L.). *Naturwissenschaften*, Vol. 93, No. 3, (March 2006), pp. 127-130. ISSN 0028-1042

Städler, E. (1984) Contact chemoreception. In*: Chemical Ecology of Insects*, Bell, W. J., Cardé, R.T. (eds), pp. (3-35), Chapman and Hall, ISSN 9780878930692, New York

Stopfer, M., Jayaramann, V., & Laurent, G. (2003). Intensity versus identity coding in an olfactory system. *Neuron*, Vol. 39, No. 6 (September 2003), pp. 991-1004. ISSN 0896-6273

Strausfeld, N.J. (1989). Cellular organization in male-specific olfactory neuropil in the moth *Manduca sexta*. In: *Dynamics and Plasticity in Neuronal Systems:* Proceedings of the 17th Göttingen Neurobiology Conference, Elsner N, Singer W, eds), p (79), Thieme, ISSN 3137380014, Stuttgart, Germany.

Sun, X.J., Tolbert, L.P., & Hildebrand, J.G. (1993). Ramification pattern and ultrastructural characteristics of the serotonin immunoreactive neuron in the antennal lobe of the moth *Manduca sexta*: a laser scanning confocal and electron microscopic study. *Journal of Comparative Neurology*, Vol. 338, No. 1 (December 1993), pp. 5-16. 1096- 9861. ISSN 1096-9861.

Thorne, N., Chromey, C., Bray, S., & Amrein, H. (2004). Taste perception and coding in *Drosophila. Current Biology,* Vol. 14, No. 12, (June 2004), 1065–1079. ISSN 0960-9822.

Tumlinson, J.H., Brennan, M.M., Doolittle, R.E., Mitchell, E.R., Brabham, A., Mazomenos. B.E., Baumhover, A.H., Jackson, D.M. (1989). Identification of a pheromone blend attractive to *Manduca sexta* (L.) males in a wind tunnel. *Archives of Insect Biochemistry and Physiology*, Vol. 10, No. 4, (April 1989), pp. 255-271. ISSN 1520-6327.

Van der Pers, J.N.C. & Den Otter, C.J. (1978). Single cell responses from olfactory receptors of small ermine moths to sex attractants. *Journal of Insect Physiology* Vol. 24, No. 4, (April 1978), pp. 337-343. ISSN 0022-1910

Vickers, N.J., Christensen, T.A., Baker, T.C., & Hildebrand, J.G. (2001). Odour-plume dynamics influence the brain's olfactory code. *Nature*, Vol. 410, No. 6827, (March 2001), pp. 466-470. ISSN 0028-0836

Visser, J.H. & Avé, D.A. (1978). General green leaf volatiles in the olfactory orientation of the Colorado beetle, *Leptinotarsa decemlineata. Entomologia Experimentalis et Applicata*, Vol. 24, No. 3, (November 1978) pp. 538-549. ISSN 0013-8703

Vosshall, L.B. (2001). The molecular logic of olfaction in *Drosophila. Chemical Senses*, Vol. 26. No. 2, (February 2001), pp. 207-213. ISSN 0379-864X

Vosshall, L.B., Amrein, H., Morozov, P.S., Rzhetsky, A., & Axel, R. (1999). A spatial map of olfactory receptor expression in the *Drosophila* antenna. *Cell,* Vol. 96, No. 5, (March 1999), pp. 725-736. ISSN 0092-8674

Wang, J.W., Wong, A.M., Flores, J., Vosshall, L.B., & Axel, R. (2003). Two-photon calcium imaging reveals an odor-evoked map of activity in the fly brain. *Cell* Vol. 112, No. 2 (January 2003), pp. 271-282. ISSN 0092-8674

Wang, Z., Singhvi, A., Kong, P., & Scott, K. (2004). Taste representations in the *Drosophila* brain. *Cell,* Vol. 117, No. 7, (June 2004). pp. 981–991. ISSN 0092-8674

Wanner, K.W. & Robertson, H.M. (2008). The gustatory receptor family in the silkworm moth *Bombyx mori* is characterized by a large expansion of a single lineage of putative bitter receptors. *Insect Molecular Biology*, Vol. 17, No. 6, (December 2008), pp. 621-629. ISSN 1365-2583

Wu, W., Anton, S., Löfstedt, C., & Hansson, B.S. (1996). Discrimination among pheromone component blends by interneurons in male antennal lobes of two populations of the turnip moth, *Agrotis segetum*. *Proceedings of the National Academy of Sciences*, Vol. 93, No. 15, (July 1996), pp. 8022-8027. ISSN 0027-8424

Yamamoto, R.T., Jenkins, R.Y., & McClusky, R.K. (1969). Factors determining the selection of plants for oviposition by the tobacco hornworm, *Manduca sexta*. *Entomologia Experimentalis et Applicata*, Vol. 12, No. 5, (December 1969), pp. 504-508. ISSN 0013- 8703

Zar, J.H. (1984). Biostatistical analysis, 2nd edition, Englewood Cliffs, ISSN 9780130779250, Prentice Hall, NJ.

CHAPTER 8

HISTOPATHOLOGICAL CHANGES IN VARIOUS ORGANS OF ADULT *CLARIAS GARIEPINUS* (BURCHELL, 1822) EXPOSED TO 4-NONYLPHENOL

1. INTRODUCTION: AN OVERVIEW

Endocrine-disrupting chemicals (EDCs) comprise both synthetic and naturally occurring substances that interfere with the normal physiological functioning in animals (Razia et al. 2006). Research indicates that exposure to both natural and synthetic estrogenic compounds might negatively impact wildlife and human health (Colborn et al. 1993). In vitro studies (Soto et al. 1992; Soto et al. 1994; Toomey et al. 1999) have validated the impact of endocrine-disrupting chemicals on tissue architecture and cellular mechanisms. Nonylphenol ethoxylates (NPEs) are endocrine-disrupting chemicals (EDCs) utilized worldwide in the manufacture of plastics, insecticides, and cleaning agents, and are found in sewage effluents internationally (Talmage, 1994). NP is identified as the principal degradation product of NPEs due to its increased resistance to biodegradation, toxicity, capacity for bioaccumulation in aquatic creatures, and estrogenic properties (Ahel et al. 1994). NP is present in surface waters, aquatic sediments, and groundwater (Bennie, 1999; Talmage, 1994) and has estrogenic properties in numerous aquatic organisms (Nimrod and Benson, 1996; Talmage, 1994; Servos, 1999).

The application of environmental toxicological studies on non-mammalian vertebrates is rapidly expanding; and for aquatic system, fish have become valuable indicator for the evaluation of the effects of noxious compounds (Khidr and Mekkawy, 2008). Histology and histopathology can be used as biomonitoring tools for health in toxicity studies (Meyers and Hendricks, 1985). Histoplathological alterations are biomarkers of effect exposure to environmental stressors, revealing alterations in physiological and biochemical function (Hinton et al. 1992). Histopathology, the study of lesions or abnormalities on cellular and tissue levels is useful tool for assessing the degree of pollution, particularly for sublethal and chronic effects (Bernet et al. 1999). More than one tissue may be studied for assessment of the biological effects of a toxicant on localized portions of certain organs and also for assessment of subsequent derangements (degradations) in tissues or cells in other locations and this allows for diagnoses of the observed changes (Adeyemo, 2008). NP has been shown to cause histopathological changes in the germ and Sertoli cells of the male eelpot (Christiansen et al. 1998). The skin of fish is continuously exposed to and in direct contact with the environment pollutants such as NP. Histological changes in skin of rainbow trout with mucosomes in goblet cells were recorded after exposure to 10 µg/l of 4-nonylphenol. Several studies demonstrated the high susceptibility of skin to environmental pollutant impacts (Burkhardt-Holm et al. 1997; Iger et al. 1995; Shephard, 1994). Burkhardt-Holm et al. (1997) hypothesized that in trout, xenobiotic estrogens might affect the skin, like natural estrogens, via the steroid receptor.

In trout species, nonylphenol was found to accumulate in the liver, gill, skin, gut, fat, and kidney tissue (Ahel et al. 1993; Coldham et al. 1998; Lewis and Lech, 1996). So that, 4- nonylphenol may affects those organs in corresponding with its impacts on reproductive ones. Most of NP studies revealed sever effects on the liver and gonads of fish tissues (Christiansen et al. 1998; Jobling et al. 1996; Lech et al. 1996) and the corresponding metabolism. The liver is important in many aspects of nutrition, including lipid and carbohydrates storage and alterations in liver structure may be useful as biomarker that indicate prior exposure to environmental stressors (Hinton and laurén, 1990). Stressors- associated alterations of hepatocytes may be found in the nucleus or cytoplasm or both (Marchand et al. 2008). Malik and Hodgson, (2002) reported that the liver plays a major role in complex enzymatic processes of thyroid hormones conversion. So, liver dysfunction and disease affects thyroid hormone metabolism. Although gills are not only the prime organs for gaseous exchange, they perform several other physiological functions including osmoregulation and excretion. Parashar and Banerjee, (2002) reported that changes in environmental parameters often damage this delicate vital organ which has direct contact with aquatic environment. Many studies demonstrated that increased concentrations of different pollutants including several heavy metals seriously damage the gills of teleostean fish (Dutta et al. 1996; Wendelaar Bonga, 1997).

The African catfish (Clarias gariepinus), an omnivorous freshwater species, is a widely appreciated delicacy across tropical Africa (Nguyen and Janssen, 2002) owing to its rapid growth rate, high stocking-density potential, strong consumer acceptance, and notable resilience to suboptimal water quality and oxygen deficiency (Adewolu et al. 2008; Akinwole and Faturoti, 2007; Karami et al. 2010). The African catfish, being a significant cultured species (Adeyemo, 2008), has been utilized in numerous foundational experimental studies (Mahmoud et al., 2009). This study extends prior research by the authors (Mekkawy et al., 2011; Mahmoud et al., 2011; Sayed et al., 2011) to assess the degree to which histopathological variations in certain organs of the adult catfish, Clarias gariepinus (Burchell, 1822), are concurrently associated with biochemical and physiological changes induced by nanoparticles, particularly concerning endocrine disruption.

2. MATERIALS AND METHODS

2.1 Specimen collection

Specimens of adult catfish *C. gariepinus* were collected from the River Nile at Assiut and then were transported to Fish Biology Laboratory of Zoology Department, Faculty of Science, Assiut University. The fish (500–1200 g) were fed on a commercial pellet diet (3% of body weight per day) and kept together in 100 l rectangular tanks containing tap water (conductivity 2000 ls/cm; pH 7.5; oxygen 88–95% saturation; temperature 27-28 ºC; photoperiod 12:12 light: dark). After 2 week acclimatization, fishes were used for the experimental setup.

2.2 4-nonylphenol

4-Nonylphenol was obtained from Sigma- Aldrich (Schnelldrof, Germany)

2.3 Experimental setup

The adapted adult fish classified into four groups (10 fish per each): control, 4-nonylphenol- treated group (for15 day/ for 0.05mg/l day), 4-nonylphenol-treated group (for 15 day/for 0.08mg/l day), and 4-nonylphenol-treated group (for 15 day/ for 0.1 h/ day). In the present study, the range of NP exposures was 0.05-0.1 mg/l and the exposure concentrations are environmentally relevant. The conditions of the experiment were as that of acclimatization with changing all the tap water and concentrations of 4-nonylphenol every day.

2.4 Hematoxylin-Eosin (HE) and Masson's Trichrome (TRI) histopathological preparations

For microscopic preparations, after 15 days, 3 surviving fish of each group were removed and dissected. Small pieces of the liver, kidneys, gills, and skin were taken and immediately fixed in 10% neutral buffered formalin. Fixed tissues were processed routinely for paraffin embedding technique. Embedded tissues were sectioned at 5-7μ in thickness and then stained with Harris' hematoxylin and eosin stain (H & E) and Masson's Trichrome (TRI) stain according to Bancroft and Steven, (1982). Sections were visualized and studied using OLYMPUS microscope model BX50F4 (Olympus optical Co., LTP. Japan).

2.5 Transmission electron microscope (TEM)

Small pieces of liver of newly scarified fish were fixed in 2 % glutaraldehyde, washed in cacodylate buffer and post-fixed in 1% osmium tetroxide. Dehydration was carried out in ascending grads of alcohol and then embedded in epon-araldite mixture. Semithin sections of liver were cut at 1µm and stained with toluidine blue for examination under a light microscope. Ultrathin sections were stained with uranyl acetate followed by lead nitrate (Johannessen, 1978). Electron micrographs were obtained using a Jeol JEM 1200 EX Transmission Electron Microscope at Electron microscope center of Assiut University.

3. RESULTS

During the experiment, the mortality rates of fish were 1.2%, 3.75%, 5.5%, and 8% for the control group and for concentrations of 0.05 mg/l, 0.08 mg/l, and 0.1 mg/l of 4-nonylphenol, respectively. Lesions were detected in the gills, epidermis, kidneys, and liver of tested fish across all exposure dosages and durations of 4-nonylphenol. The frequency and extent of changes were positively correlated with the levels of 4-nonylphenol, whereas samples from the control group exhibited normalcy in all organs throughout the experiment's period.

3.1 Histopathological changes in the gills

Histologically, the gills of the adult catfish *Clarias gariepinus* are composed of primary lamellae (pl), secondary lamellae (sl), epithelial cell (epc), mucous cell (mc), and chloride cell (chc) (Fig. 1a). The initial lesions in the gills were manifested in groups exposed to 0.05, 0.08, and 0.1 mg/l of 4-nonylphenol for 15 days (Fig. 1b, c, d, e). The anomalies include epithelial lifting, edema, deformed secondary lamellae in fishes exposed to 0.05 mg/l 4-nonylphenol (Fig. 1b) while in fishes exposed to 0.08 mg/l 4-nonylphenol, desquamation and necrosis were recorded (Fig. 1c). As Fig. (1d, e) shows gills with degeneration of cartilaginous bar malformed secondary lamellae, increase in chloride cell size and number, epithelial hyperplasia, diffusion of secondary lamellae and increase number of mucous cells in fishes exposed to 0.1 mg/l 4-nonylphenol for 15 days.

3.2 Histopathological changes in the skin

Normal structure of skin of adult catfish, *Clarias gariepinus* was shown in fig. (2a), where it consists of alarm cell (ac), mucous cell (mc) and epithelium (ep) with pigment cell (p). The fishes exposed to 0.05 mg/l of 4-nonylphenol showed enlarged alarm cell (eac), with vacuoles (va) in their skin structure (Fig. 2f). As shown in fig. (2c) other changes such as ruptured epithelial cells (repc) and enlarged mucous cells (emc) in the skin of fishes exposed to 0.08 mg/l of 4-nonylphenol for 15 days. Severe damage was recorded in fishes exposed to 0.1 mg/l of 4-nonylphenol as in Fig. (2b, d,

e) in which ruptured epithelial cells, necrotic cell (nc), granuled cells (gc), vacuoles (va) and fat cells (fc) were recorded.

3.3 Histopathological changes in the kidney

The functional units of the kidney of the control fish, *Clarias gareipinus* are nephrons which are composed of renal corpuscles (rc) and renal tubules (rt); these structures are surrounded by haemopoietic tissue (ht). The shape of the renal corpuscle is roughly spherical consisting of a double membraned capsule (Bowman's capsule) enclosing a tuft of blood capillaries (glomerulus) (g). Bowman's space; a space between the glomerulus and the capsule (Fig. 3a). Examination of kidney sections of fish exposed to 0.05 mg/l of 4-nonylphenol for 15 days revealed edema in the epithelium lining of some renal tubules (e) and some showed degeneration (d) and rupture of Bowman's capsule (r). Hypertrophy of the glomerulus (hyt) was observed with shrinkage (sh). Moreover, necrosis (n) and pyknosis (p) were observed in some renal tubules (fig. 3b). After 15 days of exposure to the 0.08 mg/l of 4-nonylphenol, similar histological changes were observed, however, proliferation in renal tubules and haemopoieatic tissue (pr) with dissociation in some tubules (di) were recorded. Dilated blood vessels (dbv) and mealnomacrophages (m) were also observed (fig. 3c). The sections of kidney in fishes exposed to 0.1 mg/l of 4-nonylphenol showed severe damage or complete degeneration with obliterated Bowman's space (obs). Masson's Trichrome stain indicated the degeneration of connective tissue and degeneration of renal tubule and glomerulus (fig. 3d).

3.4 Histopathological changes in the liver

The liver of the control fish *Clarias gariepinus* appears as a continuous mass of hepatic cells; hepatocytes (h) which cord-like pattern interrupted by blood vessels and sinusoids (bs). The cords of hepatocytes are arranged around the central vein (cv). The hepatocytes are large in size, polygonal in shape with centrally located nuclei. The hepatocytes have homogenous eosinophilic cytoplasm. The sinusoids are seen as communicating channels occupied by blood cells with Küffer cells (kc) (fig. 4a). Examination of liver sections after exposure to 0.08 mg/l of 4-nonylphenol for 15 days showed degeneration (d) in the form of disintegration in most cytoplasmic contents. Lymphatic aggregations (la), necrosis (n), pyknosis (p), were observed (Fig. 4b). Also, melanomacrophages, pyknosis and rupture of hepatocytes (r) were recorded (fig. 4c). Less damage occurred in liver sections after exposure to 0.05 mg/l of 4-nonylphenol for 15 days (Fig. 4d). As Fig. (5a, b, c, d, e) shows marked severe damage occurred in fishes exposed to 0.1 mg/l of 4-nonylphenol for 15 days. Pyknosis indicated by arrows, fat cell (fc), lymphatic infiltration indicated by arrows, pigments diffusion, aggregation of fibers around central vein and rupture of hepatocytes were recorded. Masson's Trichrome stain indicated this severe damage in liver tissue (Fig. 5f).

3.5 Electron microscope examination of hepatocytes

The fine structure of the hepatocytes shows parallel cisternae of rough endoplasmic reticulum (rer), polygonal centrally located vesicular nuclei (n) with nucleolus (nu), numerous mitochondria (m) with different shapes and sizes and Golgi complex (g) near the nucleus (Fig. 6a).

Hepatocytes of animals exposed to 0.05 mg/l of 4-nonylphenol appeared swollen or hypertrophied with dense bodies (db), karyolysis in nucleus (fig. 6b) and damaged mitochondria (dm), rarified cytoplasm (rc) and vacuoles (v) (fig. 7b). Also, degenerative changes, shrunken and indented nuclei with cytoplasmic fat droplets (fd) were observed (Fig. 8b). In fishes exposed to 0.08 mg/l of 4-nonylphenol the cytoplasm shows tiney vacuoles (cv), the nuclei appeared irregular in shape with nuclear indentation. Some hepatocytes showed signs of karylyosis (Fig. 7a, 9a). Moreover, damaged mitochondria, degenerative rough endoplasmic reticulum (drer), increase in number of lysosomes (ly) were recorded. Some nuclei showed condensation and migration of chromatin at the nuclear periphery with prominent nucleolus with some apoptotic changes in the form of nuclear envelope (Fig. 9a). Some hepatocytes appeared swollen with large rarified areas in the cytoplasm resulting in disorganization and dissociation of cellular organelles (Fig. 7a). Hepatocytes of fishes exposed to 0.1 mg/l of 4-nonylphenol showed similar changes as those exposed to 0.05 and 0.08 mg/l of 4-nonylphenol, however, degenerative changes, hypertrophied, karylyosis, blood sinusoids collapsation, apoptosis and vacuolated hepatocytes (Fig. 10, 11). Mitochondria were swollen with destructive cristae; electron dense materials appeared at the periphery of these mitochondria (em) (Fig. 9b, 10a, 11,a, b). Concentric whorled arrangement of rough endoplasmic reticulum (cw) with detached ribosomes was also observed (Fig. 11a). Other areas of the reticulum manifested as parallel cisternae with electron-lucent cytoplasm interspersed between the cisternae; additionally, circular arrangements of rough endoplasmic reticulum were observed (Fig. 10a, 11a). An augmentation in the quantity of lysosomes and lipid droplets was observed (Fig. 10b, 11b).

4. DISCUSSION

It has been reported that NP like E2 is estrogenic and affects the histology of developing immune and endocrine organs and those in direct contact with aquatic environment (Yokota *et al.*, 2001; Kang *et al.*, 2003; Seki *et al.*, 2003; Razia *et al.*, 2006). Skin and gills are highly sensitive to pollutants due to their direct contact to aquatic environment. It has been reported that the skin is sensitive to steroid hormone activity (Pottinger and Pickering, 1985). The present results showed severe damage in the skin epithelial cells and necrosis reflecting such sensitivity to NP. NP exposure of rainbow trout resulted in a specific granulation pattern of epidermal mucous cells visible as irregularly shaped and large mucosomes (Burkhardt-Holm *et al.*, 2000). The unique granulation pattern in skin of rainbow trout represents a suitable bioindicator for nonylphenol exposure (Burkhardt- Holm *et al.,* 2000). These latter authors stated that the structural alterations in the skin of the estradiol-injected trout is pointed to a physiological response such as, detached pavement cells, vacuolation of the cytoplasm and severely deformed cell nuclei at a dose of 10μgl^{-1} which is lower than those in the present study. The damage occurred in the rainbow trout skin is similar to the quantitative changes of the mucous composition induced by hormones or environmental acidifications (Balm *et al.,* 1995). Schwaiger *et al.,* (1999) reported vitellogenin induction in the liver after exposure to 10 μgl^{-1} nonylphenol.

The present results exhibited severe damage in liver tissue of *C. gariepinus* including necrosis and decrease in the cell number along with vacuolation. Similar results were recorded by Uguz *et al.* (2003) who reported a significant increase in the Küpffer cells after one week of 4-nonylphenol exposure. Hughes *et al.* (2000) have shown NP-induced cell death. Galembeck *et al.* (1998), Hughes *et al.* (2000), and Uguz *et al.* (2003) reported that the disappearance of the cell membranes could be due to the lytic activity of alkylphenols.

In the present study, the liver cell borders disappeared and nuclei became larger after two weeks of exposure to 0.1mgl^{-1} of 4-nonylphenol, this is similar to the findings of Uguz *et al.* (2003) who reported that this may be due to the increase in the DNA/RNA ratio which was been observed in carcinogenic cells induced by NP (Chiriboga *et al.,* 2000a, b). The increase in the connective tissue with regenerating hepatocytes instead of normal liver tissues recorded in the present work was similar to those of Uguz *et al.* (2003). Such changes can be interpreted as an indication of carcinogenic development in the liver (Chiriboga *et al.,* 2000a, b; Calmak, 2001). Generally, the lesions detected in cells, tissues or organs are represent an integration of cumulative effects of physiological and biochemical stressors and therefore, can be linked to the exposure and subsequent metabolism of chemical contaminants (Adeyemo, 2008).

The gills are the primary initial target of toxicity, and the cytological changes in gill morphology in fish usually occur as a result of contaminant exposure. Gills have an extensive surface area and minimal diffusion distance between dissolved O2 and blood capillary for efficient gaseous exchange. The fusion occurred in gills of fishes exposed to 4-nonylphenbol in this investigation may cause a drastic reduction in the respiratory surface area. However, very little is known about the toxic impact of 4-nonylphenol on the functional morphology of the gills. The present results indicated such toxic impacts. Increase in the number of mucous cells in gills of fishes exposed to 0.1 mgl-1 of 4- nonylphenol was recorded. It has been reported that the immediate morpho-pathological response of the gills to ambient xenobiotics is often manifested by a significant increase in the density of its mucous cells (Dutta, 1997, Hemalatha and Banerjee, 1997). The large quantity of mucous secretion acts as a defense mechanism against several toxic substances (Handy and Eddy, 1991; Mazon et al., 1999). Similar to the findings of Dutta et al. (1996), the present results included many alterations such as increase in mucous and chloride cell number and size, necrosis, rupture of epithelium, desquamation, deformed secondary lamellae and oedema.

According to Peuranen *et al.* (1994) any discontinuity of epithelial lining of the gill lead to a negative ion balance and to changes in the haematocrite and mean cellular haemoglobin values of the blood. The number of chloride cells increased in the present study and this is similar to the results of Parashar and Banerjee, (2002). They stated that the number of chloride cells in the epithelial linings of both primary lamellae and secondary lamellae of *Heteropneustes fossilis* increased significantly following exposure to lead nitrate solution. Dutta *et al.* (1996) summarized the increased number of chloride cells in the gills of fishes following exposure to a variety of toxicants.

Ethoxylate nonylphenol was found to increase ion permeability and sodium efflux in the gill epithelial cells of rainbow trout (Pärt et al., 1985). Comparable findings in the current study were observed in the histology of gills under NP stress, corroborated by the elevated NP-induced anion gap. Fish kidneys get the highest percentage of post-branchial blood, making renal diseases potential indications of environmental pollution (Cengiz, 2006). Numerous studies have employed histological features of the kidney as indicators of pollution, particularly nonylphenol (Srivastava et al., 1990; Banerjee and Bhattacharya, 1994; Ortiz et al., 2003; Cengiz, 2006). The current study identified histological alterations in the kidney following exposure to 4-nonylphenol, including necrosis, glomerular hypertrophy, degeneration and dissociation of renal tubules and Bowman's capsule, proliferation in renal tubules and hematopoietic tissue, glomerular shrinkage, pyknosis, dilated blood vessels, rupture of Bowman's capsule, and obliteration of Bowman's space. Comparable outcomes were observed in fish following exposure to several contaminants (Cengiz, 2006; Khidr and Mekkawy, 2008; Abdel-Tawab and Al-Salahy, 2009). The findings of the present investigation indicate that exposure of adult catfish, Clarias gariepinus, to sublethal dosages of 4-nonylphenol resulted in moderate to severe damage to many organs, including the gills, skin, kidneys, and liver. The detrimental effects of nanoparticles on gills, skin, kidneys, and liver were concurrently associated with significant biochemical and physiological alterations, as well as endocrine disruption (Mekkawy et al., 2011; Mahmoud et al., 2011; Sayed et al., 2011). It is determined that NP functions as both an estrogenic and non-estrogenic agent, resulting in general and particular metabolic disruptions across many pathways.

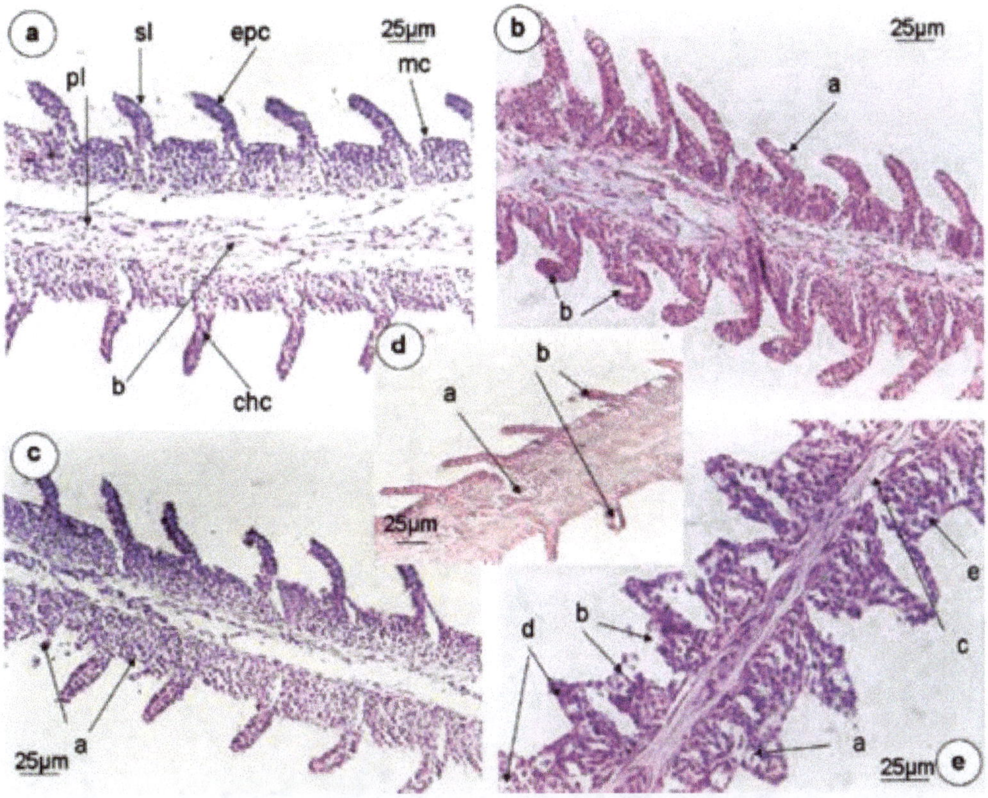

Fig. 1. (a) Gill structure of control adult fish *Clarias gariepinus*. (pl), primary lamellae; (sl) secondary lamellae; (epc) epithelial cell; (mc) mucous cell; (chc) chloride cell. (b) Gill tissue exposed to 0.05 mg/l 4-nonylphenol for 15 days showing a, epithelial lifting and oedema and b, deformed secondary lamellae. (c) Gill tissue exposed to 0.08 mg/l of 4-nonylphenol for 15 days showing a, desquamation and necrosis. (d)) Gill tissue exposed to 0.1 mg/l of 4- nonylphenol for 15 days showing a, degeneration of cartilaginous bar and b, malformed secondary lamellae. (e) Gill tissue exposed to 0.1 mg/l of 4-nonylphenol for 15 days showing a, increase in chloride cell size and number; b, epithelial hyperplasia and diffusion of secondary lamellae; c, degeneration and vacuolation of cartilaginous bar; d, desquamation and necrosis and e, increase number of mucous cells. Stain H& E. Magnification a, b, c, e (400X) and d (100X).

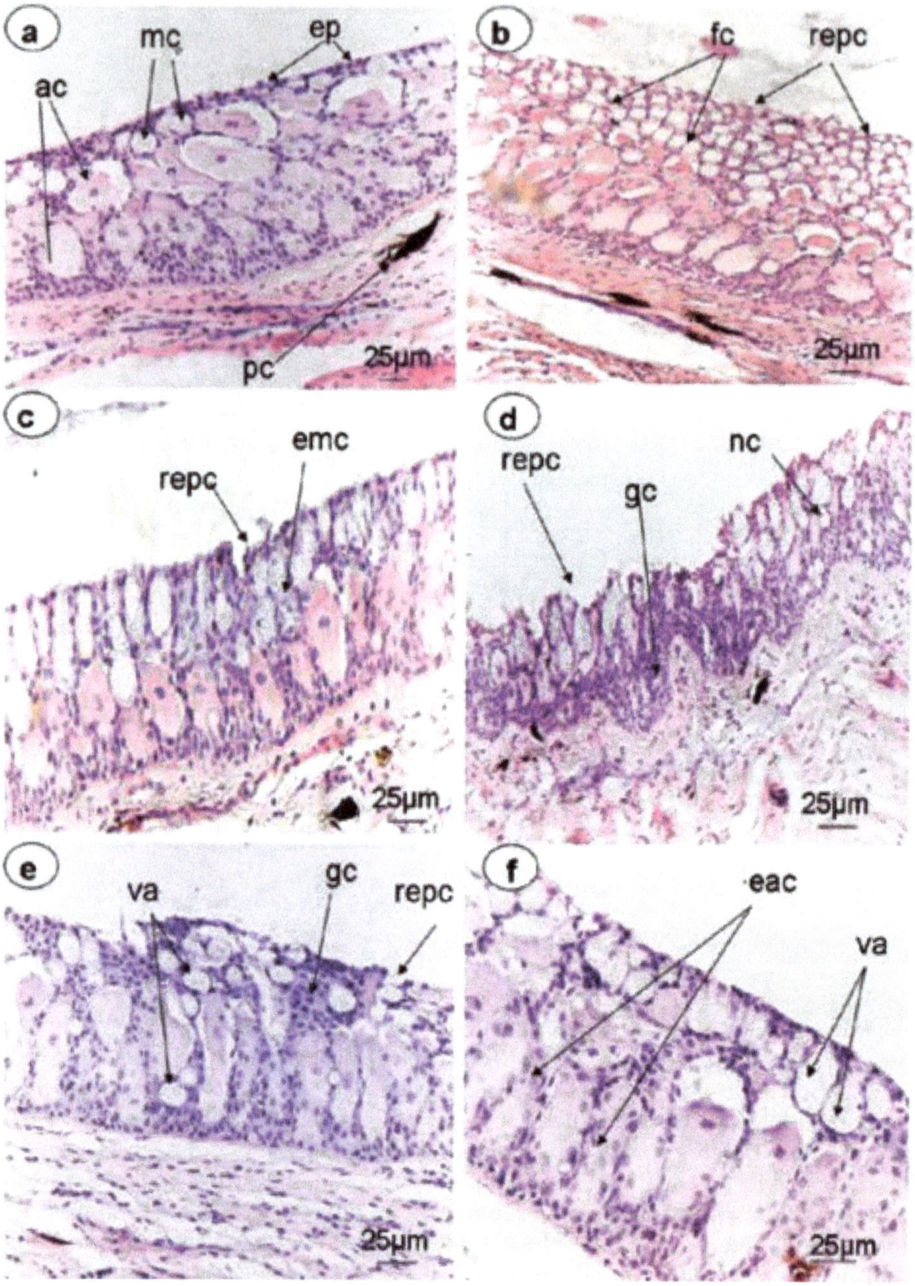

Fig. 2. Vertical sections of the skin of adult catfish *Clarias gariepinus* stained with H&E. (a) skin of control fish showing ac, alarm cell; mc, mucous cell; ep, epithelium; p, pigment cell. (b, d, e) skin of fish exposed to 0.1 mg/l of 4-nonylphenol for 15 days showing fc, fat cell; repc, ruptured epithelial cells; nc, necrotic cells; gc, granuled cells and va, vacuoles.(c) skin of exposed fish to 0.08 mg/l 4-nonylphenol for 15 days showing repc, ruptured epithelial cells and emc, enlarged mucous cells. (f) skin of exposed fish to 0.05 mg/l 4-nonylphenol for 15 days showing eac, enlarged alarm cells and va, vacuoles. Magnification a, b, c, d, e (100X) and f (400X).

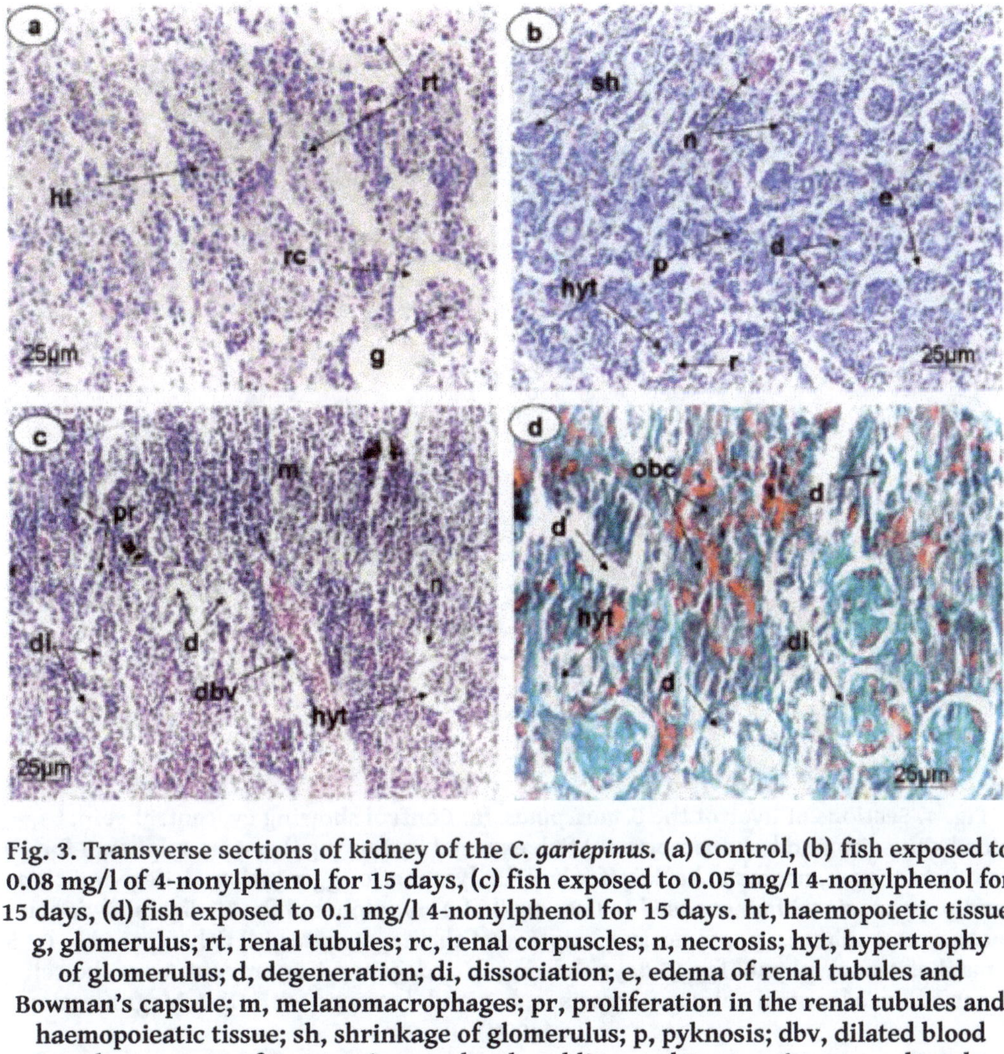

Fig. 3. Transverse sections of kidney of the *C. gariepinus.* (a) Control, (b) fish exposed to 0.08 mg/l of 4-nonylphenol for 15 days, (c) fish exposed to 0.05 mg/l 4-nonylphenol for 15 days, (d) fish exposed to 0.1 mg/l 4-nonylphenol for 15 days. ht, haemopoietic tissue; g, glomerulus; rt, renal tubules; rc, renal corpuscles; n, necrosis; hyt, hypertrophy of glomerulus; d, degeneration; di, dissociation; e, edema of renal tubules and Bowman's capsule; m, melanomacrophages; pr, proliferation in the renal tubules and haemopoieatic tissue; sh, shrinkage of glomerulus; p, pyknosis; dbv, dilated blood vessel; r, rupture of Bowman's capsule; obs, obliterated Bowman's space a, b and c Staind with H&E and d stained with masson's trichrome. Magnification a and d (400X), b and c (200X).

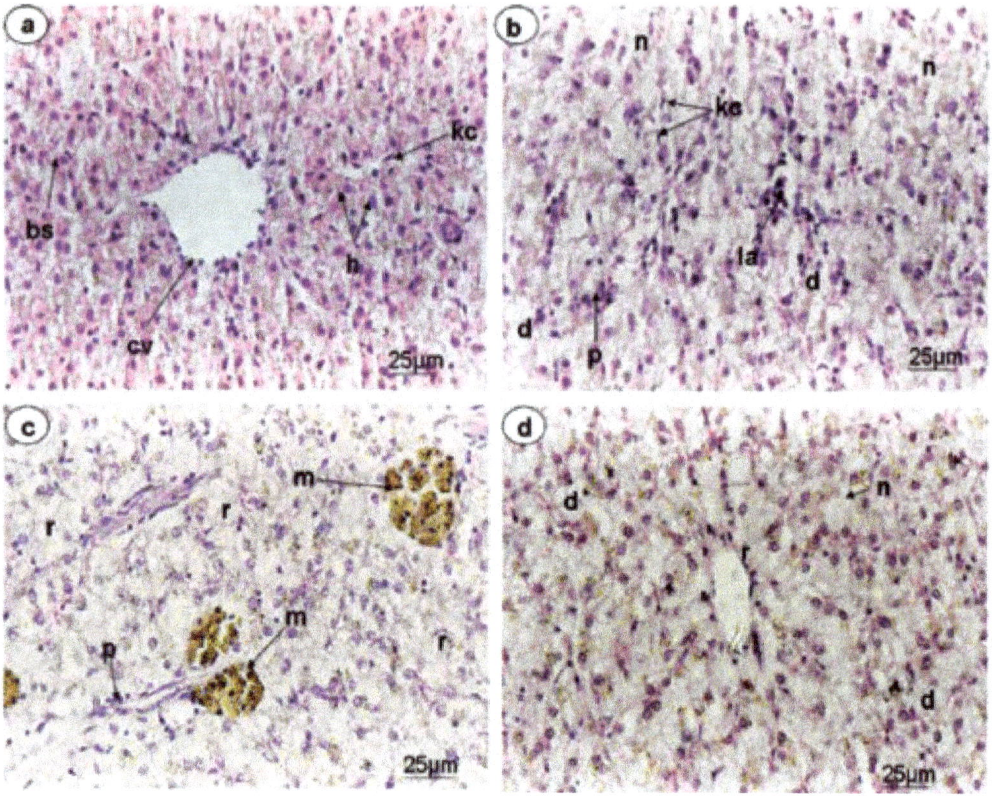

Fig. 4. Sections of liver of the *C. gariepinus.* (a) Control showing cv, central vein; bs, blood sinusoids and h, hepatocyte (b) fish exposed to 0.08 mg/l of 4-nonylphenol for 15 days showing n, necrosis; kc, küpffer cell; la, lymphatic aggregation; d, degeneration and p, pyknosis (c) fish exposed to 0.08 mg/l of 4-nonylphenol for 15 days showing m, melanomacrophages; p, pyknosis; r, rupture of the hepatocytes (d) fish exposed to 0.05 mg/l of 4-nonylphenol for 15 days showing d; d, degeneration; r, rupture of the cell membrane of central vein; n, necrosis. a, b, c and d stained with H&E, Magnification (400X).

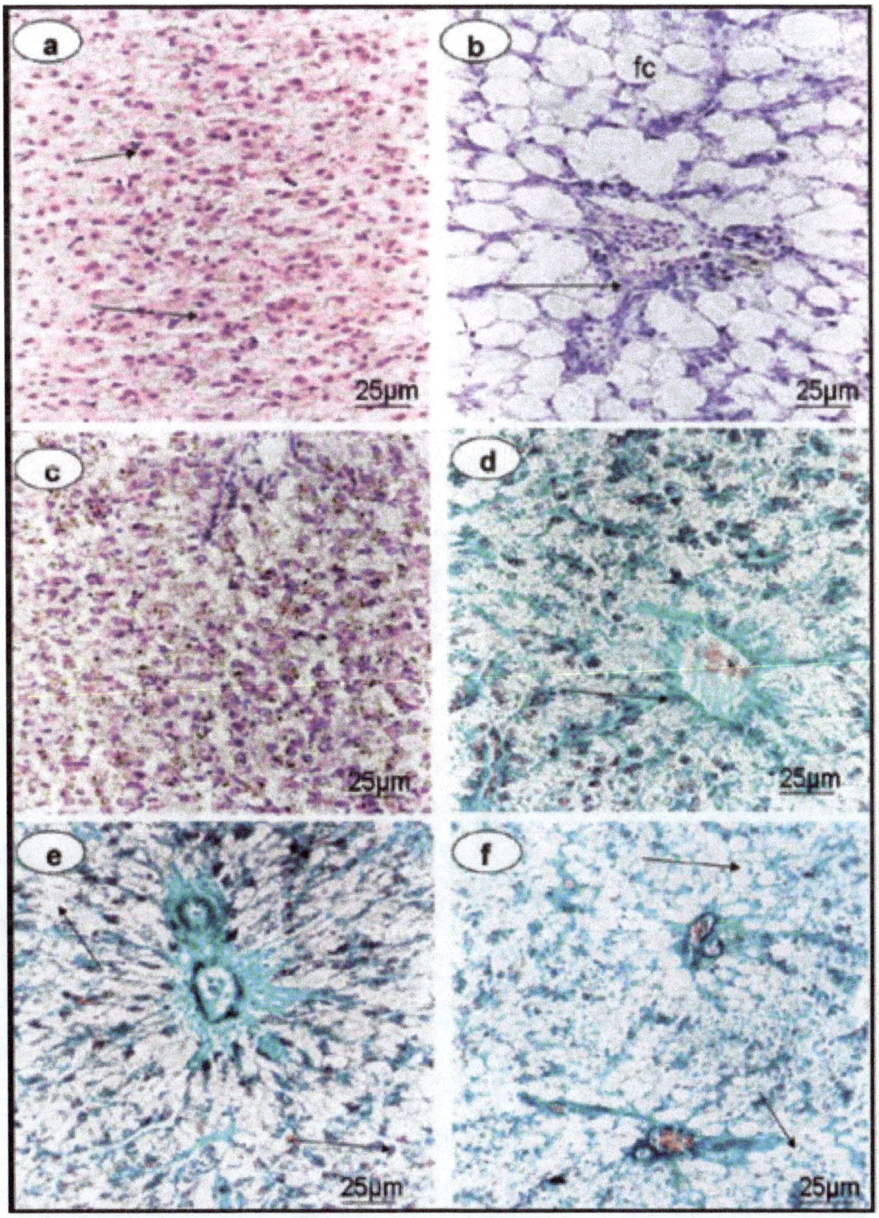

Fig. 5. Sections of liver of the *C. gariepinus.* (a) Fish exposed to 0.1 mg/l of 4-nonylphenol for 15 days showing pyknosis (arrows) (b) fish exposed to 0.1 mg/l of 4-nonylphenol for 15 days showing fc, fat cell and lymphatic infiltration (arrows) (c) fish exposed to 0.1 mg/l of 4- nonylphenol for 15 days showing pigments diffusion (d) fish exposed to 0.1 mg/l of 4- nonylphenol for 15 days showing aggregation of fibres around central vein (arrows) (e) fish exposed to 0.1 mg/l of 4-nonylphenol for 15 days showing rupture of hepatocytes and aggregation of fibres around the central vein (arrows) (f) fish exposed to 0.1 mg/l of 4- nonylphenol for 15 days showing accumulation of fats as fat cells. a, b and c staind with H&E and d, e and f stained with mssson's trichrome. Magnification (400X).

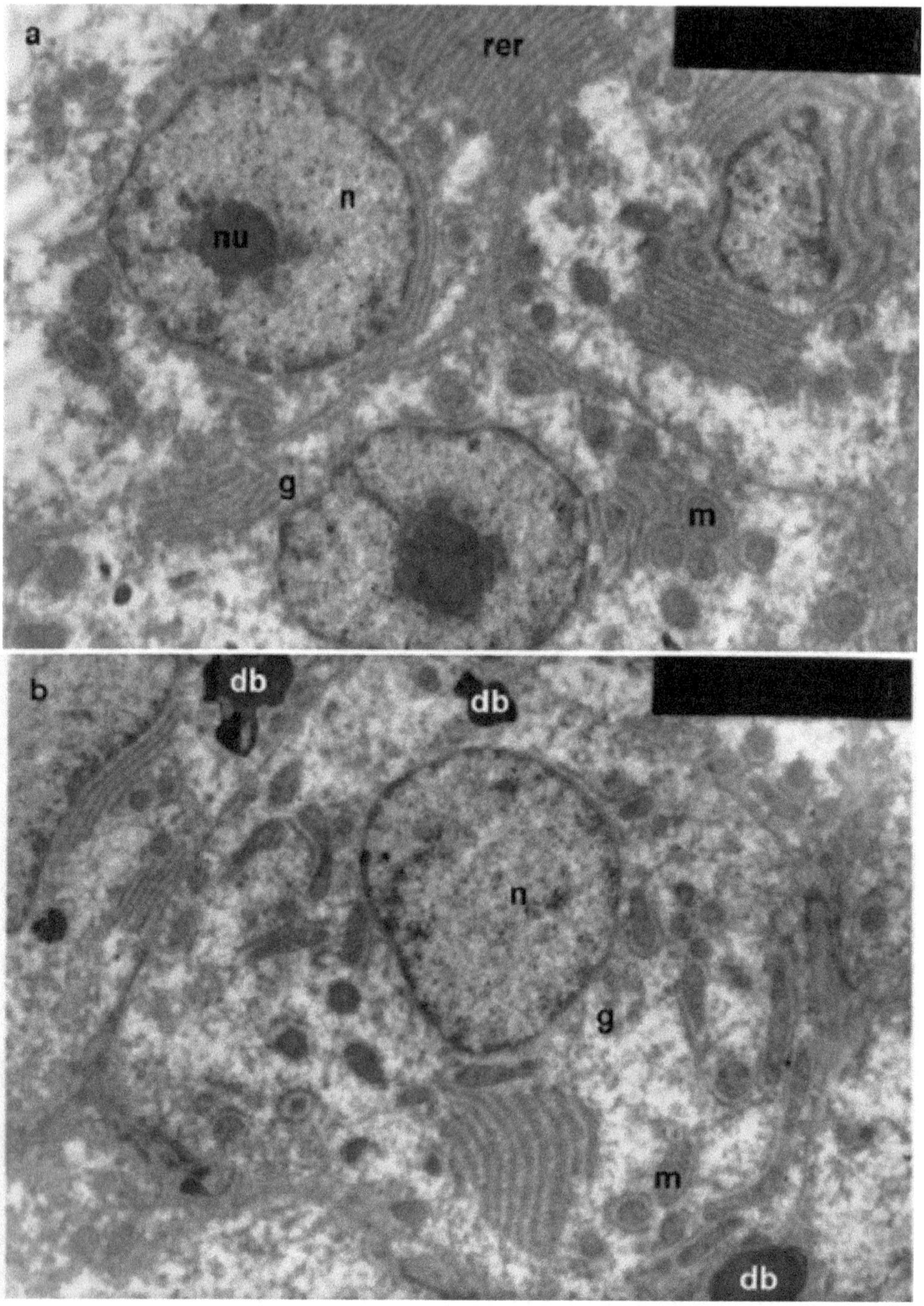

Fig. 6. Transmission electron micrograph of hepatocytes of catfish *Clarias gariepinus*. (a) control (X5000), (b) fish treated with 0.05 mg/l of 4-nonylphenol for 15 days (X5000). (n) nucleus, (nu) nuculeolus, (m) mitochondria, (rer) rough endoplasmic reticulum,(g) Golgi apparatus and (db) dense body.

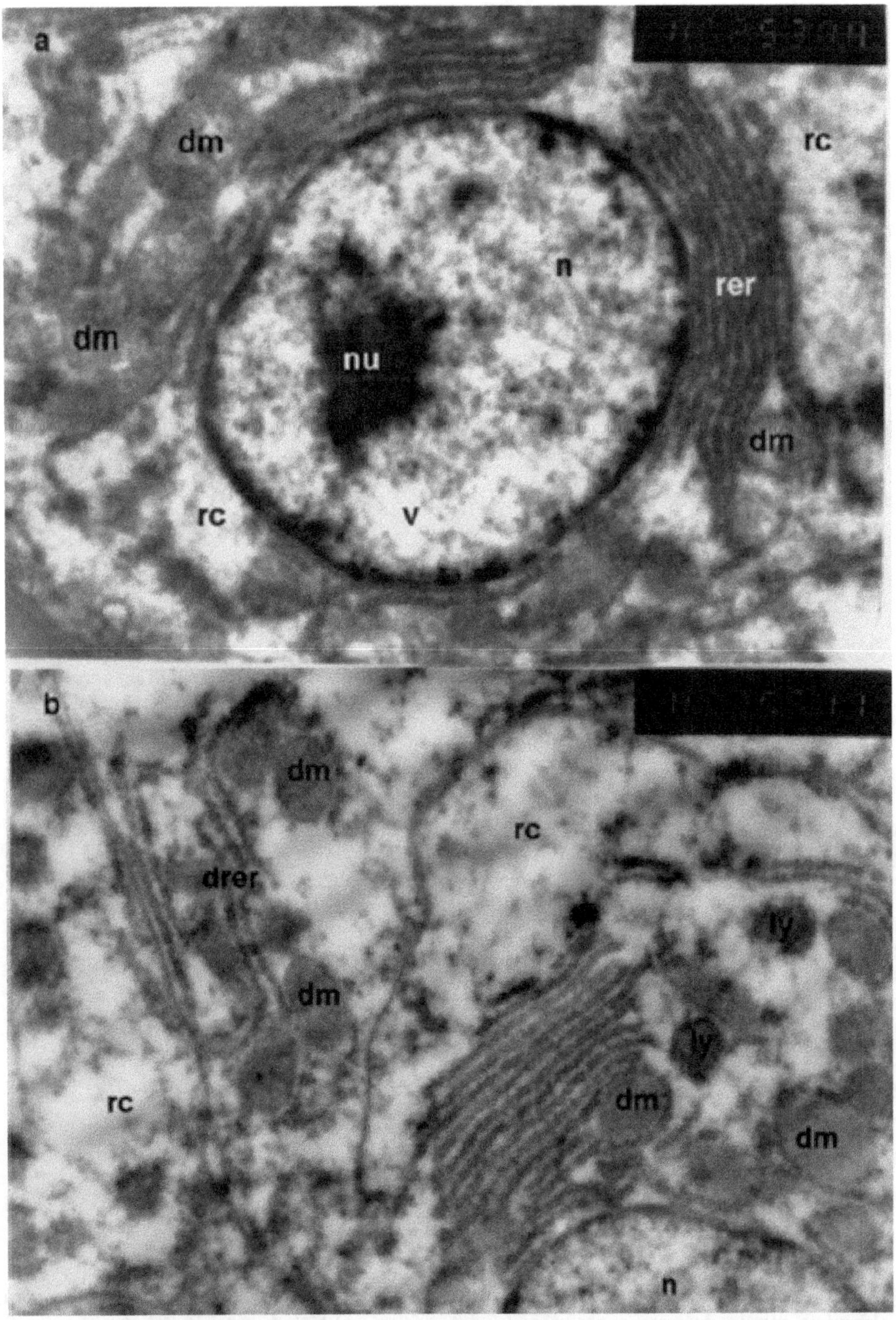

Fig. 7. Transmission electron micrograph of hepatocytes of catfish *Clarias gariepinus* showing marked degeneration of hepatocytes. (a) fish treated with 0.08 mg/l of 4-nonylphenol for 15 days (X8000), (b) fish treated with 0.05 mg/l of 4-nonylphenol for 15 days (X8000). (n) nucleus, (nu) nuculeolus, (dm) damaged mitochondria, (drer) degenerated rough endoplasmic reticulum,(rc) rarfied cytoplasm (v) vacuoles and (ly) lysosomes.

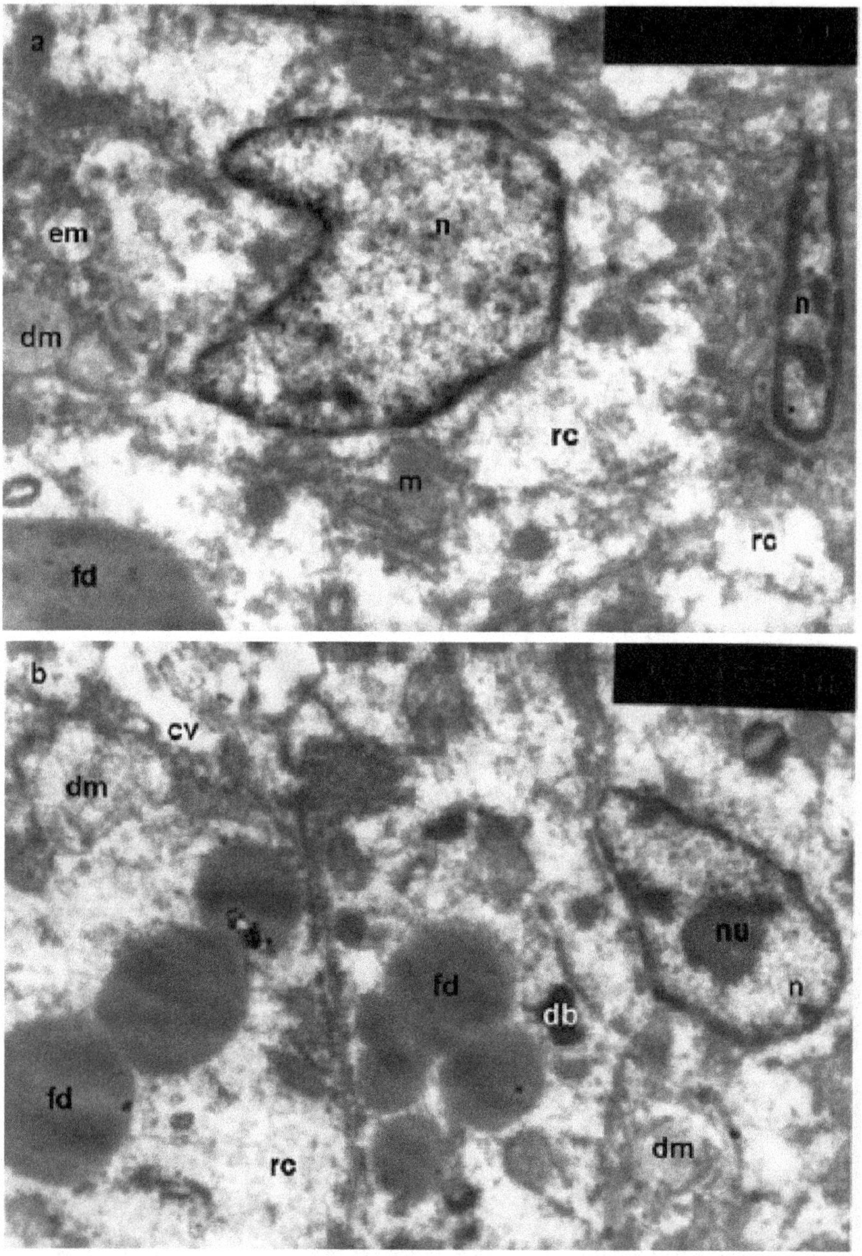

Fig. 8. Transmission electron micrograph of hepatocytes of catfish *Clarias gariepinus* showing marked degeneration of hepatocytes. (a) fish treated with 0.1 mg/l of 4-nonylphenol for 15 days (X5000), (b) fish treated with 0.05 mg/l of 4-nonylphenol for 15 days (X5000). (n) nucleus, (nu) nucleolus , (dm) damaged mitochondria, (em) empty mitochodria, (m) mitochondria, (rc) rarified cytoplasm (fd) fat drops and (db) dense body.

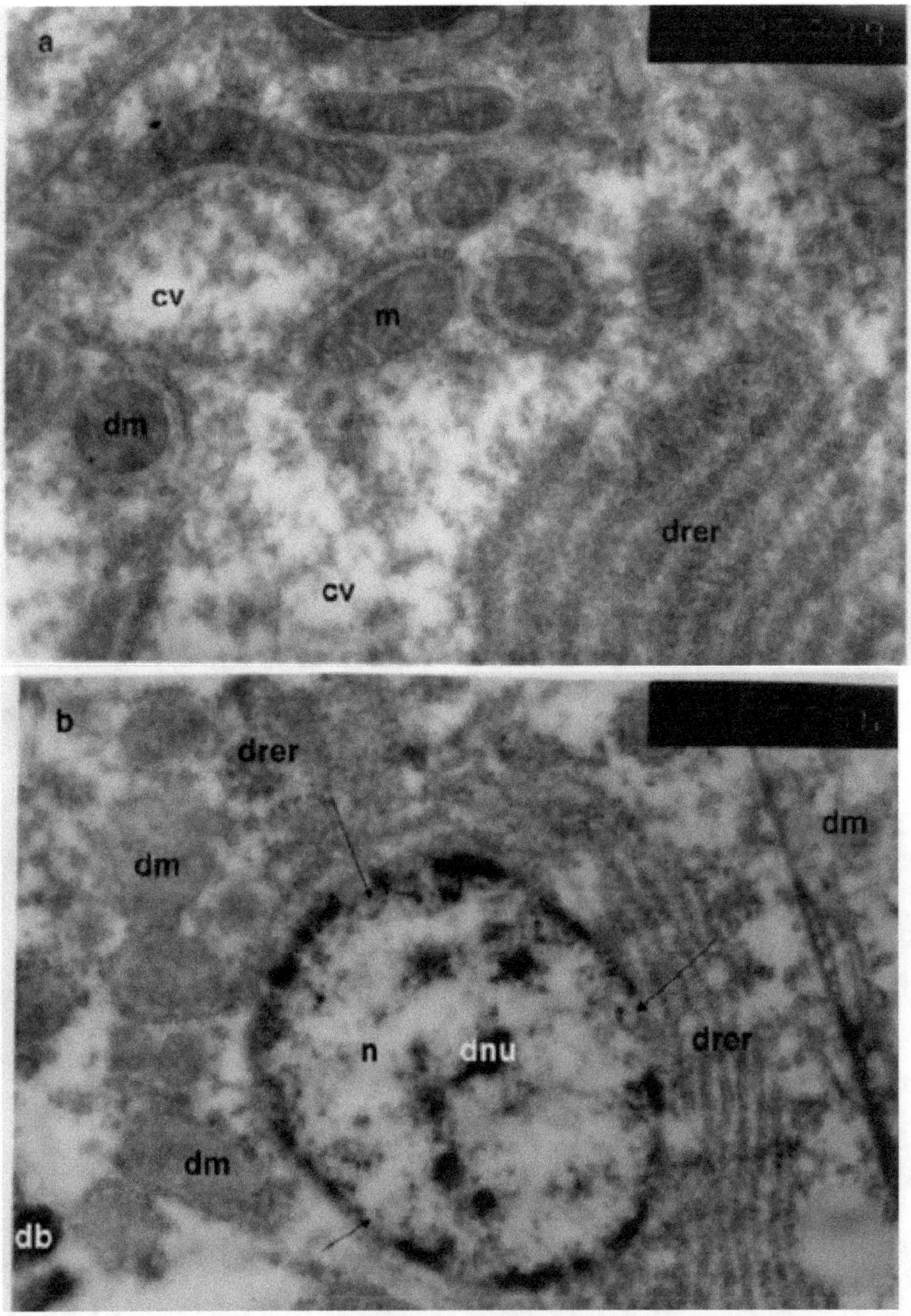

Fig. 9. Transmission electron micrograph of hepatocytes of catfish *Clarias gariepinus* showing marked degeneration of nuclear envelope (arrows) of hepatocytes. (a) fish treated with 0.08 mg/l of 4-nonylphenol for 15 days (X8000), (b) fish treated with 0.1 mg/l of 4-nonylphenol for 15 days (X8000). (n) nucleus, (nu) nuculeols, (dnu) degenerated nucleolus, (dm) damaged mitochondria, (em) empty mitochodria, (cv) cytoplasm vacuoles, (drer) degenerated rough endoplasmic reticulum and (db) dense body.

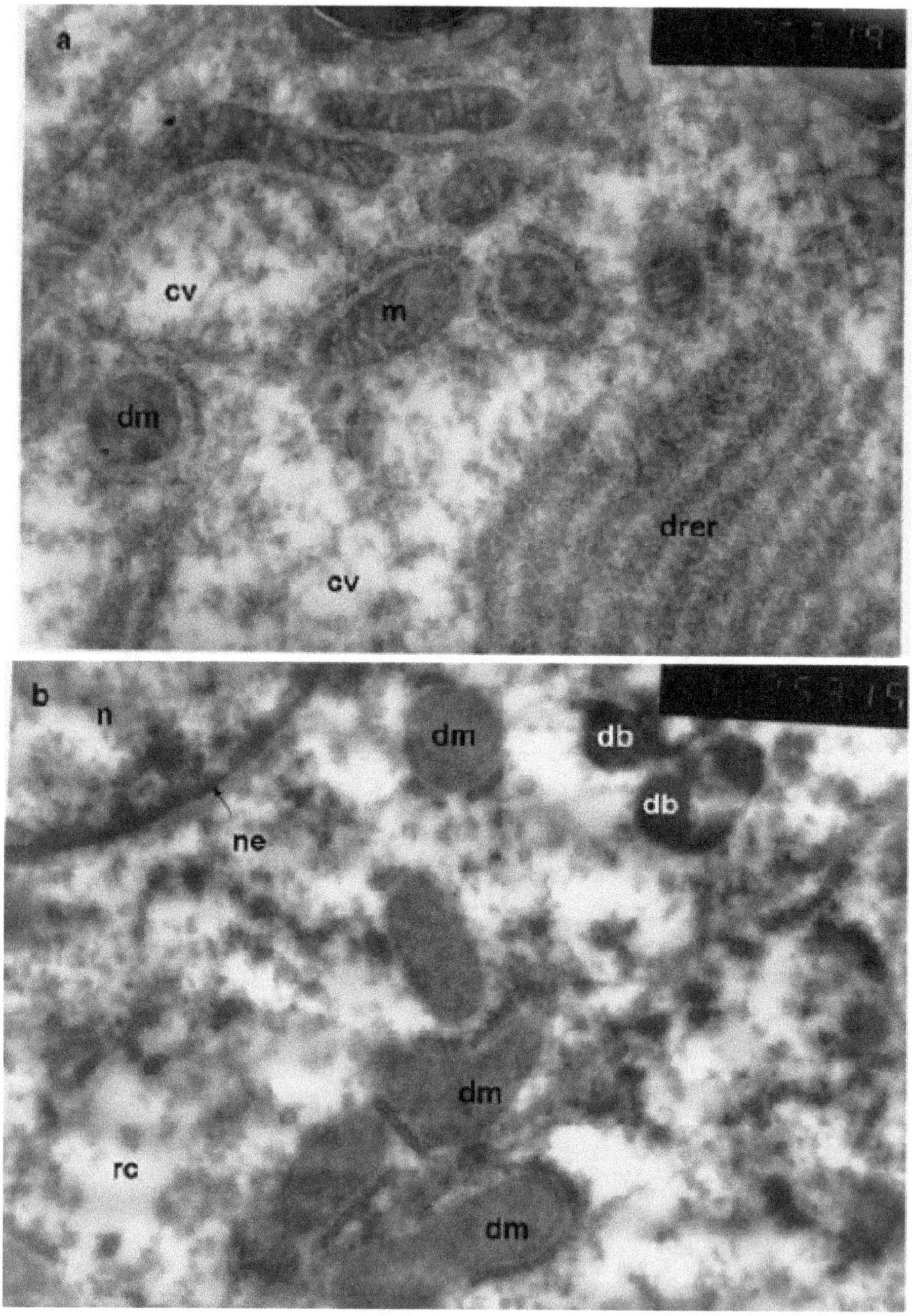

Fig. 10. Transmission electron micrograph of hepatocytes of catfish *Clarias gariepinus* showing marked degeneration of hepatocytes. (a) fish treated with 0.1 mg/l of 4-nonylphenol for 15 days (X14000), (b) fish treated with 0.1 mg/l of 4-nonylphenol for 15 days (X5000). (n) nucleus, (nu) nucleolus, (drer) degenerated rough endoplasmic reticulum, (dm) damaged mitochondria, (m) mitochodria, (rc) rarified cytoplasm, (cv) cytoplasm vacuoles, (fd) fat droplets and (db) dense body.

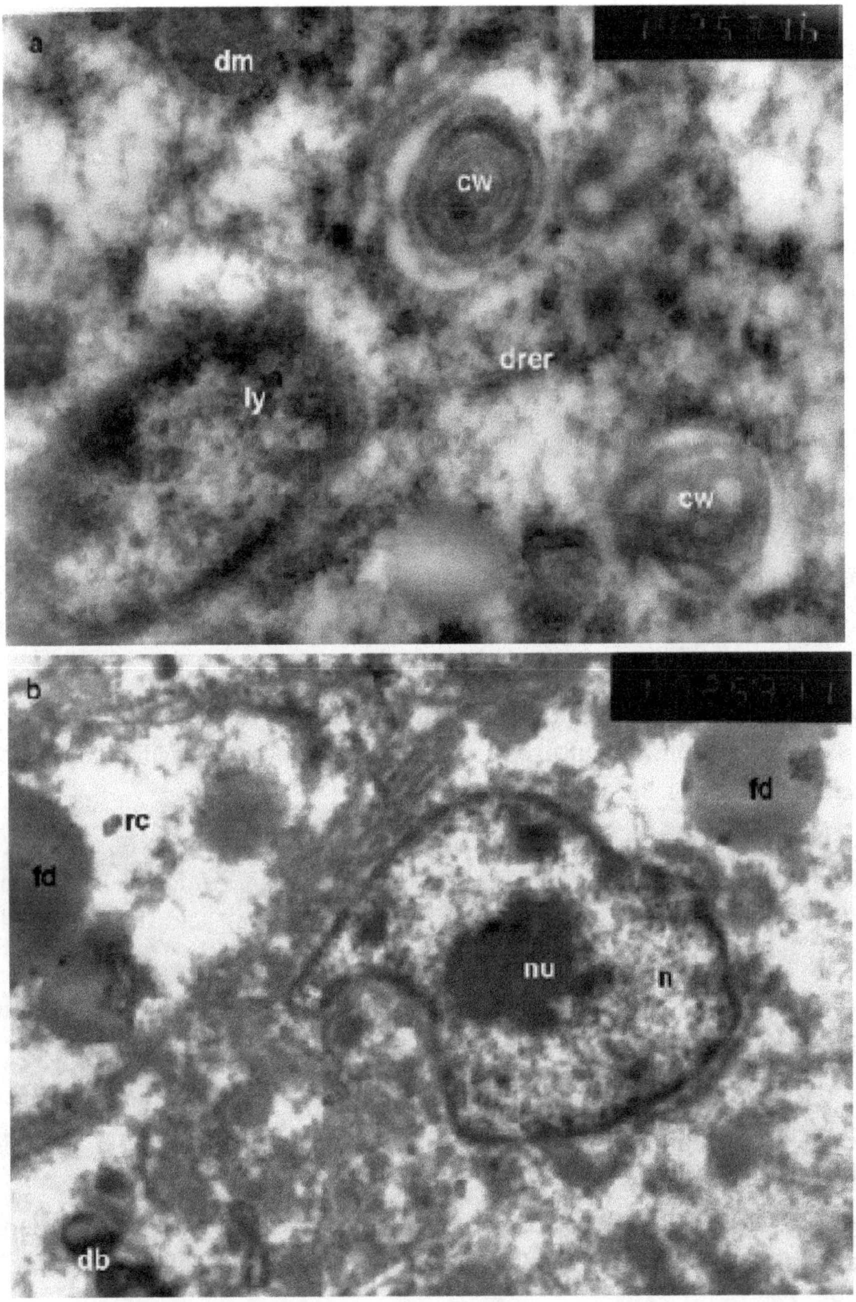

Fig. 11. Transmission electron micrograph of hepatocytes of catfish *Clarias gariepinus* showing marked degeneration of hepatocytes. (a) fish treated with 0.1 mg/l of 4-nonylphenol for 15 days (X14000), (b) fish treated with 0.1 mg/l of 4-nonylphenol for 15 days (X10000). (n) nucleus, (ne) nucleolus envelope, (drer) degenerated rough endoplasmic reticulum, (dm) damaged mitochondria, (cw) concentric whorls appearance of RER, (rc) rarified cytoplasm, (ly) lysosomes and (db) dense body.

5. REFERENCES

Abdel-Tawab, H.S. & Al-Salahy, M.B. (2010). Biocemical and ultrastructural studies of the effect of garlic juice on the liver of the fish Clarias gariepinus. Journal of Egyptian Germany Society of Zoology Vol. 60C: 39-62.

Adewolu, M.A., Adeniji, C.A. & Adejobi, A.B. (2008). Feed utilization, growth and survival of Clarias gariepinus (Burchell 1822) fingerlings cultured under different photoperiods. Aquaculture Vol. 283: 64–67.

Adeyemo, O.K. (2008). Histological Alterations Observed in the Gills and Ovaries of Clarias gariepinus exposed to environmentally relevant lead concentrations. Journal of Environmental Health Vol. 70 (9): 48-51.

Ahel, M., Giger, W. & Koch, M. (1994). Behaviour of alkylphenol polyethoxylate surfactants in the aquatic environment - I. Occurrence and transformation in sewage treatment. Water Resources Vol. 28: 1131–1142.

Ahel, M., McEvoy, J. & Giger, W. (1993). Bioaccumulation of the lipophilic metabolites of nonionic surfactants in freshwater organisms. Environmental Pollution. Vol. 79: 243-248.

Akinwole, A.O. & Faturoti, E.O. (2007). Biological performance of African Catfish (Clarias gariepinus) cultured in recirculating system in Ibadan. Aquaculture Enginering. Vol. 36: 18–23.

Balm, P.H.M., Iger, Y., Prunet, P., Pottinger, T.G. & Wendelaar Bonga, S.E. (1995). Skin ultrastructure in relation to prolactin and MSH function in rainbow trout (Oncorhynchus mykiss) exposed to environmental acidification. Cell Tissue Research. Vol. 279: 351-358.

Bancroft, J. & Stevens A, (1982). Theory and Practice of Histological Techniques, 2nd Ed. Churchill-Livingston, NY, pp 131-135.

Banerjee, S. & Bhattacharya, S. (1994). Histopathology of kidney of Channa punctatus exposed to chronic nonlethal level of elsan, mercury and ammonia. Ecotoxicology and Environmental Safety. Vol. 29 (3): 65–275.

Bennie, D.T. (1999). Review of the environmental occurrence of alkylphenols and alkylphenol ethoxylates. Water Quality Research Journal of Canada. Vol. 34: 79–122.

Bernet, D., Schmidt, H., Meier, W., Brkhardt-Holm, P. & Wahli, T. (1999). Histopathology in fish: Proposal for a protocol to assess aquatic pollution. Journal of Fish Diseases. Vol. 22: 25–34.

Burkhardt-Holm, P., Escher, M. & Meier, W. (1997). Waste water management plant effluents cause cellular alterations in the skin of brown trout Salmo trutta. Journal of Fish Biology. Vol. 50: 744-758.

Burkhardt-Holm, P., Wahli, T. & Meier, W. (2000). Nonylphenol affects the granulation pattern of epidermal mucous cells in rainbow trout, Oncorhynchus mykiss. Ecotoxicology and Environmental Safety. Vol. 46: 34–40.

Calmak, G. (2001). Spectroscopic analysis of the livers exposed to nonylphenol in rainbowtrout (Onchorynchus mykiss). MS thesis, Biology Department, Middle East Technical University, Turkey.

Cengiz, E. I. (2006). Gill and kidney histopathology in the freshwater fish Cyprinus carpio after acute exposure to deltamethrin. Environmental Toxicology Pharmacology. Vol. 22: 200–204.

Chiriboga, L., Yee, H. & Diem, M. (2000a). Infrared spectroscopy of human cells and tissue. Part VI: a comparative study of histology and infrared microspectroscopy of normal, cirrhotic, and cancerous liver tissue. Applied Spectroscopy. Vol. 54 (1): 1–8.

Chiriboga, L., Yee, H. & Diem, M. (2000b). Infrared spectroscopy of human cells and tissue. Part VII: FT-IR microspectroscopy of DNase- and RNase-treated normal, cirrhotic, and neoplastic liver tissue. Applied. Spectroscopy. Vol. 54 (4): 480–485.

Christiansen, T., Korsgaard, B. & Jespersen, A. (1998). Effects of nonylphenol and 17β- oestradiol on vitellogenin synthesis, testicular structure and cytology in male eelpout Zoarces viviparus. Journal of Experimental of Biology. Vol. 201: 179-192.

Colborn, T., Vom Saal, F.S. & Soto, A.M. (1993). Developmental effects of endocrine disrupting chemicals in wildlife and humans. Environmental Health Perspectives. Vol. 101: 378–384.

Coldham, N.G., Sivapathasundaram, S., Dave, M., Ashfleld, L.A., Pottinger, T.G., Goodall, C. & Sauer, M.J. (1998). Biotransformation, tissue distribution, and persistence of 4- nonylphenol residues in juvenile rainbow trout (Oncorhynchus mykiss). Drug Metabolism Dispososition. Vol. 26: 347-353.

Dutta, H.M. (1997). A composite approach for evaluation of the effect of pesticides on fish, In: Munshi, J. S. D., Dutta, H. M. (Eds.), Fish Morphology, Horizon of new research Science Publisher Inc., USA, pp. 249-277.

Dutta, H.M., Munshi, J.S.D., Roy, P.K., Singh, N.K., Adhikari, S. & Killius, J. (1996). Ultrastructural changes in the respiratory lamellae of the catfish, Heteropneustes fossilis after sublethal exposure to melathion. Environmental Pollution. Vol. 92: 329- 341.

Galembeck, E., Alonso, A. & Meirelles, N.C. (1998). Effects of polyoxyethylene chain length on erythrocyte hemolysis induced by poly[oxyethylene(n)nonylphenol] non-ionic surfactants. Chemico-Biollogical Interactions. Vol. 113 (2): 91–103.

Handy, R.D. & Eddy, F.B. (1991). The absence of mucous on the secondary lamellae of unstressed rainbow trout, Oncorhynchus mykiss. Journal of Fish Biology. Vol. 38: 153-155.

Hemalatha, S. & Banerjee, T.K. (1997). Histopathological analysis of sublethal toxicity of zinc chloride to the respiratory organs of the air-breathing catfish Heteropneustes fossilis (Bloch). Biological Research. Vol. 30: 11-21.

Hinton, D.E. & Lauren, D.J. (1990). Integrative histopathological approaches to detecting effects of environmental stressors on fishes. Am. Fish Soc. Sym. Vol. 8: 51–66.

Hinton, D.E., Baumann, P.C., Gardner, G.R., Hawkins, W.E., Hendricks, J.D., Murchelano, R.A. & Okihiro, M.S. (1992). Histopathological biomarkers, In: Huggett, R.J., Kimerle, R.A., Mehrle, P.M., Jr Bergman, H.L., (Eds.), Biomarkers: Biochemical, Physiological, and Histological Markers of Anthropogenic Stress, Lewis, USA, pp 155–196.

Hughes, P.J., McLellan, H., Lowes, D.A., Khan, S.Z., Bilmen, J.G., Tovey, S.C., Godfrey, R.E., Michell, R.H., Kirk, C.J. & Michelangeli, F. (2000). Estrogenic alkylphenols induce cell death by inhibiting testis endoplasmic reticulum Ca2+ pumps. Biochemical and Biophysical Research Communications. Vol. 277: 68–574.

Iger, Y., Balm, P.H., Jenner, H.A. & Wendelaar Bonga, S.E. (1995). Cortisol induces stress- related changes in the skin of rainbow trout (Oncorhynchus mykiss). General and Comparative Endocrinology. Vol. 97: 188-198.

Jobling, S., Sheahan, D., Osborne, J.A., Matthiessen, P. & Sumpter, J. P. (1996). Inhibition of testicular growth in rainbow trout (Oncorhynchus mykiss) exposed to estrogenic alkylphenolic chemicals. Environmental Toxicology and Chemistry. Vol. 15: 194-202.

Johannessen, J. (1978). Instruction and techniques in Electron Micrscopy in human medicine. Mchgraw-Hill Int. Book Co.

Kang, I.J., Yokota, H., Oshima, Y., Tsuruda, Y., Hano, T., Maeda, M., Imada, N., Tadokoro, H. & Honjo, T. (2003). Effect of 4-nonylphenol on the reproduction of Japanese medaka, Oryzias latipes, Environmental Toxicology and Chemistery. Vol. 22: 2438- 2445.

Karami, A., Christianus, A., Ishak, Z., Courtenay S. C., Syed, M. A., Noor Azlina, M. & Noorshinah H. (2010). Effect of triploidization on juvenile African catfish (Clarias gariepinus). Aquaculture International. Vol. 18: 851–858.

Khidr, M. B. & Mekkawy, I.A.A. (2008). Effect of separate and combined lead and selenium on the liver of the cichlid fish Oreochromis niloticus: ultrastructural study. Egyptian Journal of Zoology. Vol. 50: 89-119.

Lech, J.J., Lewis, S.K. & Ren, L. (1996). In vivo estrogenic activity of nonylphenol in rainbow trout. Fundam. Applied Toxicology. Vol. 30: 229-232.

Lewis, S.K. & Lech, J.J. (1996). Uptake, disposition and persistence of nonylphenol from water in rainbow trout (Oncorhynchus mykiss). Xenobiotica. Vol. 26: 813-819.

Mahmoud, U.M., Sayed, A.H. & Mekkawy, I.A. (2011). Biochemical changes of African catfish Clarias gariepinus exposed to 4-nonylphenol. African Journal of Biochemistery,11-053 (Accepted)

Malik, R. & Hodgson, H. (2002). The relationship between the thyroid gland and the liver. Quarterly Journal of Medicine. Vol. 95: 559-569.

Marchand, M.J., van Dyk, J.C., Pieterse, G.M., Barnhoorn, I.E.J. & Bornman, M.S., (2008). Histopathological Alterations in the Liver of the Sharptooth Catfish Clarias gariepinus from Polluted Aquatic Systems in South Africa. Environmental Toxicology. DOI 10.1002/tox.

Mazon, A.F., Cerqueira, C.C.C., Monteiro, E.A.S. & Fernandes, M.N. (1999). Acute copper exposure in freshwater fish, Morphological and physiological effects. In: Val, A.L., Almeida-Val, V.M.F., (Eds.), Biology of Tropical Fishes, INPA, Manaus, pp. 263-275.

Mekkawy, I.A., Mahmoud, U.M. & Sayed, A.H. (2011). Effects of 4-nonylphenol on blood cells of the African catfish Clarias gariepinus (Burchell, 1822) Tissue Cell, doi:10.1016/j.tice.2011.03.006

Meyers, T.R. & Hendricks, J.D. (1985). Histopathology. In: Loux, D.B., Dorfman, M., (Eds.), Fundamentals of Aquatic Toxicology: Methods and Applications, Hemisphere USA, pp. 283–330.

Nguyen, L.T.H. & Janssen, C.R. (2002). Embryo-larval toxicity tests with the African catfish (Clarias gariepinus): comparative sensitivity of endpoints. Archives of Environmental Contamination and Toxicology. Vol. 42: 256–262.

Nimrod, A.C. & Benson, W.H. (1996). Environmental estrogenic effects of alkylphenol ethoxylates. Critical Reviews in Toxicology. Vol. 26: 335–364.

Ortiz, J.B., De Canales, M.L.G. & Sarasquete, C. (2003). Histopathological changes induced by lindane (gamma-HCH) in various organs of fishes. Scientia Marina. Vol. 67 (1): 53–61. Parashar, R.S. & Banerjee, T.K. (2002). Toxic impact of lethal concentration of lead nitrate on the gills of air-breathing catfish Heteropneustes fossilis (Bloch). Veterinarski Arhiv. Vol. 72 (3): 167-183.

Pärt, P., Svanberg, O. & Bergstrom, E. (1985). The influence of surfactants on gill physiology and cadmium uptake in perfused rainbow trout gills. Ecotoxicology and Environmental Safety. Vol. 9: 135-144.

Peuranen, S., Vuorinen, P. J., Vuorinen, M. & Hollender, A. (1994). The effect of iron, humic acids and low pH on the gills and physiology of brown trout (Salmo trutta). Annales Zoologici Fennicii. Vol. 31: 389-396.

Pottinger, T.G. & Pickering, A.D. (1985). Changes in skin structure associated with elevated androgen levels in maturing male brown trout, Salmo trutta L. Journal of Fish Biology. Vol. 26: 745-753.

Razia, S., Maegawa, Y., Tamotsu, S. & Oishi, T. (2006). Histological changes in immune and endocrine organs of quail embryos: Exposure to estrogen and nonylphenol. Ecotoxicology and Environmental Safety. Vol. 65: 364–371.

Sayed, A.H., Mekkawy, I.A. & Mahmoud, U.M. (2011). Histopathological alterations in some organs of adults of Clarias gariepinus (Burchell, 1822) exposed to 4-nonylphenol. The 19th Conference of Egyptian-German Scocity of Zoology, Bin Sueif University, Egypt.

Schwaiger, J., Spieser, O.H., Nardy, E., Kalbfus, W., Braunbeck, Th. & Negele, R. D. (1999). Toxic effects versus endocrine disruption Does the xenoestrogen nonylphenol influence physiological functions in fish? SETAC: 9th Annual Meeting, 25-29 May 1999, Leipzig, Germany.

Seki, M., Yokota, H., Maeda, M., Tadokoro, H. & Kobayashi, K. (2003). Effect of 4-nonylphenol and 4- tert octylphenol on sex differentiation and vitellogenin induction in medaka (Oryzias Latipes), Environmental Toxicology and Chemistery. Vol. 22:1507-1516.

Servos, M.R. (1999). Review of the aquatic toxicity, estrogenic responses and bioaccumulation of alkylphenols and alkylphenol polyethoxylates. Water Qualality Research Journal of Canada. Vol. 31: 123–177.

Shephard, K.L. (1994). Functions for fish mucus. Reviews in Fish Biology and Fisheries. Vol. 4: 401-429.

Soto, A.M., Chung, K.L. & Sonnenschein, C. (1994). The pesticides endosulfan, toxaphene and dieldrin have estrogenic effects on human estrogen sensitive cells. Environmental Health Perspectives. Vol. 102: 380–385.

Soto, A.M., Lin, T.M., Justicia, H., Silvia, R.M. & Sonnenenschein, C. (1992). An 'in culture' bioassay to assess the estrogenicity of xenobiotics (E-screen), In: Colburn, T.C. (Eds.), Chemically Induced Alterations in Sexual and Functional Development; The Wildlife-human Connection. Princeton Scientific Publishing, Princeton, NJ, USA, pp. 295–309.

Srivastava, S.K., Tiwari, P.R. & Srivastav, A.K. (1990). Effects of chlorpyrifos on the kidney of freshwater catfish. Heteropneustes fossilis. Bulletin of Environmental Contamination and Toxicology. Vol. 45: 748–751.

Talmage, S.S. (1994). Environmental and Human Safety of Major Surfactants: Alcohol Ethoxylates and Alkylphenol Ethoxylates. Lewis Publishers, Boca Raton, FL.

Toomey, B.H., Monteverdi, G.H. & Di Giulia, R.T. (1999). Octylphenol induces vitellogenin production and cell death in fish hepatocytes. Environmental Toxicology and Chemistry. Vol. 18: 734–739.

Uguz, C., Iscan, M., Ergu, A., Belgin I.V. & Togan, I. (2003). The bioaccumulation of nonyphenol and its adverse effect on the liver of rainbowtrout (Onchorynchus mykiss). Environmental Research. Vol. 92: 262–270.

Wendelaar Bonga, S.E. (1997). The stress response in Fish. Physiological Reviews. Vol. 77: 591-625.

Yokota, H., Seki, M., Maeda, M., Oshima, Y., Tadokoro, H., Honjo, T. & Kobayashi, K. (2001). Life-cycle toxicity of 4-nonylphenol to medaka (Oryzias latipes), Environmental Toxicology and Chemistry. Vol. 20: 2552-2560.

CHAPTER

CELL DEATH AND OVARIAN FOLLICULAR ATRESIA IN ANIMALS

9

1. INTRODUCTION: AN OVERVIEW

The reproductive efficacy of any mammalian species can be improved by optimizing nutrition, regulating environmental variables such as photoperiod, temperature, and humidity, employing selective breeding, endocrine interventions, and implementing superior management practices (Sharma, 2000; Maillet et al., 2002; Bussiere et al., 2002; Iwata et al., 2004, 2005; Rudolf, 2007; Bhardwaj and Sharma, 2011). Despite its complexity, endocrine control is highly effective in enhancing the reproductive output of organisms. In females, follicular development and estrous cyclicity are closely interconnected (Craig et al., 2007; Sharma and Batra, 2008). Among the hundreds of oocytes and primordial follicles in the neonatal ovary, only a minuscule proportion, roughly 0.001%, are ovulated during the active reproductive lifespan of mammals (Tabarowski et al., 2005; Sharma and Bhardwaj, 2009). Follicular atresia is a prevalent occurrence that constrains the number of ovulations, hence limiting the reproductive capacity of a species. This leads to significant depletion of germ cells during many stages of development, including prenatal, neonatal, prepubertal, pubertal, estrous cycle, pregnancy, breastfeeding, and post-reproductive phases in mammals (Guraya, 1997, 1998; Sharma, 2000; Manabe et al., 2003, 2004; Bhardwaj and Sharma, 2011). Follicular atresia is a natural regulatory mechanism of fertility that can be effectively manipulated to enhance or diminish the reproductive capacity of a species by altering the frequency of atresia. Consequently, numerous research papers have been published on the morphology, histochemistry, biochemistry, and endocrinology of follicular atresia (Williams and Smith, 1993; Guraya, 1998, 1999; Monniaux, 2002; Sharma, 2003; Sharma and Batra, 2008; Sharma and Bhardwaj, 2009; Bhardwaj and Sharma, 2011). Nevertheless, the mechanism of atresia requires further exploration and analysis for its effective application in fertility regulation programs. The molecular mechanism of atresia is best elucidated by the process of apoptosis (Palumbo and Yeh, 1994; Sharma, 2000; Yu et al., 2004; Sharma and Bhardwaj, 2009). Apoptosis, a form of natural cell death, is the opposite of mitosis and plays a crucial role in the maintenance of tissue homeostasis (Collins and Lopez, 1993; Schwartzman and Cidlowski, 1993). It impacts individual dispersed cells inside live tissues without provoking an inflammatory reaction and is modulated by growth factors and hormones. Numerous studies on the morphological alterations in granulosa cells and theca-interstitial cells of follicles undergoing atresia have established that apoptosis is likely the principal mechanism facilitating cell loss during follicle degeneration (Tsafriri and Braw, 1984; Hirshfield, 1991; Tilly, 1996). The initial accounts of apoptosis, the physiological process of cell death, in the ovaries were documented in 1885 through the morphological examination of granulosa cells amid the degeneration of antral follicles in the rabbit ovary (Flemming, 1885). The process known as 'Chromatolysis' was suggested as a cause for the loss of granulosa cells, and these discoveries corresponded closely with the morphological criteria recognized as hallmarks of apoptosis (Kerr et al., 1972, 1994).

The ovarian processes governing life and death decisions entail interaction between pro-apoptotic and pro-survival molecules (Hussein, 2005). Apoptosis functions in ovarian follicles at both fetal and adult stages of life. Apoptosis is confined to oocytes throughout fetal development, although in adulthood, it is commonly detected in the granulosa cells of secondary and antral follicles (Hussein, 2005). Apoptosis, a genetically regulated process of cellular self-destruction, is an energy-dependent mechanism frequently observed during early embryonic development. Apoptosis facilitates the secure elimination of cells once they have completed their biological roles (Hussein, 2005). The ovary exemplifies efficient apoptosis during the animal's active reproductive phase. Apoptosis is fundamental to various ovarian functions and is mediated by multiple molecular mechanisms. The Bcl-2 (B-cell lymphoma 2) family, tumor necrosis factor (TNF), caspases, and transforming growth factor-β (TGF-β) proteins represent recognized pathways of apoptosis (Hussein, 2005). Apoptosis is morphologically defined by cellular shrinkage, membrane blebbing, and cytoplasmic fragmentation in oocytes (Wu et al., 2000; Chaube et al., 2005), granulosa cells (Sharma and Sawhney, 1999; Sharma, 2003; Sharma and Bhardwaj, 2007, 2009), and the corpus luteum (Sharma and Batra, 2005). The activation of a Ca^{2+}/Mg^{2+} dependent endogenous endonuclease is a distinctive biochemical occurrence in apoptosis. This enzyme cleaves genomic DNA in internucleosomal regions, producing fragments of 180-200 base pairs (bp).

The DNA fragments can be observed as a distinct ladder pattern by agarose gel electrophoresis. The occurrence of this DNA sequence in cells is regarded as a defining characteristic of apoptosis (Schwartzman and Cidlowski, 1993; Hsueh et al., 1994; Yang and Rajamahendran, 2000). Apoptosis is indicated not only by the presence of oligonucleosomes in isolated DNA but also by the distinctive morphological features of degenerating cells and the identification of fragmented DNA in individual cells in situ via the 3' end-labeling technique, specifically terminal deoxynucleotidyl transferase-mediated dUTP nick end-labeling (TUNEL) (Gavrieli et al., 1992; Palumbo and Yeh, 1994; Sharma and Bhardwaj, 2009). The initial development of mammalian ovarian follicles is inadequately comprehended. While follicle-stimulating hormone (FSH) and luteinizing hormone (LH) are crucial for follicular development from the secondary stage onward, the initiation of primordial follicle growth occurs independently of gonadotropin hormones (Tilly, 1996).

The factors that trigger the transfer of quiescent primordial follicles to the expanding follicle pool remain unidentified. Research on animals has demonstrated that many locally synthesized growth factors play a role in the complex regulation of primary and secondary follicle growth (Durlinger, 2000). Recent discoveries suggest that follicular loss due to atresia may be mitigated by the generation of new primordial follicles from germ line stem cells (Johnson et al., 2004), a proposition that remains contentious and unverified (Bristol Gould et al., 2006; Liu et al., 2007). During active reproductive life, follicles regularly exit the resting reservoir to enter the growing pool and progress through successive developmental stages, influenced by stage-specific subsets of intra-ovarian regulators and endocrine factors such as growth factors, cytokines, and gonadal steroids (Gilchrist et al., 2004; Pangas, 2007). Follicles have varying responses to stimuli that stimulate cell proliferation, growth, differentiation, and apoptosis at different embryonic stages, with only a limited number progressing to ovulation (Tilly et al., 1991; Jiang et al., 2003; Craig et al., 2007). A significant quantity of follicles experiences atresia during the late pre-antral to early antral phases. In this hormone-dependent growth phase, a selection process determines whether to permit the follicles to develop to the pre-ovulatory stage (Durlinger et al., 2000). Gonadotropins are survival factors that inhibit atresia (Uilenbroek et al., 1980; Chun et al., 1994). Furthermore, growth factors such as epidermal growth factor, transforming growth factor-α, basic fibroblast growth factor (Tilly et al., 1992), insulin-like growth factor (Chun et al., 1994), and the cytokine interleukin (IL)-1β (Chun et al., 1995) also inhibit the onset of atresia.

The catalase activity in follicular fluid and granulosa cells demonstrated a decreasing trend from healthy to atretic follicles. Catalase is a ubiquitous enzyme present in all known species. It facilitates the decomposition of hydrogen peroxide into water and molecular oxygen (Aebi, 1984; Vohra, 2002). The activity of the enzyme superoxide dismutase exhibits a decreasing trend from healthy to mildly atretic to atretic follicles in both the follicular fluid and granulosa cells. The superoxide dismutase (SOD) family of metalloenzymes serves as antioxidants, safeguarding cells from the harmful effects of the oxygen free radical superoxide (Fridovich, 1975, 1986). SOD facilitates the transformation of superoxide anion into oxygen and hydrogen peroxide. Hydrogen peroxide is subsequently transformed into oxygen and water by catalase and peroxidases. Trace elements have historically been recognized for their significant involvement in cellular homeostasis, which involves the precise regulation of cell death. Zinc, recognized as a microelement that prevents or suppresses apoptosis, has been shown in numerous studies to actively induce cellular injury and death, encompassing both apoptosis and necrosis, in both malignant and normal cells (Iitaka et al., 2001; Bae et al., 2006; Wiseman et al., 2006; Rudolf, 2007; Bhardwaj and Sharma, 2011). Zinc induces apoptosis and necrosis by disrupting calcium homeostasis through oxidative stress, leading to the activation of mitogen-activated protein kinases (MAPKs) via both p53-dependent and p53-independent pathways, or through its interaction with the cytoskeleton (Feng et al., 2002; Chen et al., 2003).

The significance of reactive oxygen species (ROS) and antioxidants for female reproductive function has, conversely, garnered limited scrutiny, despite evidence of both physiological and pathological impacts (Guerin et al., 2001; Chaudhary et al., 2004).. Yang et al. (1998) discovered elevated hydrogen peroxide levels in fragmented embryos and unfertilized oocytes, whereas Paszkowski and Clarke (1996) indicated heightened antioxidant consumption, implying enhanced reactive oxygen species (ROS) activity during the incubation of subpar quality embryos. Recent studies demonstrate a strong correlation between the death of granulosa cells and the levels of estradiol and progesterone in follicular fluid (Yu et al., 2004; Sharma and Batra, 2008; Sharma et al., 2008). The concentration of insulin-like growth factor-1 (IGF-1), rather than insulin-like growth factor-II (IGF-II), is the determining factor in the maturation or atresia of a follicle (Yu et al., 2004). The function of trace elements as pro-apoptotic or anti-apoptotic agents remains inadequately understood to date. The series of morphological and biochemical changes must be examined to salvage germ cells or prevent somatic cell loss.

A substantial population of ovarian follicles in mammals is diminished through the pervasive occurrence of follicular atresia, which constrains the number of ovulations and thereby limits the reproductive capacity of a species (Sharma, 2000; Yu et al., 2004; Tabarowski et al., 2005; Slomczynska et al., 2006; Sharma and Bhardwaj, 2009). Follicle atresia is a prevalent degenerative process wherein follicles deteriorate in structural integrity, resulting in the loss of oocytes from the ovaries, aside from the ovulation process (Guraya, 1997; Sharma, 2000). Follicular atresia reduces the frequency of ovulations, and the components of the follicle wall transform into steroidogenically active interstitial gland tissue, which aids in the endocrine regulation of ovarian physiology (Sharma and Guraya, 1992; Manabe et al., 2003). Atresia impacts follicles at every developmental stage, with significant germ cell loss occurring during early development, prenatal, neonatal, prepubertal, pubertal, estrous cycle, pregnancy, lactation, and post-reproductive phases in mammals (Guraya, 1998; Sharma, 2000;

Tabarowski et al., 2005). In early development, the primordial follicle population is most significantly impacted, while in the active reproductive phase, the incidence of atresia is highest in antral follicles (Danell, 1987; Guraya et al., 1994; Guraya, 2000; Sharma, 2000). The adjustment of atresia frequency can influence animal fertility (Sharma, 2000). Follicle atresia is defined by the presence of pyknotic granulosa cells within the intact membrana granulosa; large, free-floating granulosa cells exhibiting pyknotic nuclei; hyalinization of granulosa cells; chromolysis and enucleation of granulosa cell nuclei; loosening and delamination of the intercellular matrix; colloidal, opaque, and cloudy follicular fluid; detachment of the cumulus-oophorous complex from mural granulosa cells; emergence of red blood cells within the follicle; invasion of connective tissue fibers into the follicle; and significant accumulation of follicular fluid leading to cyst formation (Sharma, 2000; Sharma and Bhardwaj, 2009; Bhardwaj and Sharma, 2011).

2. APOPTOSIS IN NORMAL CYCLING MATURE OVARY

In prepubertal and mature mammalian ovaries the preantral atretic follicles, the atretic oocyte is characterized by shrinked ooplasmic contents surrounded by a peripheral zone of lipoidal components which is strongly sudanophilic. This zone in a few atretic oocytes is so dense that it occuludens the central mass. (Sharma *et al.*, 1992; Guraya *et al.*, 1994). Initially, pyknosis in granulosa cells was restricted only to specific area which consequently prolonged and large zone of atretic pyknotic granulosa cells was formed, which results in loosening of membrana granulosa (Sharma *et al.*, 1992). These pyknotic granulosa cells give positive test for lipids and 3-β-HSDH, indicating their active role in steroidogenesis (Guraya *et al.*, 1994). Pyknosis of granulosa cells as the first sign of atresia is evocative of the general plan and path of atretogenic changes in mammals (Guraya *et al.*, 1987; Tilly *et al.*, 1996; Burke *et al.*, 2005; Tatone *et al.*, 2008; Sharma and Bhardwaj, 2009; Bhardwaj and Sharma, 2011).

The various nuclear and ooplasmic contents are drastically affected in atretic preantral follicles. The organelles showing abnormal morphology and distribution increase in number and proceed towards disintegration and then lead to an accumulation of lipids in the oocytes and degenerating granulosa cells because of liberation of cytosolic membrane bound lipids (Byskov, 1978). In the atretic preantral follicles the granulosa cells develop histochemical characteristics of steroidogenic cells as evidenced by weak 3-β-HSDH activity (Sharma, 2000). In antral follicles, apoptosis is characterized by the presence of pyknotic nucleus in the membrana granulosa, appearance of spaces and release of fragments of pyknotic nuclei in the peripheral zone of antrum in type 1b (Sharma *et al.*, 1992; Guraya *et al.*, 1994; Sharma and Bhardwaj, 2009; Bhardwaj and Sharma,2011). The granulosa cell glyco- conjugates are changed during follicular atresia in the pig and rat ovaries as shown with lectins (Sharma and Guraya 1992; Kimura *et al.*, 1999). The levels of fibronectin, laminin, type IV collagen, proteoglycans, insulin like growth factor II/mannose 6 phosphate receptors, and matrix metaloproteinases 2 and -9 increased whereas 450 aromatase and connexin 43 decreased within the wall of granulosa cells during follicular atresia in sheep (Huet *et al.*, 1997, 1998, Sharma and Guraya 1998a, b). Guraya *et al.*, (1994) have demonstrated that the degenerative signs also appear in cumulus cells of antral atretic follicles, whereas atresia proceeds gradually, membrana granulosa thins off and theca hypertrophy enhances in antral atretic follicles (Sharma *et al.*, 1992; Guraya *et al.*, 1994).

The appearance of pyknotic granulosa cells as the first sign of atresia in goat strongly advocate the concept that phenomenon of atresia in bovine species is similar (Zimmerman *et al.*, 1987; Wezel *et al.*, 1999; Hastie and Haresign, 2006). The chromophilic pyknotic granules observed in goat atretic follicles were similar to that of DNA - positive masses observed in sheep (Hay *et al.*, 1976, Hay and Cran 1978; Zhou and Zhang, 2005). However such granular material is negligible in cow (Guraya, 1997). The delamination of mural granulosa cells from the basal lamina leads to disruption causing a decline in oxygen carrying blood transudate that decreases the metabolic pace and results in onset of atresia (Guraya, 1985, 1998; Sharma and Guraya, 1998 b, c). However, Hay *et al.*, (1976) suggested that it is because of decline in estrogen synthesis and release. In advance stage of atresia, there is a tendency towards accumulation and storage of sudanophilic lipids in the atretic oocytes (Guraya, 1973 a). The mural layers acquire more lipids as compared to cumulus cells or peripheral granulosa cells. Similar trends were reported in hamster and rabbit (Guraya and Greenwald, 1964), thus indicating granulosa heterogeneity in morphology and physiology (Parshad and Guraya 1983; Sharma and Guraya, 1990). The NADH and NADPH-dependent tetrazolium reductase activity in goat further endorse this concept. The theca interna of atretic antral follicles undergo hypertrophy and develop lipid droplets largely comprised of phospholipids which are associated with steroidogenesis in mammals (Guraya, 1973 b, 1974 a, b, 1977, 1978 a,b; Nicosia, 1980). This hypertrophied theca interna in atretic follicles of goat finally constitute conspicuous masses of interstitial gland tissue which show strong 3-β-HSDH activity (Guraya, 1971, 1973; Guraya *et al.*, 1994; Sharma and Batra, 2008). These lipid droplets are largely phospholipids and triglycerides which change to cholesterol, its esters and phospholipids, and remain associated with theca type interstitial gland tissue. Various carbohydrates, proteins and enzyme histochemistry were used as an index to indicate whether atresia has initiated or not (Guraya, 1984, 1985; Sharma and Guraya, 1992, 1997; Burke *et al.*, 2005). The follicles undergoing atresia disintegrate oocytes and granulosa cells are reabsorbed while theca interna hypertrophy and acquire morphological and histochemical characteristics of a steroidogenic tissue (Sharma and Batra, 2008).

3. STRUCTURE AND ULTRASTRUCTURE OF THE PROCESS

Intrafollicular paracrine steroid interactions rely on FSH and LH, which govern follicular development and the synthesis and release of estrogen (Guraya, 1985, 1998; Ireland, 1987; Richards et al., 1993; Gore-Langton and Armstrong, 1994; Hillier, 1994; Guthrie and Cooper, 1996; Maillet et al., 2002; Burke et al., 2005). Alterations in the production of steroid in the antral fluid are the first biochemical manifestations of atresia. (Guraya, 1985, 1998; Hillier, 1985; Greenwald and Terranova, 1988; Westhof *et al.*, 1991; Gore-Langton and Armstrong, 1994; Greenwald and Roy, 1994; Quirk *et al.*, 2006). Variations in normal and atretic follicular fluid concentration of androgens and that of progesterone are not significantly different (Hillier, 1985; Westhof *et al.*, 1991; Moor, 1977; Gore-Langton and Armstrong, 1994; Slomczynska *et al.*, 2006). Hormonal profile of normal and atretic follicles depends largely on the stage of advancement of atresia. Normal1y it is oestrogen to progesterone ratio which determines whether a follicle will mature or undergo atresia (Moor *et al.*, 1978; Ireland and Roche, 1982, 1983 a, b; Guraya, 1997; Yu *et al.*, 2004; Sharma *et al.*, 2008).

The estrogen level is highest in the preovulatory healthy follicles of pig, sheep, cow, and goat as compared to atretic ones (Eiler and Nalbandov, 1977; Moor *et al.*, 1978; Carlson *et al.*, 1981; Fortune *et al.*, 1988; Burke *et al.*, 2005; Sharma *et al.*, 2008). Proliferation, migration, differentiation and cellular death constitute the most important stages in the development and growth ovarian follicles and is actively involved in turnover of ovarian tissues. The cellular death basically involves two pathways: necrosis or apoptosis (Wyllie *et al.*, 1980; Kressel and Groscurth, 1994). Necrosis is induced as the result of injuries or environmental pathological influences and produces a series of cellular alterations that begin with the changes in cellular membrane permeability with consequent disruption of cytoplasmic structures and ensuing nuclear degeneration (Pol De *et al.*, 1997). On the other hand, apoptosis is the process of cellular self destruction which also involves active process of intracellular synthesis (Wyllie *et al.*, 1980; Cohen and Duke, 1984) and is controlled by cellular genes. In granulosa cell apoptosis, the primary events observed were the nuclear compaction, the chromatin collapses in to large irregular masses surrounded by nuclear envelope, and plasma membrane introflexes forming deep incisions that confer to the cell a very irregular appearance (Sharma and Bhardwaj 2009; Bhardwaj and Sharma, 2011). Despite this, the cellular organelles maintain their morpho-functional integrity. Subsequently the cell fragments into spheroidal subunits surrounded by membranes (apoptotic bodies) that contain portions of cytoplasm and nucleus that are finally phagocytosed by neighbouring cells or by macrophages.

The cellular death in granulosa cells therefore is distinguishable on accounts of histological and ultrastructural morphology, and pattern of changes in cell organelles and the chromatin. In necrosis, chromatin was altered at the end in a disorderly fashion, whereas during apoptosis it was precociously excised in a regular sequence. The apoptotic changes observed in the degenerating granulosa cells from goat ovarian follicles revealed diminished size, withered surface morphology and pyknosis (Hay and Cran, 1978; Hirshfield, 1983; Guraya, 1985; Kaur and Guraya, 1987; Tilly, 1996; Manabe *et al.*, 2003, 2004; Sharma and Bhardwaj, 2009; Bhardwaj and Sharma, 2011). The goat granulsoa cells resemble rat atretic granulosa cells in terms of nuclear condensation and cytoplasmic shrinkage as well as presence of apoptotic bodies (Hurwitz *et al.,* 1996).

The undulations of nuclear membrane and pinching off of the apoptotic bodies strongly advocate the concept that the apoptosis is the basic mechanism involved in the phenomenon of atresia of granulosa cells in mammals. The vacuolization of condensed chromatin material within the nucleus was the specific positive indicator of apoptosis (Figs. 1, 2). Various studies, conducted during atresia of ovine and caprine follicles, cumulatively indicate that degenerative changes are restricted only to membrana granulosa layers that lie adjacent to the antral cavity during early phases of atresia (Sharma *et al.,* 1992). The detachment or delamination of mural granulosa cells from the basal lamina in the initial phases induce atresia due to disruption of oxygen and nutritional milieu carying blood transudate to the cells (Guraya, 1985, 1998; Sharma and Guraya, 1998 b, c; Sharma, 2000, 2003). Ultrastructurally, typical apoptotic granulosa cells from the antral follicle of goat show compaction and segregation of chromatin, condensation of the cytoplasm maintaining the integrity of organelles, and subsequent fragmentation in to membrane bounded apoptotic bodies (Figs.3, 4). Necrotic cells are characterized by irregular

clumping of chromatin, swelling and dissolution of organelles, rupture of the cell membrane, and finally disintegration of all cellular components. Ultrastructural changes in degenerating granulosa cells of goat ovaries strongly support the morphological hallmarks of apoptotic cells. First, the nucleus and cytoplasm became condensed, then the condensed cells were fragmented but retained the integrity of organelles, i.e. mitochondria, and nuclear and plasma membrane. Since cell debris in atretic follicles increased in number following the appearance of apoptotic bodies and had condensed nuclei similar to apoptotic bodies, the debris is possibly comprised of degraded apoptotic bodies. The investigations on fine morphology of granulosa cells *in vivo* showed similar characteristic features as observed after gonadotrophins and steroid treatment in rat *in vitro*, thus, confirming the role of hormones in structural modification of granulosa cells at different phases of the cycle (Balboni and Zecchi, 1981; Yu *et al.*, 2004; Sharma *et al.*, 2008; Sharma and Batra, 2008).

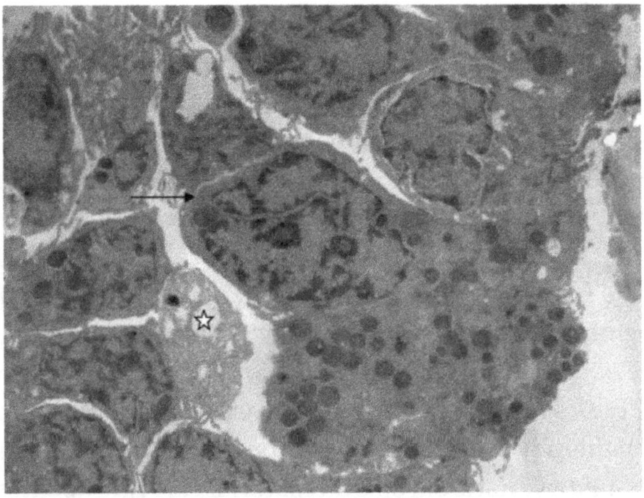

Fig. 1. Electron micrograph of apoptotic granulosa cells showing increased indentation (arrow) and vacuoles (star) of different shapes within condensed chromatin material.

The retraction of standard intercellular microvilli or interacting surfaces with the progression of atresia, alongside the emergence of lysosomal vacuoles and lipid droplets of varying sizes noted during atresia in goats, further corroborates the histological and histochemical attributes of follicular atresia in mammals (Motta, 1972; Parshad and Guraya, 1983; Sharma and Bhardwaj, 2009; Bhardwaj and Sharma, 2011). The occurrence of lysosomal vacuoles in the degenerating granulosa cells of goats during the advanced stage of atresia corroborates the findings of Sharma and Guraya (1997), who reported that lysosomal vacuoles emerge in the cytoplasm, ultimately destabilizing the plasma membrane and resulting in the expulsion of intracellular contents, rendering it more hyaline.

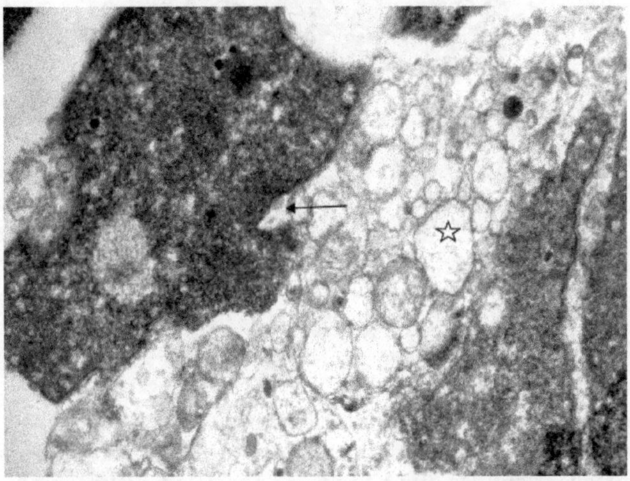

Fig. 2. Electron micrograph of apoptotic granulosa cells showing uneven wavy nuclear envelope (arrow) and increased vacuolization (star) of cytoplasm.

The presence of hyaline granulosa cells in the mural and antral layers of membrana granulosa and free floating in follicular fluid were observed in goat ovary (Sharma and Bhardwaj, 2007). The similar distribution of cells were also observed in primates and other mammals (Byskov, 1974; Balboni and Zecchi, 1981; Bill and Greenwald, 1981; Sharma and Guraya, 1992, 1997). The increased undulation and indentations of the nuclear membrane and pinching off of the nuclear fragments suggests that apoptosis is involved in initiation and execuation of cell death during atresia in goat. The increase in the frequency and dimensions of nuclear pores as well as flattening of the nuclear membrane observed in goat follicles are similar to the earlier findings on ultrastructure of apoptotic granulosa cells in rat and cow (Coucouvanis *et al.*, 1993; Grotowski *et al.*, 1997; Isobe and Yoshimura, 2000; Yang and Rajamahendran, 2000; Inoue *et al.*, 2003), thereby suggesting a common plan of apoptosis in bovine species. The membrane bound pyknotic chromatin material carrying apoptotic vesicles were observed lying within the cytoplasm during the advanced stages of atresia which endorse the concept that apoptotic bodies are formed from condensed chromatin material packed in small vacuoles limited by the nuclear membrane.

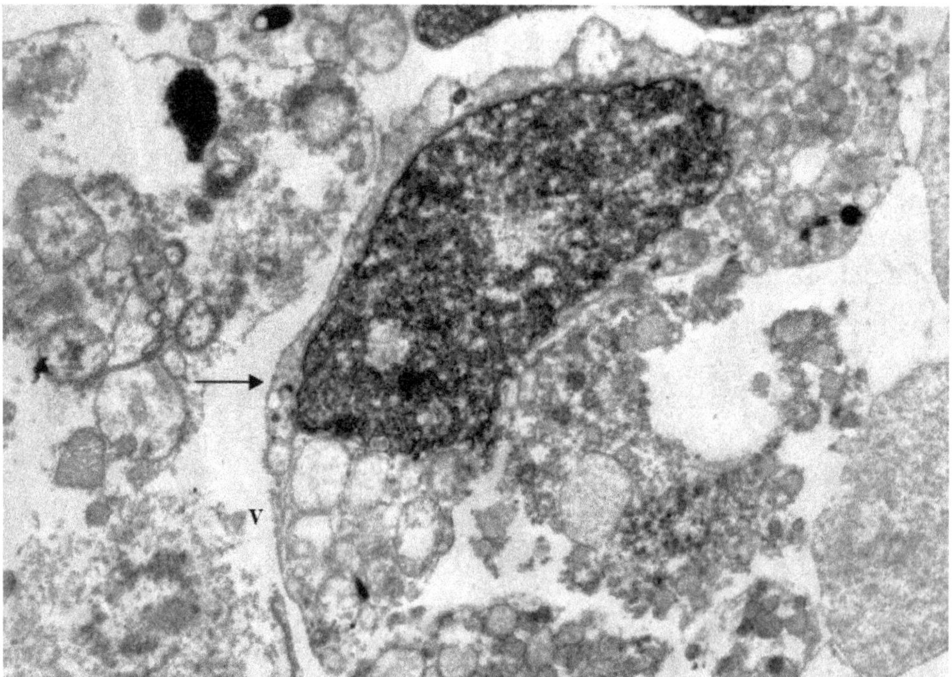

Fig. 3. Electron micrograph of apoptotic granulosa cells showing the pyknotic chromatin material adhering to the periphery of nuclear membrane (arrow) and vacuoles (v) of different shapes and sizes were observed within condensed chromatin material.

In a few cells, the presence of condensed cytoplasm in contrast to hyaline one was possibly due to the differential functional impairment of the cytoplasmic membrane. Sharma and Guraya (1992) have reported changes in glycoconjugates and carbohydrates of atretic granulosa cells in rat and have postulated that changes in histochemical mapping of negatively charged moieties induces uncoupling of membrane interactions subsequently leading to impairment of membrane permeability characteristics that finally lead to atresia or cell death due to apoptosis. The alterations in acidic phospholipids phosphatidyl serine content that acts as apoptosis inducing agent (Krishnamurthy *et al.,* 2000), modulates the membrane chemistry leading to a change in its permeability to water molecules. The cell becomes larger and hyaline if the permeability is enhanced whereas the contents become pyknotic and electron dense if the permeability decreases. Recent studies have demonstrated that apoptosis involves cleaving of DNA in several animal species. Internucleosomal DNA fragmentation has been considered to be characteristic of apoptosis and is one of the earliest event (Schwartzman and Cidlowski, 1993; Okamura *et al.*, 2001; Hastie and Haresign, 2006).

In addition to the detection of oligonucleosomes in extracted DNA, the occurrence of apoptosis may also be inferred from the characteristic morphological appearance of degenerating cells, together with the detection of fragment DNA in single cells *in situ* using TUNEL (Gavrieli *et al.,* 1992; Palumbo and Yeh, 1994; Bristol and Gould *et al.,* 2006; Sharma and Bhardwaj, 2009; Bhardwaj and Sharma,2011). Using *in situ* 3' end labeling (TUNEL), which can detect apoptosis precisely at the single cell level without disruption of the tissue morphology (Gavrieli *et al.,* 1992; Palumbo and Yeh, 1994; Liu *et al.*, 2007; Sharma and Bhardwaj, 2009), the specific morphological features of granulosa cell death in follicular atresia (nuclear pyknosis, karyorrhexis, and formation of apoptotic bodies) can be related to the physiological process of apoptosis. The relationship is supported by a combination of biochemical, classic histological evidences, and *in situ* histochemical localization of DNA fragmentation. Different cellular details were observed in atretic and healthy follicles classified by morphological criteria, including cells with a single shrunken and dense nucleus (pyknotic appearance) and cells with marginated chromatin and/or nuclear fragmentation. According to Lussier *et al.,* (1987), non atretic follicles should have intact and normal granulosa layers with the mean pyknotic index per class varying from 0.13 percent to 0.67 percent. However, in another study in cows (Ireland and Roche,1983), pyknotic cells were observed in the granulosa cell layer in 30-60 percent of estrogen-active large follicles.

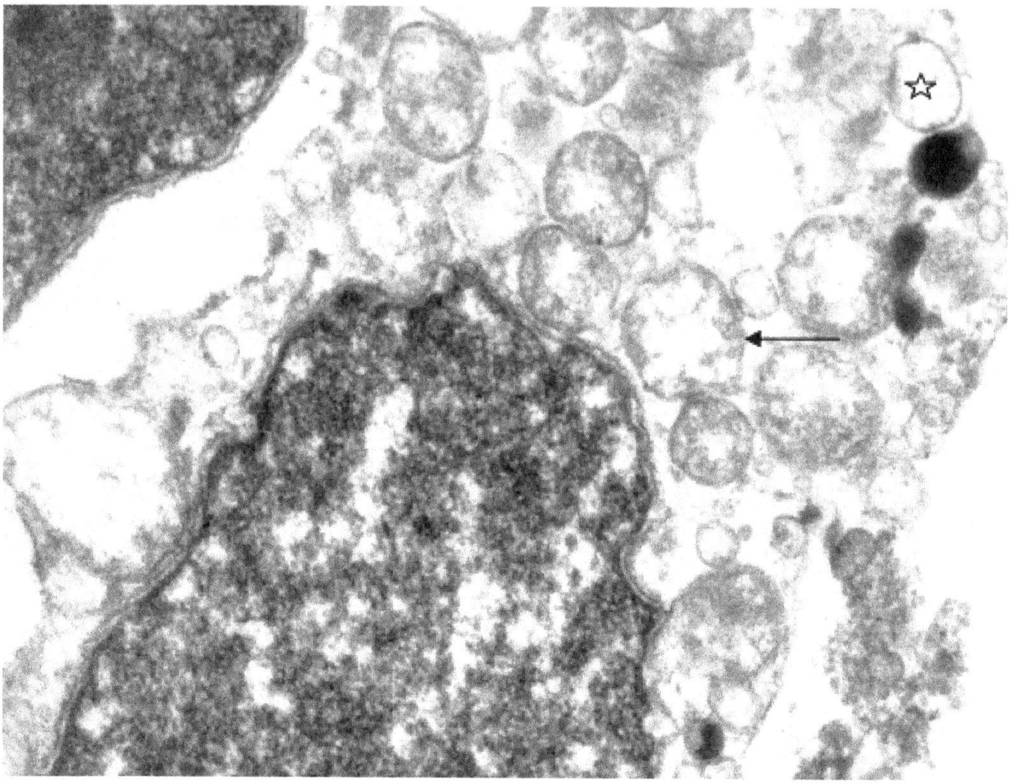

Fig. 4. Electron micrograph of granulosa cells revealing vacuolated cytoplasm (star) and mitochondria (arrow).

Thus, the mere presence of pyknotic cells in the granulosa cell layer does not imply that they are atretic. However, the morphological and biochemical results strongly indicate that apoptosis may occur to a certain level during normal follicle growth and development and that apoptotic death of granulosa cells may be detectable before other morphological and biochemical signs of degeneration in goats. Alkaline phosphatase activity in follicular fluid and granulosa cells exhibited a declining trend from healthy to slightly atretic and atretic follicles.

The biochemical estimation of alkaline phosphatase endorse the earlier histochemical mapping of alkaline phsophatase activity opining the possible role of alkaline phsophatase in active transport of nutrients and secretary material across the membrane (Verma and Guraya, 1968; Sangha and Guraya, 1988/89; Sharma, 2000). The association of alkaline phosphatase positive sites with theca interna indicates the involvement of this enzyme in steroid metabolism and transport (Britenecker *et al.*, 1978; Gilchrist *et al.*, 2004). The decline in levels of alkaline phosphatase in follicular fluid and granulosa cells of atretic follicles may be tangibly due to increased vascularity and changed morphology and biochemistry of granulosa cells for steroid hormone synthesis (Guiseppe, 1983; Gilchrist *et al.*, 2004; Pangas, 2007; Tatone *et al.*,

2008). Acid phosphatase activity in follicular fluid and granulosa cells exhibited an increasing trend from healthy to slightly atretic to atretic follicles. The increase in Golgi complex and lysosomes in atretic follicles/cells is possibly attributable to the rise in acid phosphatase activity.

The ultrastructural modifications associated with the lysosomal accumulation during atresia which is further increased due to luteinization wherein chief protein synthesizing cells transmutate to steroidogenic cells may be attributable to the increase in acid phosphatase activity in the granulosa cells (Dorrington *et al.* 1975; Armstrong and Dorrington, 1976; Sangha and Guraya, 1988/89). The degenerative/transformative changes involved in reshaping of ovarian subcellular components that facilitate differentiation during follicular development while bringing about lysis and formation of apoptotic vesicles may be responsible for the rise in lysosomal activity (Sangha and Guraya, 1988/89; Sharma, 2000). The increase in acid phosphatase enzyme activity observed in the follicular fluid and granulosa cells of atretic follicles may also be related to some mechanism for the secretion of steroids (Sawyer *et al.,* 1979; Dimino and Elfont, 1980; Pangas, 2007).

It has been reported that acid phosphatase activity is higher in active and regressing corpora lutea, provides a lurking possibility that during follicular atresia the rise in acid phosphatase activity may not be associated exclusively with regression but also with the formation of interstitial gland tissue. Catalase is generally associated with superoxide dismutase, constituting a reciprocally protective set, while catalase is inhibited by oxyradicals, and SOD is inhibited by H_2O_2 (Lapluye, 1990). If H2O2 produced by SOD, action on oxyradicals is not removed immediately it will react with super oxide radicals (Haber-Weiss reaction) giving rise to highly reactive hydroxyl radicals (Michiels et al., 1994). However, with the increase in H_2O_2 concentration, the catalase contribution for its degradation concomitantly increased (Verkek and Jond Kind, 1992). Michiels et al., (1991) reported a 30 percent increase in survival when catalase was injected in combination with SOD, whereas the survival was only 21 percent and 4 percent for SOD and catalase respectively, when independently injected in human fibroblasts. Singh and Pandey (1994) reported an increased catalase activity in the ovary of metaoestrous rats Pari pasu with a decline in H_2O_2 production in the mitochondria and microsomal fraction. In addition to its effects on oxygen free radical metabolism, SOD has been shown to influence cell functions by increasing the levels of the second messenger cGMP (Ignarro et al., 1987; Schmidt et al., 1993; Burke et al., 2005).

There are three known forms of SOD with specific subcellular and extracellular distributions (Ravindranath and Fridovich, 1975; Crouch et al., 1984; Redmond et al., 1984; Tibell et al., 1987). The manganese-associated form of SOD is localized in mitochondria of cells, whereas the copper/zinc associated form is found in the cytoplasm. Furthermore, there is an extracellular form of SOD that is secreted from cells. All three forms of SOD are expressed in the ovary (Laloraya et al., 1988; Shiotani et al., 1991; Sato et al., 1992; Hesla et al., 1992, Tilly and Tilly, 1995) and the pattern of expression appears to be related to gonadotropin induced follicular development and luteal steroidogenesis and regression (Laloraya et al., 1988; Hesla et al., 1992; Sato et al., 1992; Tilly and Tilly, 1995; Pangas, 2007). Furthermore, SOD effectively blocks gonadotropin-induced ovulation (Miyazaki et al., 1991).

The activity of enzyme glutathione peroxidase in follicular fluid and granulosa cells shows a declining trend from healthy to slightly atretic to atretic follicles. It has been reported that glutathione peroxidase (GPx) catalyzes the breakdown of H_2O_2 with much more affinity than the catalase (Gul *et al.,* 2000). GPx is reported to provide significant protection against both lipid peroxide and H2O2 (Halliwel and Gutteridge, 1985). Land and Verdetti (1989) observed a reduction in GPx levels with advancing age in the kidneys and liver of rats, whereas Imre et al. (1984) and Hazelton and Lang (1985) noted a decline in GPx activity with age in the liver, kidneys, and several other tissues of mice. GPx activity was observed to be stable in the caudate putamen and temporal cortex, but it diminished in the substantia nigra and thalamus in rats (Benzi et al., 1989).

4. CONCISE SUMMARY

Therefore, comprehending the molecular machinery of apoptosis is essential, as programmed cell death plays a role in both health and illness, triggered by many physiological and pathological stimuli. Furthermore, the extensive role of apoptosis in disease pathophysiology allows for therapeutic intervention at several checkpoints. Consequently, comprehending the mechanisms of apoptosis and other forms of programmed cell death at the molecular level offers profound insights into numerous disease processes and may consequently affect therapeutic strategies.

5. REFERENCES

Aebi, H. (1984). Catalase *in vitro.* In: Methods in Enzymology (I. Packer, ed.). Vol. 105, Academic Press, NY, pp. 121-126.

Armstrong, D.T., and Dorrington, J.H. (1976). Androgen augment FSH-induced progesterone secretion by cultured rat granulosa cells. *Endocrinol.* 99, 1411-1414.

Bae, S.N., Lee, Y.S., Kim, M.Y., Kim, J.D., Park, L.O. (2006). Antiproliferative and apoptotic effects of zinc-citrate compound on human epithelial ovarian cancer cell line, OVCAR-3. *Gynecol. Oncol.* 103, 127-136.

Balboni, G.C., and Zecchi, S. (1981). On the structural changes of granulosa cells cultured *in vitro.* Histochemical, ultrastructural and stereological observations. *Acta Anat.* 110, 136-145.

Benzi, G., Marzatico, F., Pastoris, O., Villa, R.F. (1989). Relationship between ageing, drug treatment and the cerebral enzymatic antioxidant system. *Exp. Gerontol.* 24, 469- 479.

Bhardwaj, J.K. and Sharma, R.K. (2011). Changes in trace elements during follicular atresia in goat (Capra hircus) ovary. *Biol Trace Elem Res* 140, 291-298.

Bhardwaj, J.K. and Sharma, R.K. (2011). Scanning electron microscopic changes in granulosa cells during follicular atresia in caprine ovary. *J. Scanning.* 33, 21-24.

Bill, C.H., and Greenwald, G.S. (1981). Acute gonadotropin deprivation. A model for the study of follicular atresia. *Biol Reprod.* 24, 913-921.

Brietenecker, G., Friedrich, F., Kemeter, P. (1978). Further investigation on the maturation and degeneration of human ovarian follicles and their oocytes. *Fertil. Steril.* 29, 336- 341.

Bristol-Gould, S.K., Kreeger, P.K., Selkirk, C.G., Kilen, S.M., Mayo, K.E., Shea, L.D., Woodruff, T.K. (2006). Fate of the initial follicle pool: empirical and mathematical evidence supporting its sufficiency for adult fertility. *Dev. Biol.* 298, 149-154.

Burke, C.R., Cardenas, H., Mussard, M.L., Day, M.L. (2005). Histological and steroidogenic changes in dominant ovarian follicles during oestradiol-induced atresia in heifers- *Reproduction.* 129, 611-620.

Bussiere, F.I., Zimowska, W., Gueux, E., Rayssiguier, Y., Mazur, A. (2002). Stress protein expression cDNA array study supports activation of neutrophils during acute magnesium deficiency in rats. *Magnes. Res.* 15, 37-42.

Byskov, A.G. (1978). *The Vertebrate Ovary*: Follicular Atresia. In: Jones, R.E. (ed.). Plenum press, New York, pp. 533-562.

Chaube, S.K., Prasad, P.V., Thakur, S.C., and Srivastava, T.G. (2005). Hydrogen peroxide modulates meiotic cell cycle and induces morphological features characteristics of apoptosis in rat oocytes cultured *in vitro. Apoptosis.* 10, 863-875.

Chaudhary, A., Sharma, R.K., Saini, K. (2004). Antioxidant effect of Mandukparni on cerebellum and spinal cord of ageing rats. *J. Tiss. Res.* 4, 113-115.

Chen, W., Wang, Z., Zhang, Y. (2003). The effect of zinc on the apoptosis of cultured human retinal pigment epithelial cells. *J. Huazhong Univ. Sci Technology. Med Sci.* 23, 414- 417.

Chun, S.Y., Eisenhauer, K.M., Kubo, M., Hsueh, A.J. (1995). Interleukin-1β suppresses apoptosis in rat ovarian follicles by increasing nitric oxide production. *Endocrinology.* 136, 3120-3127.

Cohen, J.J., and Duke, R.C. (1984). Glucocorticoid activation of a calcium dependent endonuclease in thymocyte nuclei leads to cell death. *J. Immunol.* 132, 38-42.

Coucouvanis, E.C., Sherwood, S.W., Carswell-Crumpton, C., Spack, E.G., Jones, P.P. (1993). Evidence that the mechanism of prenatal germ cell death in mouse is apoptosis. *Exp. Cell Res.* 209, 238-247.

Craig, J., Orisaka, M., Wang, H., Orisaka, S., Thompson, W., Zhu, C., Kotsuji, F., Tsang, B.K. (2007). Gonadotropin and intra-ovarian signals regulating follicle development and atresia: the delicate balance between life and death. *Front Biosci.* 12, 3628-3639.

Crouch, R.K., Gandy, S., Patrick, J., Reynolds, S., Buse, M.G., Simson, J.A. (1984). Localization of copper-zinc superoxide dismutase in the endocrine pancreas. *Exp Mol Pathol.* 41, 377-383.

Danell, B. (1987). 'Oestrous behaviour, ovarian morphology and cyclical variations in follicular system and endocrine pattern in water buffalo heifers'. Ph.D. Thesis, pp. 1-124. Swedish Univ. Agricultural Sci., Uppsala, Sweden.

Dimino, M.J., and Elfont, E.A. (1980). The role of lysosomes in ovarian physiology. In: *Biology of the Ovary.* (Eds.) Motta, P.M. and Hafez, E.S.E., Nijhoff, The Hague/ Boston/London ed. Chap. XV, 196-201.

Dorrington, J.H., Moon, Y.S., Armstrong, D.T. (1975). Estradiol 17 β biosynthesis in cultured granulosa cells from hypophysectomized immature rats. Stimulation by follicle- stimulating hormone. *Endocrinol.* 97, 1328-1331.

Durlinger, A.L.L., Krammer, P., Karels, B., Grootegoed, J.A., Vilenbroek, J., Themmen, A.P.N. (2000). Apoptotic and proliferative changes during induced atresia of pre- ovulatory follicles in the rat. *Hum. Reprod.* 15, 2504-2511.

Feng, P., Li, T.L., Guan, Z.X., Franklin, R.B., Costello, L.C. (2002). Direct effect of zinc on mitochondrial apoptogenesis in prostate cells. *Prostate.* 52, 311-318.

Flemming, W. (1885). Uber die bildung von richtungsfiguren in sauge thiereiern beim untergang graafscher follikel. *Arch Anat Entwickl.* 221-244.

Fridovich, I. (1975). Superoxide dismutases. *Annu Rev Biochem.* 44, 147-159.

Fridovich, I. (1986). Biological effects of the superoxide radical. *Arch Biochem Biophys.* 247, 1- 11.

Gavrieli, Y., Sherman, Y., Bensasson, S.A. (1992). Identification of programmed cell death *in situ* via specific labeling of nuclear DNA fragmentation. *J. Cell. Biol.* 119, 493-501.

Gilchrist, R.B., Ritter, L.J. Armstrong, D.T. (2004). Oocyte-somatic cell interactions during follicle development in mammals. *Anim. Reprod. Sci.* 82-83, 431-446.

Gore-Langton, R.E., and Armstrong, D.T. (1994). Follicular steroidogenesis and its control. *The physiology of Reproduction,* Vol. I, 2nd Edn., Knobil, E and Neil, J.D., Raven Press, New York, pp. 571-628.

Greenwald, G.S. and Terranova, P.F. (1988). Follicular selection and its control In: *The Physiology of Reproduction.* (Eds.) Knobil, E. and Neil, J.D. Vol. 1, 387-446. Raven Press, New York.

Greenwald, G.S., and Roy, S.K. (1994). Follicular development and its control. *The Physiology of Reproduction.* Knobil, E. and Neil, J.D. Raven Press, New York, pp. 629-724.

Grotowski, W., Lecybyl, R., Warenik-Szymankiewicz, A., Trzeciak, W.H. (1997). The role of apoptosis in granulosa cells in follicular atresia. *Ginekol. Pol.* 68, 317-326.

Guerin, P., El Mouatassim, S., Menezo, Y. (2001). Oxidative stress and protection against reactive oxygen species in the pre-implantation embryo and its surroundings. *Hum. Reprod. Update.* 7, 175-189.

Guiseppe, C.B. (1983). Structural changes: Ovulation and luteal phase. In: *The Ovary.* (Ed.) Serra, G.B. Raven Press, New York, Vol. I, Chap-VIII, 123-142.

Gul, M., Kutay, F.Z., Temocin, S., Hannien, O. (2000). Cellular and Clinical implications of glutathione. *I.J. Exp. Biol.* 38, 325-634.

Guraya, S.S. (1971). Morphology, histochemistry and biochemistry of human ovarian compartments and steroid hormone synthesis. *Physiol. Rev.* 51, 785-807.

Guraya, S.S. (1973a). Morphology, histochemistry and biochemistry of follicular growth and atresia. *In: Proceedings of the Symposium-Ovogenesis-Folliculogenesis.* Nouzilly, *Ann. Biol. Anim. Biochem. Biophys.* 13, 229-240.

Guraya, S.S. (1973b). Interstitial gland tissue of mammalian ovary. *Acta endocr.* 72, 1-27.

Guraya, S.S. (1974a). Morphology, histochemistry, and biochemistry of human oogenesis and ovulation. *Int. Rev. Cytol.* 37, 121-151.

Guraya, S.S. (1974b). Gonadotrophins and functions of granulosa and thecal cells *in vivo* and *in vitro.* Gonadotrophins and Gonadal Functions. (Ed.) Moudgal N.R. Academic press, New York, pp. 280-337.

Guraya, S.S. (1985). Biology of Ovarian Follicles in Mammals. Springer Verlag, Heidelberg- Berlin, New York.

Guraya, S.S. (1997). Comparative biology of corpus luteum cellular and molecular regulatory mechanisms. In: *frontiers in Environmental and Metabolic Endocrinology* (Ed.) Maitra, S.K. pp. 31-58, University of Burdwan, India.

Guraya, S.S. (1998). Cellular and Molecular Biology of Gonadal Development and Maturation in Mammals: Fundamentals and Biomedical Implications. Narosa Publishing House, New Delhi.

Guraya, S.S. (1998). Cellular and Molecular Biology of Gonadal Development and Maturation in Mammals : Fundamentals and Biomedical Implications. Narosa Publishing House, New Delhi.

Guraya, S.S. (2000). Comparative Cellular and Molecular Biology of Ovary in Mammals: Fundamentals and Applied Aspects: Science Publishers, INC, PO Box 699, Enfield, USA.

Guraya, S.S., and Greenwald, G.S. (1964). A comparative histochemical study of interstitial tissue and follicular atresia in the mammalian ovary. *Anat. Rec.* 149, 411-434.

Guthrie, H.D., and Cooper, B.S. (1996). Follicular atresia, follicular fluid hormones, and circulating hormones during the mid luteal phase of the estrous cycle in pigs. *Biol. Reprod.* 55, 543-547.

Halliwell, B., and Gutteridge, J.M.C. (1985). Lipid peroxidation: A radical chain reaction. In B. Halliwell and J.M.C. Gutteridge eds: *Free radicals in biology and medicine,* Clarend press, Oxford.

Hastie, P.M., and Haresign, W. (2006). Expression of mRNAs encoding insulin-like growth factor (IGF) ligands, IGF receptors and IGF binding proteins during follicular growth and atresia in the ovine ovary throughout the oestrous cycle. *Anim. Reprod. Sci.* 92, 284-299.

Hay, M.F., and Cran, D.G. (1978). Differential response of the components of sheep Graafian follicle to atresia. *Annl. Biol. Anim. Biochem. Biophys.* 18, 453-460.

Hay, M.F., Cran, D.G., and Moor, R.M. (1976). Structural changes occurring during atresia in sheep ovarian follicles. *Cell Tissue Res.* 169, 515-529.

Hazeleton, G.A., and Lang, C.A. (1985). Glutathione reductase and peroxidase activity in aging mouse. *Mech. Aging Dev.* 29, 71.

Hesla, J.S., Miyazaki, T., Dasko, L.M., Wallach, E.E., Dharmarajan, A.M. (1992). Superoxide dismutase activity, lipid peroxide production and corpus luteum steroidogenesis during natural luteolysis and regression induced by oestradiol deprivation of the ovary in pseudopregnant rabbits. *J. Reprod Fertil.* 95, 915-924.

Hillier, S.G. (1985). Sex steroid metabolism and follicular development in the ovaries. *Oxford Rev. Reprod. Biol.* 7, 168-222.

Hillier, S.G. (1994). Hormonal control of folliculogenesis and luteinization. *Molecular Biology of Female Reproductive System.* Academic Press, San Diego, pp. 1-37.

Hirshfield, A.N. (1991). Development of follicles in the mammalian ovary. *Int. Rev. Cytol.* 124, 43-101.

Ho, J.S., Gargano, M., Cao, J., Bronson, R.T., Heimler, I., Hutz, R.J. (1998). Reduced fertility in female mice lacking copper-zinc superoxide dismutase. *J. Biol. Chem.* 273, 7765- 7769.

Hsueh, A.J.W., Billig, H., Tsafriri, A. (1994). Ovarian follicle atresia: a hormonally controlled apoptotic process. *Endocr Rev.* 15, 707-724.

Huet, C., Monget, P., Pisselet, C., Hennequet, C., Locatelli, A., Monniaux, D. (1998). Chronology of events accompanying follicular atresia in hypophysectomized ewes. Changes in levels of steroidogenic enzymes, connexin 43, insulin-like growth factor II/mannose 6 phosphate receptor, extracellular and matrix metaloproteinases. *Biol. Reprod.* 58, 175-185.

Huet, C., Monget, P., Pisselet, C., Monniaux, D. (1997). Changes in extracellular matrix components and steroidogenic enzymes during growth and atresia of antral ovarian follicles in sheep. *Biol. Reprod.* 56, 1025-1034.

Hughes, F.M.J.R., and Gorospe, W.C. (1991). Biochemical identification of apoptosis (Programmed cell death) in granulosa cells: evidence for a potential mechanism underlying follicular atresia. *Endocrinology.* 129, 2415-2422.

Hurwitz, A., Ruutiainen-Altman, K., Marzella, L., Botero, L., Dushnik, M., Adashi, E.Y. (1996). Follicular atresia as an apoptotic process: atresia-associated increased in the ovarian expression of the putative apoptotic marker sulfated glycoprotein-2, *J. Soc. Gynecol. Investig.* 3, 199-208.

Hussein, M.R., (2005). Apoptosis in the ovary: molecular mechanisms. *Hum. Reprod.* 11, 162- 178.

Ignarro, L.J., Byrns, R.E., Buga, G.M., Wood, K.S. (1987). Endothelium derived relaxing factor from pulmonary artery and vein possesses pharmacologic and chemical properties identical to those of nitric oxide radical. *Circ Res.* 61, 866-879.

Iitaka, M., Kakinuma, S., Fujimaki, S., Posuta, I., Fujita, T., Yamanaka, K., Wada, S., Katayama, S. (2001). Induction of apoptosis and necrosis by zinc in human thyroid cancer cell lines. *J. Endocrinol.* 169, 417-424.

Imre, S., Toth, F., Fachet, J. (1984). Superoxide dismutase, catalase and lipid peroxidation in liver mice of different ages. *Mech. Aging Dev.* 28, 297.

Inoue, N., Manabe, N., Matsui, T., Maeda, A., Nakagawa, S., Wada, S., Miyamoto, H. (2003). Role of tumor necrosis factor-related apoptosis-inducing ligand signaling pathway in granulosa cell apoptosis during atresia in pig ovaries. *J. Reprod. Dev.* 49, 313-321.

Ireland, J.J. (1987). Control of follicular growth and development. *J. Reprod. Fertil.* 34, 39-54.

Ireland, J.J., and Roche, J.F. (1982). Development of antral follicles in cattle after prostaglandin-induced luteolysis. Changes in serum, hormones, steroids in follicular fluid and gonadotrophin receptors. *Endocrinology.* 111, 2077-2086.

Ireland, J.J., and Roche, J.F. (1983). Growth and differentiation of large antral follicles after spontaneous luteolysis in heifers: changes in concentration of hormones in follicular fluid and specific binding of gonadotropins to follicles. *J. Anim. Sci.* 57, 157-167.

Ireland, J.J., and Roche, J.F. (1983a). Development of nonovulatory antral follicles after spontaneous luteolysis in heifers: changes in the concentration of hormones in follicular fluid and specific binding of gonadotrophins to follicles. *J. Anim. Sci.* 57, 157-167.

Ireland, J.J., and Roche, J.F. (1983b). Development of nonovulatory antral follicles in heifer: changes in the steroids in follicular fluid and receptors for gonadotrophins. *Endocrinology.* 112, 150-156.

Isobe, N., and Yoshimura, Y. (2000). Localization of apoptotic cells in the cystic ovarian follicles of cows: a DNA-end labeling histochemical study. *Theriogenology.* 53, 897-904.

Iwata, H., Hashimoto, S., Ohota, M., Kimura, K., Shibano, K., Miyake, M. (2004). Effects of follicle size and electrolytes and glucose in maturation medium on nuclear maturation and developmental competence of bovine oocytes. *Reproduction.* 127, 159-164.

Iwata, H., Hayashi, T., Sato, H., Kimura, K., Kuwayama, T., Manju, Y. (2005). Modification of ovary stock solution with magnesium and raffinose improves the developmental competence of oocytes after long preservation. *Zygote.* 13, 303-308.

Jiang, J.Y., Cheung, C.K., Wang, Y., Tsang, B.K. (2003). Regulation of cell death and cell survival gene expression during ovarian follicular development and atresia. *Front Biosci.* 8, 222-237.

Johnson, N.C., Dan, H.C., Cheng, J.Q., Kruk, P.A. (2004). BRCAI 185detAG mutation inhibits Akt-dependent, IAP-mediated caspase-3 inactivation in human ovarian surface epithelial cells. *Exp. Cell Res.* 298, 9-18.

Kaur, P., and Guraya, S.S. (1987). Ovarian characteristics of the Indian mole rat *Bandicota bongalensis. Proc. Ind. Natl Acad Sci.* 96, 667.

Kerr, J.F., Winterford, C.M., and Harman, B.V. (1994). Apoptosis. Its significance in cancer and cancer therapy. *Cancer.* 73, 2013-26.

Kerr, J.F.R., Wyllie, A.H., Currie, A.R. (1972). Apoptosis: a basic biological phenomenon with wide ranging implications in tissue kinetics. *Br. J. Cancer.* 26, 239-257.

Kimura, Y., Manabe, N., Nishihara, S., Matsushita, H., Tajima, C., Wada, S., Miyamoto, H. (1999). Up-regulation of the alpha 2,6-sialyl- transferase messenger ribonucleic acid increased glycoconjugation containing alpha 2,6-linked sialic acid residues in granulosa cells during follicular atresia of porcine ovaries. *Biol. Reprod.* 60, 1475- 1482.

Kressel, M., Groscurth, P. (1994). Distinction of apoptotic and necrotic cell death by *in situ* labeling of fragmented DNA. *Cell Tiss. Res.* 278, 549-556.

Krishnamurthy, K.V., Krishnaraj, R., Chozhavendan, R., Samuel, C.F. (2000). The program of cell death in plants and animals – A comparison. *Curr. Sci.* 79, 1169-1181.

Laloraya, M., Kumar, G.P., Laloraya, M.M. (1988). Changes in the levels of superoxide anion radical and superoxide dismutase during the estrous cycle of *Rattus norvegicus* and induction of superoxide dismutase in rat ovary by lutropin. *Biochem Biophys Res Commun.* 157, 146-153.

Lapluye, G. (1990). SOD mimicking properties of copper (II) complexes: health side effects. In: Emerit I., Packer, L., Auclair, C. Antioxidants in therapy and preventive medicine. Plenum Press, New York, 59.

Liu, Y., Wu, C., Lyu, Q., Yang, D., Albertini, D.F., Keefe, D.L., Liu, L. (2007). Germline stem cells and neo-oogenesis in the adult human ovary. *Dev. Biol.*, 306, 112-120.

Lussier, J.G., Matton, P., Dufour, J.J., (1987). Growth rates of follicles in the ovary of the cow. *J. Reprod. Fertil.* 81, 301-307.

Maillet, G., Breard, E., Benhaim, A., Leymarie, P., Feral, C. (2002). Hormonal regulation of apoptosis in rabbit granulosa cells *in vitro:* evaluation by flow cytometric detection of plasma membrane phosphatidylserine externalization. *Reproduction.* 123, 243- 251.

Manabe, N., Inoue, N., Miyano, T., Sakamaki, K., Sugimoto, M., Miyamoto, H. (2003). Ovarian follicle selection in mammalian ovaries : regulatory mechanism of granulosa cell apoptosis during follicular atresia. In : *The Ovary.* Leung PK, Adashi E (Eds.), Academic press/Elsevier Science Publishers, Amsterdam, pp. 369-385.

Manabe, N., Inoue, N., Miyano, T., Sakamaki, K., Sugimoto, M., Miyamoto, H. (2003). Ovarian follicle selection in mammalian ovaries : regulatory mechanism of granulosa cell apoptosis during follicular atresia. In : *The Ovary.* Leung PK, Adashi E (Eds.), Academic press/Elsevier Science Publishers, Amsterdam, pp.

369-385.

Matzuk, M.M., Dionne, L., Guo, Q., Kumar, T.R., Lebovitz, R.M. (1998). Ovarian function in superoxide dismutase 1 and 2 knockout mice. *Endocrinology*. 139, 4008-4011.

Michiels, C., Raes, M., Haubion, A., Remacle, J. (1991). Association of antioxidant system in the protection of human fibroblasts against oxygen derived free radicals. *Free Rad. Res. Comm.* 14, 323.

Michiels, C., Raes, M., Joussaint, O., Remacle, J. (1994). Importance of Se-glutathione peroxidase, catalase and Cu/Zn-SOD for cell survival against oxidative stress. *Free Rad. Biol. Med.* 17, 235.

Monniaux, D., (2002). Oocyte apoptosis and evolution of ovarian reserve. *Gynecol Obstet Fertil.* 30, 822-826.

Moor, R.M. (1977). Sites of steroid production in ovine Graafian follicles in culture. *J. Endocrinol.* 73, 143-150.

Moor, R.M., Hay, M.F., Dott, H.M., and Cran, D.G. (1978). Macroscopic identification and steroidogenic functions of atretic follicles in sheep. *Endocrinology.* 77, 309-318.

Motta, P. (1972). Histochemical evidence of early atretic follicles in different mammals. *J. Cell Biol.* 55, 18.

Nicosia, S.V. (1980). Endocrine Physiopathology of the ovary. Tozzini RI, Reeves G. and Pineda R.L. (Eds.) Amsterdam, New York, pp. 43-62.

Okamura, Y., Miyamoto, A., Manabe, N., Tanaka, N., Okamura, H., Fukomoto, M. (2001). Protein tyrosine kinase expression in the porcine ovary. *Mol Human Reprod.* 7, 723- 729.

Osman, P. (1985). Rate and Course of atresia during follicular development in the adult cycling rat. *J. Reprod. Fertil.* 73, 261-270.

Palumbo, A., and Yeh, J. (1994). *In situ* localization of apoptosis in rat ovary during follicular atresia. *Biol Reprod.* 51, 888-895.

Pangas, S.A. (2007). Growth factors in ovarian development. *Semin Reprod Med.* 25, 225-234. Parshad, V.R., and Guraya, S.S. (1983). Histochemical distribution of tetrazolium reductases, dehydrogenases and lipids in the follicular wall of normal and atretic follicles in the ovary of the Indian gerbil, *Tatera indica* (Muridea: Rodentia). *Proc. Ind. Acad. Sci.* 92, 121-128.

Paszkowski, T., Clarke, R.N. (1996). Antioxidant capacity of preimplantation embryo culture medium declines following the incubation of poor quality embryos. *Hum Reprod.* 11, 2493-2495.

Paszkowski, T., Traub, A.I., Robinson, S.Y., McMaster, D. (1995). Selenium dependent glutathione peroxidase activity in human follicular fluid. *Clin. Chim. Acta.* 236, 173- 180.

Pol De, A., Vaccina, F., Forabosco, A., Cavazzuti, E., Marzono, L. (1997). Apoptosis of germ cells during human prenatal oogenesis. *Hum Reprod.* 12, 2235-2241.

Quirk, S.M., Cowan, R.G., and Harman, R.M. (2006). The susceptibility of granulosa cells to apoptosis is influenced by oestradiol and the cell cycle. *J. Endocrinol.* 189, 441-453.

Rajakoshi, E. (1960). The ovarian follicular system in sexually mature heifers with special reference to seasonal, cyclical and left right variations. *Acta endocr.* 34, 7-68

Ravindranath, S.D., and Fridovich, I. (1975). Isolation and Characterization of a manganese- containing superoxide dismutase from yeast. *J. Biol Chem.* 250, 6107-6112.

Redmond, T.M., Duke, E.J., Coles, W.H., Simson, J.A., Crouch, R.K. (1984). Localization of corneal superoxide dismutase by biochemical and histocytochemical techniques. *Exp. Eye Res.* 38, 369-378.

Richards, J.S., Sirois, J., Natraj, V., Morris, J.K., Fiotzpatrick, S.L., Clemens, J.W. (1993). Molecular regulation of genes involved in ovulation and luteinization. Ovarian cell interaction. Genes to physiology. Hsueh AJW and Schomberg DW (eds.), W. Springs-Verlag, New York, pp. 125-133.

Rudolf, E. (2007). Depletion of ATP and oxidative stress underlie zinc-induced cell injury. *Acta Medica.* 50, 43-49.

Sangha, G.K., and Guraya, S.S. (1988/89). Histochemical changes in acid and alkaline phosphatase activities in the growing follicles and corpora lutea of the rat ovary. *Acta Morphol. Neerl. Scand.* 26, 43-49.

Sato, T., Irie, S., Krajeueski, S., Reed, J.C. (1994). Cloning and sequencing of a cDNA encoding the rat Bcl-2 protein. *Gene.* 140, 291-292.

Sawyer, H.R., Abel, J.H., McClellan, M.C., Schwitz, M., Niswender, G.D. (1979). Secretory granules and progesterone secretion by ovine corpora lutea *in vitro. Endocrinol.* 104. 476-486.

Schmidt, H.H., Lohmann, S.M., Walter, U. (1993). The nitric oxide and cGMP signal transduction system: regulation and mechanism of action. *Biochim Biophys Acta.* 1178, 153-175.

Schwartzman, R.A., and Cidlowski, J.A. (1993). Apoptosis: the biochemistry and molecular biology of programmed cell death. *Endocr Rev.* 14, 133-151.

Sharma, R.K. (2000). Follicular atresia in goat: A review. *Ind. J. Anim. Sci.* 70, 1035-1046. Sharma, R.K. (2003). Structural analysis of cumulus and corona cells of goat antral follicles: possible functional significance. *Indian J. Anim. Sci.* 73, 28-32.

Sharma, R.K. and Batra, S. (2005). Ultrastructure of the regressing corpus luteum in the goat ovary. *Indian J. Anim. Sci.* 75, 936-937.

Sharma, R.K. and Bhardwaj, J.K. (2009). Ultrastructural characterization of apoptotic granulosa cells in caprine ovary. *J. Microsc.* 236, 236-242.

Sharma, R.K., and Batra, S. (2008). Changes in the steroidogenic cells of the ovaries in small ruminants. *Indian J. Anim. Sci.,* 78, 584-596.

Sharma, R.K., and Bhardwaj, J.K. (2007). Granulosa cell apoptosis *in situ* in caprine ovary. *J. Cell Tissue Res.* 7, 1111-1114.

Sharma, R.K., and Bhardwaj, J.K. (2009). *In situ* evaluation of granulosa cells during apoptosis in caprine ovary. *Int. J. Integ Biol.* 5, 58-61.

Sharma, R.K., and Guraya, S.S. (1990). Granulosa heterogeneity: A histo-chemical lectin staining and scanning electron microscopic study on *Rattus rattus* ovary. *Acta Embr. Morph. Exp.* 11, 107-129.

Sharma, R.K., and Guraya, S.S. (1992). Lectin staining studies on follicular atresia in house rat (*Rattus rattus*). *Acta Morphol. Hung.* 40, 25-34.

Sharma, R.K., and Guraya, S.S. (1992). Lectin staining studies on follicular atresia in house rat (*Rattus rattus*). *Acta Morphol. Hung.* 40, 25-34.

Sharma, R.K., and Guraya, S.S. (1998a). Distribution, histochemistry and biochemistry of carbohydrates in the mammalian oocytes during folliculogenesis. *J. Agri. Rev.* 19, 73-85.

Sharma, R.K., and Guraya, S.S. (1998b). Carbohydrate histochemistry of atretic follicles in house rat ovary. *J. Anim. Morphol. Physiol.* 45, 112-117.

Sharma, R.K., and Guraya, S.S. (1998c). Simultaneously histochemical and lectin staining studies and electron microsopic observations on interstitial gland tissue in the rat ovary. *J. Anim. Morphol. Physiol.* 45, 112-117.

Sharma, R.K., Khajuria, M., and Guraya, S.S. (1992). Morphology of normal and atretic follicles of goat during anoestrous. *Int. J. Anim. Sci.* 6, 81-85.

Sharma, R.K., Sharma, M.B., and Bhardwaj, J.K. (2008). Gonadotropins titre and apoptosis in caprine granulosa cells (Abstract). In: Proceedings of International conference on *Molecular and Clinical aspects of gonadal and nongonadal actions of gonadotropins.* Feb. 7-9. AIIMS, New Delhi, p. 51.

Shiotani, M., Noda, Y., Narimoto, K., Imai, K., Mori, T., Fujimoto, K., Ogawa, K. (1991). Immunohistochemical localization of superoxide dismutase in the human ovary. *Hum Reprod.* 6, 1349-1353.

Singh, D., and Pandey, R.S. (1994). Changes in the catalase activity and hydrogen peroxide production in the rat ovary during estrous cycle. XVI. *Int. Cong. Biochem. Mol. Bio.* 364.

Slomczynska, M., Tabarowski, Z., Duda, M., Burek, M., Knapczyk, K. (2006). Androgen receptor in early apoptotic follicles in the porcine ovary at pregnancy. *Folia Histochem. Cytobiol.* 44, 185-188.

Tabarowski, Z., Szaltys, M., Bik, M., Slomczynska, M. (2005). Atresia of large ovarian follicles of the rat. *Folia Histochem. Cytobiol.* 43, 43-55.

Tatone, C., Amicarelli, F., Carbone, M.C., Monteleone, P., Caserta, D., Marci, R., Artini, P.G., Piomboni, P., Focarelli, R., (2008). Cellular and molecular aspects of ovarian follicle ageing. *Human Reprod.* 14, 131-142.

Tibell, L., Hjalmarsson, K., Edlund, T., Skogman, G., Engstrom, A., Marklund, S.C. (1987). Expression of human extracellular superoxide dismutase in Chinese hamster ovary cells and characterization of the product. *Proc Natl Acad Sci USA.* 84, 6634-6638.

Tilly, J.L. (1996). Apoptosis and ovarian function. *Rev Reprod.* 1, 162-172.

Tilly, J.L. (1996). The molecular basis of ovarian cell death during germ cell attrition, follicular atresia, and luteolysis. *Front Biosci.* 1, d1-11.

Tilly, J.L., and Tilly, K.I. (1995). Inhibitors of oxidative stress mimic the ability of follicle- stimulating hormone to suppress apoptosis in cultured rat ovarian follicles. *Endocrinology.* 136, 242-252.

Tilly, J.L., Kowalski, K.I., Johnson, A.L., Huseh, A.J.W. (1991). Involvement of apoptosis in ovarian follicular atresia and postovulatory regression. *Endocrinology.* 129, 2799- 2801.

Tsafriri, A. and Braw, R. (1984). Experimental approaches to atresia in mammals. *Oxford Review in Reproductive Biology.* 6, 226-265.

Uilenbroek, J.M.J., Wouterson, P.J., Van der Schoot, P. (1980). Atresia of preovulatory follicles: gonadotropin binding and steroidogenic activity. *Biol. Reprod.* 23, 219-229.

Verkek, A., and Jondkind, J.F. (1992). Vascular cells under peroxide induced stress: a balance study on *in vitro* peroxide handling by vascular endothelial and smooth muscle cells. *Free Rad. Res. Comm.* 17, 121.

Verma, S.K., and Guraya, S.S. 1968. The localization and functional significance of alkaline phosphatase in the vertebrate ovary. *Experientia.* 24, 398-399.

Vohra, B.P.S., James, T.J., Sharma, S.P., Kansal, V.K., Chaudhary, A., Gupta, S.K., (2002). Dark neurons in the ageing cerebellum: their mode of formation and effect of Maharishi Amrit Kalash. *Biotechnology,* 3, 347-354.

Westhof, G., Westhof, K.F., Braendle, W.L., Dizerega, G.S. (1991). Differential steroid secretion and gonadotrophin response by individual tertiary porcine follicles *in vitro.* Possible physiological role of atretic follicles. *Biol. Reprod.* 44, 461-468.

Wezel, I.L.V., Dharmarajan, A.M., Lavranos, T.C., Rodgers, R.J. (1999). Evidence for alternative pathways of granulosa cell death in healthy and slightly atretic bovine antral follicles. *Endocrinology.* 140, 2602-2612.

Williams, G.T., and Smith, C.A. (1993). Molecular regulation of apoptosis: genetic controls on cell death. *Cell.* 74, 777-779.

Wiseman, D.A., Wells, S.M., Wilham, J., Hubbard, M., Welker, J.E., Boack, S.M. (2006). Endothelial response to stress from exogeneous Zn^{+2} resembles that of NO-mediated nitrosative stress, and is protected by MT-1 overexpression. *Am. J. Physiol Cell Physiol.* 291, 555-568.

Wu, Ji., Zhang, L. and Wang, X. (2000). Maturation and apoptosis of human oocytes *in vitro* are age related. *Fertil steril.* 74, 1137-1141.

Wyllie, A.H., Kerr, J.F.R., Currie, A.R. (1980). Cell death: the significance of apoptosis. *Int. Rev. Cytol.* 68, 251-306.

Yang, H.W., Hwang, K.J., Kwon, H.C., Kim, H.S., Choi, K.W., Oh, K.S. (1998). Detection of reactive oxygen species (ROS) and apoptosis in human fragmented embryos. *Hum. Reprod.* 13, 998-1002.

Yang, M.Y., and Rajamahendran, R. (2000). Morphological and biochemical identification of apoptosis in small, medium, and large bovine follicles and the effects of follicle- stimulating hormone and insulin-like growth factor-1 on spontaneous apoptosis in cultured bovine granulosa cells. *Biol Reprod.* 2000, 62, 1209-1217.

Yu, Y.S., Sui, H.S., Han, Z.B., Li, W., Luo, M.J., Tan, J.H. (2004). Apoptosis in granulosa cells during follicular atresia : relationship with steroids and insulin like growth factors. *Cell Research.* 14, 341-346.

Zhou, H., and Zhang, Y. (2005). Effect of growth factors on *in vitro* development of caprine preantral follicle oocytes. *Ann. Reprod. Sci.* 90, 265-272.

W. (1987). *In vitro* steroid secretion of tertiary atretic bovine follicles in a superfusion system correlated to their histological features. *Human Reprod.* 2, 457- 461.

GLOSSARY

Abaptation: The process by which an organism is fitted to its environment as a consequence of the characters it inherits, which have been filtered by natural selection in previous environments. Because present environments seldom differ greatly from recent past environments, abaptive fitness can resemble adaptation. In this sense, however, adaptation appears to imply advance planning, or design, which is misleading.

Abduction: Movement away from the midventral axis of the body.

Abductor muscle: A muscle that draws a structure (e.g. a limb) away from the centre line of the body.

Ablogenesis: The development of living organisms from nonliving matter, as in the origin of life on Earth, or in the concept of spontaneous generation which was once held to account for the origin of life but which modern understanding of evolutionary processes has rendered outdated.

Ablotic: Nonliving; devoid of life.

Abomasum: In Ruminantia, the fourth and final region of the specialized stomach, corresponding to the stomach in other mammals and the zone in which digestion proceeds with the usual mammalian digestive enzymes.

Aboral: Away from the mouth; on the opposite side of the body from the mouth in animals that lack clearcut dorsal and ventral surfaces.

Acanthizidae: A family of birds all of which build domed nests. They have thin, pointed bills with basal bristles, are insectivorous, and vary from being strictly arboreal (fairy warblers) to groundfeeding (scrub Wrens). The sexes are usually similar. There are 17 genera, comprising 63 species. (*Acanthiza* (thornbills) are sometimes placed in the Sylviidae or Maluridae.) They are found mainly in Australia and New Guinea.

Acanthobdeilida (phylum Annelida, class Hirudinea): An order of parasitic worms that lack an anterior sucker. Chaetae occur in the anterior region. There is one genus, *Acanthobdella*; it occurs only on salmon.

Acanthocephala (thornyheaded worms): A phylum of bilaterally symmetrical, pseudocoelomate, wormlike organisms most of which do not have an excretory system; all lack a gut. A retractable proboscis is present, covered with the recurred spines that give the animals their common name. All acanthocephalans are endoparasitic, living in two or three marine or terrestrial vertebrates during their development. There are about 600 species.

Acanthoclinidae (subclass Actinopterygil, order Perciformes): Family of small (about 8 cm long), agile fish that inhabit tropical waters of the IndoPacific region. They have pelvic fins reduced to one spine and two soft rays.

Acanthodil: Class of primitive, fossil fish that had a true bony skeleton, a heterocercal tail fin, a persistent notochord, ganoid scales, and stout spines in front of the fins. The acanthodians lived from the Silurian Period to the Permian Period and may be related to ancestors of the more modern bony fish.

Acanthopterygil: Superorder of bony fish that includes all the spinyfinned fish (e.g. Perciformes and Beryciformes).

Acanthuridae: Large family of slabsided fish that have a sharp spine on each side of the tailbase (hence the name 'surgeonfish'). The spine is retractable in some species. Many species exhibit striking colour patterns. There are about 80 species, found in ail tropical seas; many feed on organisms that cover the surface or rocks, or on seagrasses.

Acarinum: A small pouch in the abdomen of Old World carpenter, bees of the genus *Xylocopa* (subgenus *Koptortosoma*), which provide protection for symbiotic mites of the genus *Dinogamasus*.

Acceleration: Evolution that occurs by increasing the rate of ontogenetic (*see* ONTOGENY) development, so that further stages can be added before growth is completed. This form of heterochrony was proposed by E. H. Haeckel as one of the principal modes of evolution.

Accipitridae (hawks, eagles, buzzards, kites, Old World vultures; class Aves, order Falconiformes) The largest family of birds of prey, containing many cosmopolitan genera, especially *Accipiter* (sparrowhawks and goshawks), *Buteo* (buzzards or hawks), Circus (harriers), *Elanus* (kites), *Haliaeetus* (fish eagles), and *Pernis* (honey buzzards), which make up nearly half the family. Other genera are more restricted in range, and more than half are monotypic. *Aquila chrysaetos* (golden eagle) has a Holarctic distribution (there are 10 species of *Aquila*, found worldwide except for S. America and Malaysia). Hawk eagles (10 species of *Spizaetus*) occur widely in low latitude forests. Accipiters are generally carnivorous, preying on snarls, insects, fish, mammals, reptiles, and birds; but one species, *Gypohierax angolensis* (palmnut vulture), feeds on oilpalm husks. Vultures (seven species of *Gyps*) have reduced talons which aid walking, inhabit open plains, mountains, and forests, feed on carrion, and nest in trees and on crags. There are 63 genera in the family, with more than 200 species.

Acetylcholine (ACh): An acetyl ester of choline that is involved in synaptic (*see* SYNAPSE) transmission between nerve cells. It is released from vesicles by the presynaptic neurone and diffuses across the synaptic deft where it interacts with specific receptors to produce a local depolarization of the postsynaptic membrane, thus enabling the transmission of nerve impulses.

Acid: According to the BrønstedLowry theory, a substance that hi solution liberates hydrogen ions or protons. The Lewis theory states that it is a substance that acts as an electronpair acceptor. An acid reacts with a base to give a salt and water (neutralization), and has a pH of less than 7.

Acipenseridae: A family of large, fairly sluggish, bottomfeeding fish that have a low-slung mouth, toothless jaws, four barbels in front of the upper jaw, and rows of large plates along the body. They are found in Europe and N. America. Most of the 26 species are marine but ascend rivers to spawn. A few species, e.g. the sterlet (*Acipenser ruthenus*), live wholly in fresh water. Sturgeons are very fertile: large females of the common sturgeon (*Acipenser sturio*) produce up to 6000000 eggs per year (the roe of ripe females after processing is renowned as caviar).

Acochlidiacea: (class Gastropoda, subclass Opisthobranchia) An order of molluscs in which individuals are very small and benthic, living between sand grains. The visceral sac is thin and spiculate, and is often much longer than the foot and markedly separate from it. There is no shell. These gastropods are deposit feeders. According to different authorities the order contains three families or only one.

Actinoceratida: (class Cephalopoda, subclass Nautiloidea) A large order of predominantly orthoconic cephalopods that have large siphuncles. Large amounts of siphuncular and cameral deposits are present. Most are assumed to have been nektonic. The actinoceratids ranged in age from Lower Ordovician to Lower Carboniferous.

Active immunity: Resistance to a disease that is acquired by an animal as the result of the production of antibodies in response to antigens produced by the disease organism whilst inside the host animal.

Active site: Part of an enzyme molecule, the conformation of which is such that it binds to the substrate or substrates to form an enzymesubstrate complex. The conformation is not absolute and may alter according to reaction conditions.

Active transport: The transport of substances across a membrane against a concentration gradient. Such processes require energy, the source often being the hydrolysis of adenosine triphosphate (ATP). Protein or lipoprotein carrier

Adaptation: A characteristic or trait that enhances an organism's ability to survive and reproduce in its environment.

Adrenal gland: In vertebrates, an organ that secretes certain hormones. Many vertebrates possess multiple adrenal glands, but in mammals there is one gland close to each kidney. In tetrapods, each gland consists of a central medulla and an outer cortex. The medulla secretes adrenalin and noradrenalin, hormones needed when the animal is in an excited state and must engage in strenuous activity (e.g. fighting or fleeing). The *cortex* secretes sex hormones and other hormones concerned with regulating the water and salt balances of the body.

Aerotaxis: A change in direction of locomotion, in a motile organism or cell, made in response to a change in the concentration of oxygen in its immediate environment.

Agamic generation (asexual generation): In the life cycle of some species (e.g. many members of the Cynipidae) a generation that comprises parthenogenetically reproducing females that are genetically diploid. The unfertilized eggs laid may be male (having undergone meiosis) or female (not having undergone meiosis). In Cynipidae, agamic females usually emerge from stout galls designed to overwinter or to survive a period when resources are scarce.

Amiiformes: A family with only one living representative, *Amia calva*, a fairly large (up to 6 kg) carnivorous fish that has a long dorsal fin with only soft rays, a gular plate, a heterocercal tail, and a lunglike swimbladder. The male guards the nest and larvae.

Ammonoidea: A subclass of cephalopods which generally have planispirally, tightly coiled, septate shells (although some are coiled loosely or spirally). The protoconch is globular; the shells maybe either involute or evolute. Some forms have marked ventral keels; ribs and nodes may also be present. The siphuncle is variable but mainly ventral in position. Sutures are often very complex. Cameral deposits are absent. The Ammonoidea were probably tetrabranchiate cephalopods. They constitute the largest cephalopod subclass, with 163 families, including the ammonites, in which the suture lines form very complex patterns; the ceratites, in which part of the suture line is frilled; and the goniatites, which have relatively simple suture lines. They range in age from Devonian to Upper Cretaceous. All members are now extinct.

Apomorphic: Applied to an evolutionarily advanced character state. The long neck of the giraffe is apomorphic; the short neck of its ancestor is plesiomorphic. Apomorphic features are those possessed by a group of biological organisms that distinguish those organisms from others descended from the same ancestor. The term is taken from the Greek *apo*, 'from' or 'away', and *morphe*, 'form', and means 'new-featured'. It refers to 'derived' characters which have appeared during the course of evolution. Apomorphic features may be autapomorphic (uniquely derived) or synapomorphic (sharedderived). Apomorphic is the opposite of plesiomorphic.

Apostomatida: An order of ciliate protozoa in which mature forms have spirally arranged body cilia. The organisms live as parasites or commensals and often have two hosts, one of which is commonly a marine crustacean (e.g. *Foettinga* lives inside the gastrovascular cavity of sea anemones; motile young leave the sea

Asexual reproduction: Reproduction involving a single parent that produces offspring genetically identical to itself (e.g., budding, fission).

Asymmetry: Lack of symmetry in the body structure of an organism (e.g., sponges).

Autotomy: The ability of some animals to deliberately shed or lose a body part (e.g., lizards shedding their tails).

Backcross: A cross of an F_1 hybrid or heterozygote with an individual whose genotype is identical to that of one or other of the two parental individuals. Matings involving a hybrid genotype and a recessive parental genotype are used in genetic analyses to determine linkage and crossover values.

Bagridae: A large family of freshwater catfish that have a large, adipose dorsal fin, small eyes, and four pairs of barbels around the mouth. The family is distributed widely in tropical Africa and Asia.

Balaenidae (order Cetacea, suborder Mysticeti): A family that comprises the right and bowhead whales. The body is robust, the snout blunt, the throat ungrooved. A dorsal fin is present in some species. The flippers are broad and short. The posterior margin of the tail is notched at the midline. The cervical vertebrae are fused. The baleen plates are long and narrow, and left and right rows are separated anteriorly. The animals are dark grey or black and feed on plankton.

Balanced polymorphism: A genetic polymorphism that is stable, and is maintained in a population by natural selection, because the heterozygotes for particular

Balanldae (acorn barnacles; order Thoracica, suborder Balanomorpha) Family of radially symmetrical, balanomorph barnacles in which the third pair of cerci resemble the second rather than the fourth, and in which there are no caudal appendages. The family includes the familiar barnacles exposed at low tide, of the genus *Balanus*, which has been divided into 10 subgenera.

Basal body: A structure located in the cytoplasm of cells that is essential for the formation of cilia and flagella and from which they project. Normally it is composed of nine sets of microtubules, each set arranged in triplets, embedded in a dense, granular matrix.

Basal metabolic rate (BMR): The minimum metabolic rate needed to sustain the life of an organism that is in an environment at a temperature equal to its own. The BMR is measured in organisms in a postabsorbtive, resting state.

Base analogue: A purine or pyrimidine base that differs slightly in structure from the normal base, but that because of its similarity to that base may act as a mutagen when incorporated into DNA. Once in place, these bases, which have pairing properties unlike those of the bases they replace, can produce mutations by causing insertions of incorrect nucleotides opposite them during replication. Though the original base analogue exists only in a single strand, it can cause a nucleotidepair substitution that is replicated in all DNA copies descended from that original strand. An example is 5bromouracil (5BU), an analogue of thymine that has bromine

Bilateral symmetry: Body symmetry in which an organism can be divided into two identical halves along a central axis (e.g., humans, insects).

Bilirubin: A goldenyellow bile pigment that is derived from the degradation of porphyrins, in particular haem compounds, by cells of the reticuloendothelial system, notably those of the liver.

Binomial nomenclature: The system of naming species with two Latin names, genus and species (e.g., *Homo sapiens* for humans).

Biochemical oxygen demand (biological oxygen demand, BOD) An indicator of the pollution of water by organic matter where the decomposition of the organic matter by microorganisms depletes the water of dissolved oxygen. It is measured as the mass (mg) of oxygen per litre of water that is taken up when a sample of water containing a known amount of oxygen is kept in darkness at 20°C, usually for five days (when the result is described as the BOD_5).

Biogenesis: The principle that a living organism can arise only from another living organism, a principle contrasting with concepts such as that of the spontaneous generation of living from nonliving matter. The term is currently more often used to refer to the formation from or by living organisms of any substance (e.g. *coal*, chalk, or chemicals).

Biogenetic law: The early *stages* of development in animal species resemble one another, the species diverging more and more as development proceeds. The law was formulated by the embryologist *E. K.* von Baer (17921876).

Biological clock: An endogenous, physiological mechanism, whose exact nature has not been determined, that keeps time independently of external events, enabling organisms to determine and to respond to dally, lunar, seasonal, and other periodicities. Its existence has been inferred from the observation of organisms which retain rhythmic activity under constant conditions. **bioluminescence** The production by living organisms of light without heat. Bioluminescence is a property of many types of organism (e.g. certain (mostly marine) bacteria, dinoflagellates, and fireflies).

Biosphere: The global ecological system, including all living organisms and their interactions with the environment.

Blattodea (cockroaches, roaches; phylum Arthropoda, class Insecta): Order of medium to large, exopterygote insects in which the fore wings are modified as rather strongly sclerotized tegmina. Wings are absent in some groups. The legs are gressorial, and the body is dorsoventrally compressed. Auditory organs are absent. Females have a small, concealed ovipositor, and eggs are laid in an ootheca. Most cockroaches are nocturnal, and during the day are usually found hiding in crevices. Burrowing and cavedwelling species are known, and some species are even amphibious. Cockroaches may be adapted to desert conditions, or live in tropical rain forest. It is thought that most cockroaches are omnivorous. Of the approximately 3500 species known, some are of economic importance as domestic pests.

Blood clotting: A response to the injury of a blood vessel in which the escape of liquid blood is prevented by its conversion into a solid mass, or dot, composed of fibres of fibrin in which blood cells are enmeshed. In vertebrates, a soluble protein, fibrinogen, found in the blood plasma is acted upon by the enzyme thrombin. As a result, a negatively charged peptide is split off from the fibrinogen molecule, leaving monomeric fibrin; this is capable of rapid polymerization to produce the tangle of insoluble threads that forms a clot. Active thrombin is formed from inactive prothrombin, also found in blood plasma, by a very complex process involving a lipoprotein factor liberated by rupturing blood platelets, a protein plasma thromboplastin component (PTC), antihaemophilic globulin (AHG), calcium ions, etc.

Bostrichidae (branch borer, limb borer, shothole borer, class Insecta, order Coleoptera): Family of black or brown beetles, 320mm long, with the head tucked under a rounded, rugose thorax. The elytra are sharply declined at the rear, with several projecting, horizontal spines. The antennae have a loose club. Larvae are curved, white, and fleshy, with a tiny head and enlarged thorax. They bore into felled or dying trees. There are many generations per year. There are 434 species, largely tropical.

Brachypterous: Applied to insects in which both pairs of wings are reduced.

Braulidae (bee lice; order Diptera, suborder Cyclorrapha) Family of abnormal flies which lack wings and resemble members of the Hippoboscidae. The antennae are set in lateral grooves, and the eyes are reduced in size. The thorax appears to be joined to the abdomen. The tarsi are adapted to cling to the bodies of Hymenoptera, especially bees. Bee lice eggs are laid near honeyor pollenstorage cells inside a bee nest, and the larvae feed on these substances and pupate within the nest of the host. The pupa is unique among the Cyclorrapha in being enclosed in the unmodified last (final instar) larval cuticle. Only five species have been described so far.

Bryozoa (Ectoprocta; mossanimals): A phylum of small, aquatic, colonial animals, related to the Brachiopoda, many of which possess a welldeveloped, calcite skeleton which comprises microscopic, boxlike divisions, each housing an individual animal possessing ciliated tentacles and a coelom. Food is collected by the tentacles which surround the mouth and are borne on the lophophore. Reproduction takes place by asexual budding, and by the release of larvae which give rise to new colonies. Bryozoans have occurred from the Ordovician to the present day. Fossilized branched colonies are common in some rocks. They were important reef builders in the Phanerozoic, and underwent several great adaptive radiations.

Caecotrophy: The passing of food through the alimentary canal twice. Rabbits and some other small mammals take soft faecal pellets directly from the anus at night and store them in the stomach to be mixed with food taken during the day, probably obtaining metabolites essential for digestion that are produced in the caecum. The animals will die if they are prevented from ingesting these pellets.

Calciferol: A compound, designated vitamin D_2, obtained through the ultraviolet irradiation of ergosterol.

Campodeiform: Applied to a type of larval form found in Neuroptera, Trichoptera, Strepsiptera, and some Coleoptera. The campodeiform larva is predacious, and has a well sclerotized body, flattened dorsoventrally, with six legs.

Capsomere: The protein 'building block' of a virus capsid.

Carbohydrate: A generic term for molecules that have a basic empirical formula of $C_x(H_2O)_y$ where *x* and *y* are variable numbers. They are normally classified as mono, oligo, or polysaccharides, depending upon whether they are single sugars, shortchain molecules, or large polymers respectively.

Carboniferous: The penultimate period of the Palaeozoic Era, from about 360286 Ma ago, preceded by the Devonian and followed by the Permian. In Europe the lower part of the period is termed the Dinantian. It is divided into two stages and is characterized by marine limestones with a rich coralbrachiopod fauna. In contrast the upper part, the Silesian, which is subdivided into three stages, is noted for the deposition of terrestrial and freshwater sediments. The vast forests of the Upper Carboniferous gave rise to the rich coal measures of S. Wales, England, Scotland, and many other areas worldwide. N. American geologists subdivide the Carboniferous into two periods or subperiods. Of these the lower is named the Mississippian and is the equivalent of the Dinantian stages (the Tournaisian and Visean). The upper period, the Pennsylvanian, is the equivalent of the Silesian.

Cardiac cycle: The series of muscular contractions which squeeze blood through the heart. In Amphibia and higher vertebrates, blood ispumped through a double circulatory system. The right side of the heart pumps blood through the pulmonary circuit to reach the lungs, the left side pumps oxygenated blood through the systemic circuit which

Cardinal tooth: A large hinge tooth, present in some bivalves, which is immediately below the umbo. More than one may be present on each valve.

Carnivore: An organism that primarily eats other animals.

Cartilage: In vertebrates, flexible skeletal tissue formed from groups of rounded cells lying in a matrix containing collagen fibres. It forms most of the skeleton of

Cell differentiation: The process by which descendants of a single cell produce structural and functional specializations and maintain these during the course of the life of that individual. Specializations arise during early development of the individual by differential DNARNA transcription; within a species, particular cells differentiate into particular forms, each with their own functions,

Centrifugal speciation: The principle that new species are likely to arise towards the centre of the range of the present species, rather than at the periphery. In practice, it is very often observed that primitive species are located on the edges of the distribution of a speciesgroup or genus.

Centrifuge: An apparatus used for the separation of substances by the application of centrifugal force: this is generated by highspeed rotation of a vessel containing tubes filled with a fluid in which the substances are suspended but not dissolved. Separation occurs because different substances have different rates of sedimentation according to their molecular size and shape.

Centriole: A hollow, cylindrical structure, normally one of a pair lying at right angles to one another, adjacent to the nuclear envelope in animal cells, and composed of nine sets of microtubules, each set arranged in triplets. Centrioles are thought to be organizers of microtubular structures in these cells, and during cell division a pair is found at each pole of the mitotic spindle.

Cepaea: A genus of snails, including *C. hortensis* and *C. nemoralis*, that show visible polymorphism for banding, shell colour, and shell height. They have formed the subject of extensive studies in ecological genetics, where it has been shown that different selective processes relating to various factors, including differential predation and differential response to sunlight, are responsible for the maintenance of the different morphs in the population.

Cephalaspidea (Bullomorpha; phylum Mollusca, class Gastropoda): An order of opisthobranch gastropods in which the mantle cavity is reasonably well developed.

The bubbleshaped shell is internal and very small in some genera. A wide foot and head shield are present. All are hermaphrodites.

Cere: In birds, an area of skin at the *base* of the upper mandible, surrounding the nostrils, that is found in Falconiformes, in which it is bare, in Psittaciformes, in which it is usually feathered, and in Columbidae, in which it forms a swollen flap above the nostrils.

Cerebellum: One of the three major parts of the brain. The cerebellum forms from a dorsal outgrowh from the metencephalon region of the brain stem and is involved in coordination and regulation of motor activities, such as balance or escape movements. The grey matter of the cerebellum receives impulses from various other coordination centres as well as from tendon stretch receptors and acoustic areas of the myelencephalon.

Chemlosmotic theory: A theory concerning oxidative phosphorylation in which it is proposed that the electrontransport chain is arranged such that it generates an energyrich proton gradient across the inner membrane of a mitochondrion, and electrons cross the membrane by a mechanism reminiscent of that in which molecules of solvent cross a semipermeable membrane in osmosis. The energy is then used to drive the phosphorylation of ADP through a membranebound ATPase.

Chilodontidae (head stander, subclass Actinopterygii, order Cypriniformes) A small family of freshwater fish that have a single dorsal fin and an adipose fin, a small mouth, and a forked tail fin. These fish often swim in a headdown position. They occur in southern Australia.

Chilopoda (centipedes: phylum Arthropoda, subphylum Atelocerata) A class of uniramous arthropods that have segmented bodies with one pair of legs to each segment. Some centipedes are adapted to burrowing in soil but most run on the surface. All are believed to be predators. The head bears long antennae and beneath the mouthparts there is a large pair of claws (maxillipeds, forming part of the first body segment) that inject poison. Below the mandibles a pair, of maxillae forms a functional lip, overlain by a second pair of maxillae. The name. Chilopoda, is derived from the Greek *cheilos*, lip, and *pod*, fool There are about 3000 species, found worldwide, and grouped in two subclasses. (Epimorpha and Anamorpha) and four principal orders: Geophilomorpha; Scolopendromorpha; Lithobiomorpha; and Scutigeromorpha.

Chinchillidae (chinchillas, vizcachas; order Rodentia, suborder Hystricomorpha) A family of rodents which shelter in crevices or burrow's. The hind limbs are long, with three or four digits, the fore limbs short with four or five digits. The eyes are large, the ears rounded, and in some species large. The tail is long, and in some species bushy. They are found in southern S. America. There are six species in three genera: *Chinchilla; Lagidium;* and *Lagostomus*. The Chilean chinchilla (*C. laniger*), an animal about 23cm long .excluding the tail, is raised in captivity for its fur, which has been prized at least since Inca times; in the past chinchillas have also been a source of meat.

Chioragogen tissue: Tissue composed of brown or greenish cells, located in the intestinal wall or heart of Annelida, that is an important centre of metabolism and the synthesis of haemoglobin, and that may also have an excretory function.

Chloride cells: Cells located in the gills of teleosts which pump sodium and chloride ions out into the sea against a concentration gradient. To reduce osmotic dehydration due to hyposmotic body fluid, teleosts drink large quantities of sea water, absorbing in the gut sodium and chloride ions which serve to raise the osmotic pressure in the blood and cause water to flow in from the gut. Subsequently the excess of ions absorbed are pumped out of the body via the chloride cells.

Choline: A basic, nitrogenous, organic compound that functions as a methylgroup donor in some phospholipids and in acetylcholine. Although it has no known

Cholinergic: Applied to nerve endings that secrete the neurotransmitter acetylcholine into the synapse on the arrival of a nerve impulse. Cholinergic nerve endings are characteristic of the parasympathetic nerve system, although preganglionic synapses of the sympathetic nerve system also use acetylcholine.

Circulatory system: The system that moves blood and other fluids throughout an organism's body to transport nutrients, gases, and waste.

Cladistics: A method of classification based on the evolutionary relationships among species.

Class: A taxonomic rank below phylum and above order (e.g., Mammalia, Aves).

Cocoon: A protective casing spun by some animals (e.g., caterpillars) during metamorphosis.

Coevolurtion: The complementary evolution of closely associated species. The interlocking adaptations of many flowering plants and their pollinating insects provide some striking examples of coevolution. In a broader sense, predator-prey relationships also involve coevolution (when an evolutionary advance in the predator triggers an evolutionary response in the prey). **cohesion species concept** A proposal, made by Alan R. Templeton in 1989, to broaden the biological species concept by including asexual organisms and' downplaying interbreeding in sexually reproducing organisms. A species, in this view, is a population, or series of populations, that has genetic or demographic cohesion.

Competitive inhibition: The reversible inhibition of the activity of an enzyme, brought about by the presence of an inhibitor molecule structurally similar to the normal substrate.

Complementary genes: Mutant alleles at different loci which complement one another to give a wildtype phenotype. Dominant complementarity occurs where the dominant alleles of two or more genes are required for the expression of a particular trait. Recessive complementarity is the case of suppression of a particular trait by the dominant allele of either gene, so that only the homozygous double recessive displays the trait.

Complementary resources: Two or more resources that can substitute for one another and, when taken together, augment one another, so that the consumer requires less of them when taken together than when taken separately.

Complementation map: A gene map in which each mutation is represented by a line or 'bar' that overlaps the bars for other mutations with which it will not complement. Noncomplementing mutants are represented by overlapping, continuous lines. Each bar probably, represents the region of the polypeptide that is distorted by an aminoacid substitution. Complementation maps are usually linear, and the positions of mutants on the complementation and genetic maps usually accord.

Dactylopteridae: (flying gurnard; subclass Actinopterygii, order Dactylopteriformes) A small family of marine, tropical to subtropical fish which have a somewhat bizarre profile. The very large pectoral fins are winglike, the fin tips reaching the tail fin when folded. Two free spines preceding the first dorsal fin are well forward on the back. The pelvic fins are used by the fish to `walk' on the seabottom, and the pectoral `wings' are used to glide over it. There are about four species, widely distributed in coastal areas. The fairly common *Dactylopterus volitans* grows to 35cm.

Daphaenosittidae: (Neosittidae; sittellas; class Aves, order Passeriformes) A family of small, greybrown, birds that have black and white markings on their heads, white-tipped tails, pale rumps, and coloured wing patches. They have thin, slightly hooked bills, long wings, short tails, and short legs with long toes. They are found in forests where they climb up and down trees, feed on insects, and nest in trees. There are two genera, with three species, found in Australia and New Guinea.

Dasyuromorphia: In classifications that divide the Marsupialia into a number of orders, one of those orders, containing only Dasyuroidea.

Delphinoidea (order Cetacea, suborder Odontoceti): A superfamily that comprises the families Delphinidae, Phocoenidae, Stenidae, and Monodontidae, a large group of small cetaceans with teeth in both jaws and usually a prominent dorsal fin. Most are active predators, but *Sousa* (the white or longbacked dolphin), of the coasts and rivers of the Old World tropics, is herbivorous.

Dermestidae (skin beetles, hide beetles, tallow beetles, museum beetles; class Insecta, order Coleoptera) Family of black or brown, small beetles, often with a pattern of fine hairs or scales, in which the pronotum is strongly narrowed toward the head, and the antennae are clubbed. Dermestidae occur in flowers, feeding on pollen. The larvae are subcylindrical, and covered with long hair. *Anthrenus* larvae (woolly bears or carpet beetles) attack woollen carpets, furs, and dried museum specimens. *Dermestes lardarius* larvae (bacon beetles or larder beetles) attack furs, hides, and stored meat and cheese. There are 731 species, occurring worldwide.

Desmosome: An oval or circular junction between two cells where their respective cell membranes run parallel to one another 3050nm apart. Immediately within each membrane there is a plaque of electrondense material and behind this a less dense area of microfilaments. The space between the membranes contains a

Desmostylia: An extinct order of subungulates known from Miocene and Pliocene rocks that resembled the hippopotamus in their general proportions and probably followed, like the hippopotamus, an amphibious way of life, though in shallow coastal waters. It seems that they shared a common descent with the Sirenia and Proboscidea, from early African subungulates.

Desmostylus hesperus: A species of subungulate mammals from the midTertiary (Upper Miocene to Lower Pliocene) of western N. America and eastern Asia that lived in shallow coastal waters. It was an early sirenianlike animal, although the body was large (of hippopotamus size) with massive limbs. The skull was elongate, and there were 'tusks' on both upper and lower jaws.

Diapause: A temporary cessation that occurs in the growth and development of an insect or a mammal. Insects can enter the diapause state as eggs, larvae, pupae, or as an adult; mammals, only as blastocysts. Diapause is frequently associated with seasonal environments, the insect entering it during the adverse period, and breaking from it when more favourable conditions return.

Diaphragm: In mammals, a transverse partition that separates the thoracic and abdominal cavities. When at rest the diaphragm is arched up into the thorax; its flattening increases the volume of the thorax, thus allowing the lungs to expand and so playing an important part in breathing. It is composed partly of muscle and partly of tendon, and is covered by a serous membrane.

Digestive system: The system responsible for the breakdown of food into nutrients that can be absorbed into the body.

Diploblastic: Animals with two germ layers (ectoderm and endoderm) during embryonic development (e.g., cnidarians).

Disruptive selection: Selection that changes the frequency of alleles in a divergent manner, leading to the fixation of alternative alleles in members of the population. The result after several generations of selection should be two divergent phenotypic extremes within the population; this process has been thought to provide a possible mechanism for sym

DNA fingerprinting: A technique that is used to compare DNA from different sources or to identify a genome by comparison with a known standard.

Domestication: The process by which humans selectively breed animals for particular traits or behaviors.

Echiurida (spoon worms): A phylum of coelomate (*see* COELOM), marine, wormlike animals in which mature individuals lack metameric segmentation. The body consists of a sausageshaped, cylindrical trunk and an anterior, muscular proboscis that cannot be retracted (unlike the introvert of Sipunculida which Echiurida otherwise resemble in size and shape). Parapodia are absent in most forms. All have a pharynx. The nervous system is similar to that of Annelida, but there are no special sense organs. Males and females occur. They are exclusively marine, and most are burrowers and detritus feeders.

Ecological isolation: Separation of groups of organism as a result of changes in their ecology or in the environment in which they live. This is one of the processes leading to speciation, since there will be a restriction in the movement of genes between groups thus separated, and changes in gene frequencies may occur due to local selection or drift until eventually the groups may be so divergent that reproductive barriers exist.

Ecology: The study of the interactions between organisms and their environment.

Ecosystem (ecological system): A term first used by A. G. Tansley in 1935 to describe a natural unit that consists of living and nonliving parts, interacting to form a stable system. Fundamental concepts include the flow of energy via food chains and food webs, and the cycling of nutrients biogeochemically. Ecosystem principles can be applied at all scales—thus principles that apply e.g. to an ephemeral pond apply equally to a lake, an ocean, or the whole planet. In Soviet and central European literature 'biogeocoenosis' describes the same concept.

Effect hypothesis: A model proposed in 1980 by the palaeontologist Elisabeth Vrba to account for evolutionary trends. She proposed that a species, occupying a restricted ecological niche, would continually give rise to daughter species by punctuated equilibrium. These new species would have a variety of characteristics; but because of the features of the particular ecological niche, only species that possessed a particular suite of characters would survive; the surviving species would speciate in their turn, with the same result, and at each level the lineage appears to be 'pushed' further and further in a given direction.

Effective population size: The average number of individuals in a population that actually contribute genes to succeeding generations. This number is generally rather lower than the observed, censused, population size, being reduced by the following factors: (*a*) a higher proportion of one sex may mate; (*b*) some individuals will pass on more genes by having more offspring in a lifetime than others; (*c*) any severe past reduction in population size may result in the random loss of particular

Endemism: The situation in which a species or other taxonomic group is restricted to a particular geographic region, due to factors such as isolation or response to soil or climatic conditions. Such a taxon is said to be endemic to that region. The size of the region in this context will usually depend on the status of the taxon: thus a family will be endemic to a much larger area than a species, all other things being equal.

Endomitosis: The doubling of the chromosomes within a nucleus that does not divide. The doubling may be repeated a number of times in a single nucleus. It occurs mainly in insects, where different adult tissues have specific polyploid numbers; it also occurs in some vertebrate tissues.

Endopterygota (class Insecta, subclass Pterygota) A division of insects in which the wings develop internally until the final moult, when they undergo a pupal stage and complete metamorphosis.

Endothermic: Animals that can regulate their internal body temperature (e.g., mammals, birds).

Environmental resistance factors: Those aspects of an environment which constrain the growth of a population and establish the maximum number of individuals that can be sustained. Such factors include the availability of essential re sources (e.g. food and water), predation, disease, the accumulation of toxic metabolic wastes, and, in some species, behavioural changes due to stress caused by overcrowding.

Enzyme: A molecule, wholly or largely protein, produced by a living cell, that acts as a biological catalyst. Enzymes are present in all living organisms, and through their high degree of specificity exert close control over cellular metabolism.

Eutrophication: The nutrient enrichment (usually by nitrates and phosphates) of an aquatic ecosystem, such that the productivity of the system ceases to be limited by the availability of nutrients. An increase in photosynthetic activity is often followed by a depletion of dissolved oxygen as plants die and are decomposed by aerobic organisms. Deoxygenation has an adverse effect on the aquatic animal life.

Evolutionary species concept: Asuggestion made by G. G. Simpson for adapting the biological species concept for a palaeontological context 'a species is a lineage with its own evolutionary role and tendencies'.

Excretion: The process of removing metabolic waste products from the body (e.g., urine, feces).

Exoskeleton: An external skeleton, typically made of chitin or calcium carbonate, that provides structural support (e.g., arthropods).

Fatty acid: A longchained, predominantly unbranched, carboxylic acid: it may be saturated or unsaturated. The side chains usually have an odd number (commonly 15 or 17) of carbon atoms to which hydrogen atoms may be attached; if all available hydrogen sites are occupied the fatty acid is said to be saturated, if one or more sites is vacant it is unsaturated, and if two or more sites are vacant it is polyunsaturated. Fatty acids have the general formula R—COOH, where R is hydrogen or a group of hydrogen and carbon atoms; saturated fatty acids have the general formula $C_nH_{2n+1}COOH$.

Fauna: The animal life of a particular region or time period.

Feather: A keratinous outgrowth of the skin of birds that is highly modified for the purposes of flight, insulation, and display. Feathers can be divided into distinct types: contour feathers, down feathers, intermediate feathers, filoplumes, powder down, and bristles. They are pigmented, iridescent colours being due to scattered light from specially structured feathers. Worn feathers are annually renewed by moulting.

Feral: From the Latin fetus. 'wild', an adjective applied to a wild or undomesticated organism. In particular, the term is applied to wild strains of an otherwise domesticated species or to an organism that has reverted to a wild condition following escape from captivity, etc. Some authors make these distinctions: wild species, subject to natural selection only; domestic species, subject to selection by humans: and feral species, formerly domestic species which are now, as escapees, subject once again to natural selectiom

Fermentation: A metabolic process that produces energy in the absence of oxygen, common in some invertebrates and microbes.

Fertilization: The union of two gametes to produce a zygote, which occurs during sexual reproduction. Fertilization involves the fusion of two haploid nuclei (*see* HAPLOID NUMBER), containing genetic material from two distinct individuals (crossfertilization) or from one individual (serffertilization). The resulting zygote then develops into a new individual. Most aquatic animals, e.g. echinoderms, achieve fertilization externally, gametes uniting outside the body of the parents. Some other animals, particularly terrestrial species, have internal fertilization, with the union of gametes inside the female.

Fission: A type of asexual reproduction in which an organism splits into two or more parts, each capable of growing into a new individual (e.g., bacteria).

Flehmen: Behaviour, probably with a sexual significance, exhibited by felids, goats, and some Other mammals when examining a scent mark. Having sniffed the mark intensively, the animal raises its head with its mouth partly open and upper lip slightly drawn back, stares fixedly to its front, and breathes slowly. Sniffing followed by Flehmen may Occur more than once at a single encounter With a scent mark.

Food begging: Juvenile behaviour that may endure into adulthood (e.g. forming part of some courtship rituals), in which young that are dependent on their parents for food stimulate adults to feed them.

Fossil: The remains of a onceliving organism, generally those that lived prior to the end of the last glacial period, i.e. fossils are older than 10000 years. They include skeletons, tracks, impressions, trails, borings, and casts. Fossils are usually found in hard rock, but not always: e.g. woolly mammoths living 20000 years ago were recovered from the frozen tundra of Siberia.

Gastropoda (gastropods; phylum Mollusca) A class of molluscs, of asymmetrical form, including snails and slugs, which have a true head, an unsegmented body, and a broad, flat foot. When present, the shell is in one piece and spirally coiled, at least in young stages. The mantle cavity and visceral mass have undergone

Gavialidae (gharial, gavial; subclass Archosauria, order Crocodilia) A monospecific family (*Gavialis gangeticus*) comprising the Indian gharial (gavial) which is found in the Ganges, Indus, and Brahmaputra river systems. It is a fisheater, with an elongate, narrow snout widening at the nostrils. Its teeth are all similar in shape and size. *Tomistoma schlegeli* (false gharial) is sometimes included in this family, but it is more probably a member of the Crocodylidae which has developed a long, narrow snout.

Gekkonidae (geckos; order Squamata, suborder Sauria) A family of insectivorous, mainly nocturnal lizards in which the eye has a vertical pupil and spectacle (no eyelids). The skin is soft With sparse horny tubercles. The toes usually have transverse rows of hooked lamellae for adhesion. The tail is used as a fat store, and is capable of autotomy. *Gekko gecko* (tokay gecko or great house gecko) of southeast Asia and Indonesia is one of the largest geckos, growing up to 35cm long, and is strong and aggressive, feeding on small reptiles, birds, and mammals as well as insects; its common name refers to its call of 'tokay'. There are more than 400 species of geckos, distributed widely in warm latitudes. One species, the common house gekko (*Hemidactylus bibronii*), has been spread all around the tropical world by human agency, mainly as an insecteater, but also simply for its attractiveness.

Gel filtration: A columnchromatographic technique that normally employs as a stationary phase polymeric carbohydrategel beads of controlled size and porosity. Mixture components are separated on the basis of their sizes and rates of diffusion into the beads. Smaller molecules tend to diffuse more, rapidly into the beads, thereby leaving the mainstream of solvent and so becoming retarded with respect to larger molecules. This method can also be used to determine the molecular weight of an unknown substance.

Gempylidae (snake mackerel, gemfish; subclass Actinopterygii, order Perciformes): A small family of marine, coastal to oceanic fish that have a very elongate, sometimes ribbonlike body, a large mouth provided with strong teeth, and long, low first dorsal and short second dorsal fins. Secondary small finlets may occur anterior to the forked tail fin. Several species are of commercial importance. There are about 20 species, found in tropical to temperate waters, worldwide.

Gene flow: The movement of genes within an interbreeding group that results from mating and gene exchange with immigrant individuals. Such an exchange of genes may occur in one direction or both.

Gene flow: The transfer of genetic material between populations, leading to genetic diversity.

Gene frequency: The number of loci at which a particular allele is found, divided by the total number of loci at which it could occur, for a given population, expressed as a proportion or percentage.

Gene: The fundamental physical unit of heredity. It occupies a fixed chromosomal locus, and when transcribed has a specific effect upon the phenotype. It may mutate, and so yield various allelic forms. A gene comprises a segment of DNA (in some viruses it is RNA) coding for one function or several related functions. The DNA is usually situated in threadlike chromosomes, together with protein, within the nucleus; in bacteria and viruses, though, the chromosomes comprise simply a long thread of DNA. The part of a gene that functions as one unit is called a cistron.

Genetic code: The set of correspondences between base (nucleotide pair) triplets in DNA and amino acids in protein. These base triplets carry the genetic information for protein synthesis. For example, the triplet CAA (cytosine, adenine, adenine) codes for valine. The code is universal, but degenerate, in that certain

Genetic distance: A measure of genetic similarity and evolutionary relationship (e.g. between two races of the same species), based on the frequency of a number of

Genetic polymorphism: The occurrence in a population of two or more genotypes in frequencies that cannot be accounted for by recurrent mutation. Such occurrences are generally longterm. Genetic polymorphisms may be balanced (such that allele frequencies are in equilibrium with one another at a given locus), or transient (such that a mutation is spreading through the population in a constant direction). In the former case, the different alleles may be maintained by different environmental conditions (in space or time), one being favoured under one set of circumstances and another under another set; or a heterozygous genotype may be in some way superior to the genotypes that are homozygous at that locus (called a 'heterozygous advantage'). It has been pointed out, however, that in many populations polymorphism is so high that to account for it all by natural selection would entail an impossible genetic load: thus, a good deal of polymorphism would be due to chance increase or decrease of neutral alleles.

Genus: A taxonomic rank above species and below family; species within a genus are closely related (e.g., *Panthera* for lions and tigers).

Gills: Respiratory organs that allow aquatic organisms to extract oxygen from water.

Gregarious: Describes species that live in groups or herds.

Habitat: The natural environment where an organism lives.

Habituation: A decrease in behavioural responsiveness that occurs when a stimulus is repeated frequently with neither reward nor punishment. The process involves learning to ignore insignificant stimuli and should not be confused with accommodation.

Haemochorial: In a eutherian placenta, applied to the condition in which the relationship between the embryo and its mother is very intimate, the embryonic villi being bathed in the maternal bloodstream.

Haemocoel: A body cavity, containing blood, which has the same embryonic origins as the bloodvascular system (i.e. it is not a coelom). Functionally, it replaces the coelom as the main body cavity and hydrostatic skeleton but it does not contain gametes.

Halictidae (sweat bees; order Hymenoptera, suborder Apocrita) Large, cosmopolitan family of mining bees, of great interest to behaviourists because the family encompasses all the grades of sociality. Like those of the Colletidae, the brood cells are lined with macrocyclic lactones secreted by the Dufour's gland. Small species of *Lasioglossum* are attracted to human perspiration and hence the name 'sweat bees'.

Hectocotylized arm: In male Cephalopoda, a specialized tentacle used as an intromittent organ for the transfer of sperm. In some species of Octopoda, during copulation the tentacle is inserted into the mantle cavity and may break off; the detached hectocotylus was originally thought to be a parasitic worm, and was described as *Hectocotylus* (by Cuvier).

Heliornlthidae: A family of small to large, olivebrown or brown birds, which have a long, thin head, and broadly lobed toes. They inhabit the edges of rivers and lakes, feed on insects, molluscs, and small aquatic invertebrates, and nest low in bushes. They swim, dive, and run well, but rarely fly. There are three monotypic genera, found in Central and S. America, Africa, and southern Asia.

Hemiptera (true bugs; class Insecta, subclass Pterygota) Order of insects in whose life cycle there is no true pupal stage, and which never have an eleventh abdominal segment or cerci. The mouthparts are modified into a rostrum or, in nonfeeding stages of some species, are completely lacking. There are typically two pairs of wings, of which the anterior pair are usually wholly or partly toughened to protect the membranous posterior pair over which they are folded at rest. There are two suborders, Homoptera and Heteroptera.

Hermatypic: Applied to corals that contain zooxanthellae and are reefforming. Modern scleractinian hermatypic corals are characterized by the presence of vast numbers of symbiotic zooxanthellae in their endodermal tissue. They live in waters of normal marine salinity, at depths of up to 90m, in temperatures above 18ºC, and grow vigorously in strong sunlight.

Heterogametic sex: The sex that has sex chromosomes that differ in morphology (e.g. XY), and that therefore produces two different kinds of gametes with respect to the sex chromosome. The term may also refer to the possession of an unpaired sex chromosome (a single Xchromosome). In mammals and most other animals, the male is the heterogametic sex, usually with equal numbers of different gametes X and Y, although some men, for example, have XXY or even XXXY. In birds, reptiles, some fish and amphibians, and Lepidoptera, the female is the heterogametic sex.

Heterotroph: An organism that cannot make its own food and must consume other organisms for energy (e.g., animals, fungi).

Heterozygote: A diploid or polyploid individual that has alleles at at least one locus. Its phenotype is often identical to that of an individual that has one of these alleles in the homozygous state. For example, the phenotype of A_1A_2 may be identical to that of A_1A_1, but different from that of A_2A_2. If so, that allele (A_1) is said to be dominant over the other (A_2), which is said to be recessive. When there exist only two alleles of a gene and when one of these is dominant over the other, one may use a lowercase letter for the recessive allele (e.g. one may write *Aa* rather than A_1A_2). However, alleles are not always dominant one over another; and often there are more than two alleles at a given locus, especially in polyploids. Because it has two or more different alleles at a given locus, a heterozygote does not breed true.

Hibernation: A state of dormancy in which animals conserve energy and reduce metabolic activity during cold weather.

Holoplankton: Zooplankton organisms that are planktonic throughout their life cycles.

Homalorhagida: An order of worms in which the head is covered on retraction by a series of body plates on the third segment.

Homing: The return by an animal to a particular site that is used for breeding or sleeping. The term may apply to the return of an animal to its nest after foraging, or to seasonal migrations between breeding and feeding grounds.

Homocercal tail: In fish, a tail in which the last vertebra adjoins modified bony elements (the urostyle and hypurals) which support the fin rays in such a manner that the tail fin outwardly appears symmetrical. The homocercal tail fin is typical of the higher bony fish.

Homologous structures: Body parts that have a similar structure due to common ancestry, though they may serve different functions (e.g., the forelimbs of vertebrates).

Homology: The fundamental similarity of a particular structure in different organisms, which is assumed to be due to descent from a common ancestor. Two organs sharing a similar position, similar histological appearance, and a similar embryonic development are said to be homologous organs irrespective of their superficial appearance and function in the adult.

In vitro: Literally 'in glass', but applied more generally to studies on living material that are performed outside the living organism from which the material is derived. Examples include the use of a perfused organ, tissue cultures, cell homogenates, and subcellular fractions.

Ingestion: The process of taking in food or nutrients.

Instinct: A complex, inherited behavior that is typically performed without prior learning (e.g., migratory patterns).

Invasive species: A non-native species that causes ecological harm to a new environment.

Invertebrate: An animal lacking a backbone (e.g., insects, mollusks).

Island biogeography: Basically, the relationship between area and species number, as an equilibrium between immigration and extinction, on islands; a study that is possible because islands are numerous and their biotas are often small enough to be quantified. The subject also has great relevance to the continents, where plant and animal communities are effectively reduced to islands in a sea of cultivation or urbanization.

Island biotas: The plants and animals of oceanic islands; because of their isolation, such biotas normally include numerous endemic taxa. Island biotas are generally fragile, unbalanced in that they lack plants and animals with poor transoceanic dispersal capacities, and vulnerable to disturbance by human activities and introduced species.

Jellyfish: A type of cnidarian, known for their gelatinous, umbrella-shaped bodies and tentacles.

Juvenile hormone (neotenin) A hormone, released by the corpora allata, which regulates the form of larvae after each moult. Juvenile hormone brings about the retention of juvenile characteristics in developing insects, but has other functions in adults (e.g. it may be required for the proper development of reproductive organs and for the deposition of yolk in eggs).

Juvenile: An immature organism, typically before it reaches sexual maturity.

Karyotype: The entire chromosomal complement of an individual or cell, which may be observed during mitotic metaphase.

Keratin: A group of fibrous scleroproteins that usually contain large amounts of sulphur as cystine and which form the structural bases of hair, wool, nails, and other epidermal structures in animals.

Kidney An organ that may have arisen in freshwater animals for the purpose of voiding excess water and that is concerned in modern forms with excretion and/or the retention of water.

Kin selection: A form of natural selection where individuals may increase their genetic success by helping relatives survive and reproduce.

kinesis: The phenomenon in which a motile organism or cell changes its rate of locomotion (or frequency of turning) in response to the intensity of a particular stimulus. The direction of locomotion remains random and is unrelated to the direction of the stimulus.

Kingdom: The highest taxonomic rank, grouping organisms based on major structural characteristics (e.g., Animalia, Plantae).

Labellum: In Diptera, one of a pair of fleshy lobes into which the proboscis is expanded distally. It is often used for imbibing surface liquids

Labildae (earwigs; subclass Pterygota, order Dermaptera) Cosmopolitan family of earwigs, all species of which are winged. Labiids are generally small (less than 1.5 cm long) and unremarkable in appearance.

Labridae (wrasse; subclass Actinopterygii, order Perciformes) A very large family of marine fish which are very diversified in shape and size. Usually a wrasse has a rather robust body, long and low dorsal and anal fins, and a tail fin that is rounded or notched, but never forked. The thick. protrusible lips and well developed canine-like teeth give the fish a distinctive profile. Most species are brightly coloured, but, within a single species the colour pattern may differ according to age, sex, and season. In a number of species females turn into males as they grow older. Wrasses range in size from the 10 cm *Labroides dimidiatus* (cleaner wrasse), to the huge *Cheilinus undulatus* (giant maori wrasse) which grows to 2.3 m. There are at least 400 species, found worldwide in tropical to temperate waters.

Lactase: The enzyme that metabolizes lactose. It is present in the stomach of neonatal mammals, but usually ceases to be secreted after weaning, hence milk is indigestible to most mammalian species after this time. In humans, it is a polymorphism: lactase is secreted throughout life in most people of north European and north Indian origin, but its frequency declines towards the tropics and most tropical, East Asian, Australian Aboriginal, Pacific, and Native American peoples cannot digest milk.

Lactic acid: A threecarbon hydroxyacid, formed as the major metabolic product of certain bacteria, and also from pyruvic acid in animal cells when glycolysis occurs under anaerobic conditions.

Lampbrush chromosome: One of the large chromosomes found in the eggs (primary oocytes) of amphibians, with paired loops which extend from most of their

Land bridge: Aconnection between two land masses, especially continents, that allows the migration of plants and animals from one land mass to the other. Before the widespread acceptance of continental drift, the existence of former land bridges was often invoked to explain faunal and floral similarities between continents that are now widely separated.

Larva: An early, immature form of an organism that undergoes metamorphosis to become an adult (e.g., a caterpillar before becoming a butterfly).

Lateral line: A sensory organ in fish and aquatic animals used to detect vibrations and changes in water pressure.

Lipofuscin (age pigment): The darkpigmented material that accumulates in the lysosomes of older animal cells. It may be partly responsible for the ageing of such cells.

Lipoic acid A compound that functions as a coenzyme in the oxidative decarboxylation of pyruvic acid to alphaketoglutaric acid and of the latter to succinic acid. It is sometimes classed as a vitamin as it is an essential requirement for some microorganisms.

Lipoprotein A watersoluble, conjugated protein in which the prosthetic group is a lipid. Lipoproteins transport lipids in the blood and lymph from the small intestine to the liver and from the liver to fat deposits.

Living fossll: A member of a living species of animals that are almost identical to species known from the fossil record (not the recent fossil record); i.e. that have changed very little over a long period. For example, *Latimeria chalumnae* (coelacanth) is dose to late Palaeozoic genera; *Xiphosura* species (horseshoe crab) are close to Middle Palaeozoic genera; and *Tapirus pinchaque* (Andean tapir) is dose to a Miocene species. Note, however, that none of these species is identical to any fossil taxon; evolution has been slow, but not absent.

Lymph: A colourless fluid, similar to blood plasma, that consists mainly of salts and proteins in water and suspended fats (whose presence and amount varies according to food intake). It drains from spaces between cells into a network of vessels (the lymphatic system) that convey it to the bloodstream which it enters close to the heart. Lymph nodes, occurring at intervals in the lymphatic system (in humans especially in the neck, armpits, and groin), falter out bacteria and particles of foreign matter and produce lymphocytes

Lymphatic system: Part of the circulatory system that helps in immune response and the removal of excess fluids from tissues.

Lysozyme: A bacteriolytic enzyme that is found in a number of biological fluids, including tears, saliva, and egg white. It facilitates the hydrolysis of structural polysaccharides in the cell walls of bacteria.

Macro: A prefix meaning 'large' or 'long', from the Greek *makros*, 'large' or 'long'.

Macromolecule: A molecule that has a high molecular weight, often a polymer.

Macula: In the retina of many vertebrates that lack a fovea, an area of relatively acute sensitivity to light stimulation and so of relatively acute vision. In the human retina the macula lutea is a similarly sensitive area, containing a yellow pigment, that surrounds the fovea.

Madreporite: In Echinodermata, a sievelike (it may be perforated by up to 250 pores), buttonshaped process on the aboral surface of the body; through its opening the watervascular system is connected to the water outside.

Major gene: A gene that has pronounced phenotypic effects, in contrast to a modifer gene, which modifies the phenotypic expression of another gene.

Malacopterygii: A name now little used, describing the group (superorder) comprising the 'lower' bony fish (Osteichthyes), characterized by the possession of only soft-rayed fins (no spines), often single dorsal and adipose fins, both dorsal and pelvic fins located roughly midway along the back and belly profile, and the

Malpighi, Marcelio (162894): An ltalian physiologist who is credited with having been the first person to study the structure of plants and animals by means of a microscope. He was the first person to observe directly the blood coursing through vessels on the surface of the lung, and he described the structure of secreting glands (*see* MALPLGHIAN TUBULES). He also attempted to study the fine structure of the brain.

Mammal: A warm-blooded vertebrate animal characterized by having hair or fur and producing milk for their young (e.g., humans, whales).

Mammary gland: In female mammals, a gland on the ventral surface which produces milk. It is probably derived from modified sweat glands. It gives the Mammalia its name.

***Mammuthus* (mammoth):** A genus of Pleistocene elephants that were adapted to steppe and tundra habitats. The tusks were elongated and strongly curved, and the skull was shorter and higher than that of other elephants. The woolly mammoth was adapted to Arctic environments. The largest mammoth, indeed the largest probiscidean of all time, was *Mammuthus armeniacus* of Eurasia, which stood about 4.5 m at the shoulder.

Marginal value theorem: A mathematical rule, proposed by E. L. Charnov in 1976, according to which the optimum time a foraging animal remains in a patch is defined in terms of the rate at which the forager is extracting energy at the time it leaves (the marginal value of the patch). The optimum foraging strategy is to abandon each patch when the rate of energy extraction from it falls to a certain level, this level being the same for all patches. The theorem predicts that foragers will remain a shorter time in patches with little food than in patches with more food, patches will be abandoned more quickly when they are close together than when they are scattered, and patches will be abandoned more quickly in an area of abundance than in a poorer area.

Markrecapture technique: A technique for estimating the population density of more elusive or mobile animals. A sample of the population is captured, marked, and released. Assuming that these marked individuals become randomly distributed through the wild population, and that subsequent trapping is random, any new sample should contain a representative proportion of marked to unmarked individuals. From this, the size of the population may be estimated, most simply by multiplying the number in the first sample by the number in the second sample, and dividing the product by the number of marked individuals in the recaptured sample. This calculation is appropriate only when the population is fairly static or where changes (due to migration, natality, or mortality) are known. It may also be distorted if marked individuals become more vulnerable to predators and if particular individuals become 'trapaddicted' or 'trapshy'. When a population is fluctuating rapidly, as is common in insect populations, more sophisticated indices, which allow for the probabilities of change, are preferable.

Marsupium: A pouch (from the Greek *marsupion*, 'pouch') on the surface of the body of an adult animal in which its young are held securely. Marsupia occur in some invertebrates (e.g. some Isopoda), some Anura, but most notably in the Marsupialia, which derive their name from the possession of marsupia, although not all species have them. In those female marsupials that do, and also in echidnas (*see* Tachyglossidae), the marsupium is a fold of skin, supported by epipubic bones, containing the mammary glands into which the eggs of echidnas and young of marsupials are placed.

Meianism: In an animal population, *the occurrence* of individuals that are black owing to the presence in excess of the pigment melanin.

Meiofauna: That part of the microfauna which inhabits algae, rock fissures, and the superficial layers of the muddy seabottom. The term is often used synonymously with interstitial fauna.

Membrane: A sheetlike structure, 710 nm wide, that forms the boundary between a cell and its environment and also between various compartments within the cell. It is composed mainly of lipids, proteins, and some carbohydrates, the structural arrangement of which is still subject to speculation. Membranes function as selective barriers and also as a structural base for enzymes, which may form an integral part of the membrane itself.

Metamorphosis: The process of an organism undergoing a complete or incomplete change in form from its juvenile to adult stage (e.g., tadpole to frog).

Metaspecies: Ancestral species. They are not monophyletic (in the strictest sense) so do not, and cannot, conform to the canon that taxa must always be monophyletic.

Metencephalon: The first region of the rhombencephalon in the brain. The metencephalon gives rise to a dorsal outgrowth, the cerebellum, which constitutes a major brain region.

Microbiocoenosis: The decomposer organisms that form part of the biocoenosis in a biogeocoenosis.

Microevolution: Evolutionary change within species, which results from the differential survival of the constituent individuals in response to natural selection. The genetic variability on which selection operates arises from mutation and sexual reshuffling of gene combinations in each generation.

Migration: The seasonal movement of animals from one region to another, often for breeding or feeding purposes (e.g., bird migration).

Mitochondrial DNA (mtDNA) Circular DNA that is found in mitochondria. It is entirely independent of nuclear DNA and, with very few exceptions, is transmitted from females to their offspring with no contribution from the male parent. Mitochondrial-DNA codes for specific RNA components of ribosomes that are unique to those organelles. It also codes for some of the respiratory enzymes found in mitochondria.

Mobulidae (manta, devil ray; subclass Elasmobranchii, order Rajiformes): A small family of marine rays most of which are large, with very large pectoral fins spread out like wings and pointed at the tips. A small dorsal fin is located above the base of the whiplike tall, which has no tail fin. The two flaplike appendages extending forward on each side of the mouth, which assist in the uptake of food, are characteristic of the family. Perhaps the largest of the devil rays is *Manta*

Modern synthesis (neoDarwinism): The fusion of Mendelian genetics and Darwin's natural selection to produce a more comprehensive evolutionary theory than either can offer separately. Darwin showed that evolution involves selection interacting with variation within populations, Mendel that the bases of this variation are discrete units of heredity (genes). A further synthesis (the synthetic theory) has been achieved in recent years with the incorporation of knowledge of evolution at the molecular level.

Nematode: A type of roundworm, often parasitic, found in many environments.

Niche: The specific role and position of an organism within an ecosystem, including its interactions with other species and its environment.

Nocturnal: Active during the night and resting during the day (e.g., owls, bats).

Omnivore: An organism that eats both plants and animals (e.g., humans, bears).

Organism: An individual living being, from single-celled organisms to complex multicellular life forms.

Oviparous: Animals that lay eggs, with the young developing outside the mother's body (e.g., birds, reptiles).

Parasite: An organism that lives on or inside another organism (the host) and benefits at the host's expense (e.g., tapeworms).

Phylum: A major taxonomic rank below kingdom and above class, grouping organisms with similar body plans (e.g., Arthropoda, Chordata).

Predation: The act of one organism hunting and consuming another organism for food.

Prokaryote: A single-celled organism that lacks a nucleus and other membrane-bound organelles (e.g., bacteria).

Quarantine: The practice of isolating animals or humans to prevent the spread of disease.

Radial symmetry: A body plan where an organism can be divided into similar halves along multiple planes (e.g., starfish, jellyfish).

Removal of free calcium from the blood, e.g. by adding citrate or oxalate, inhibits the formation of thrombin.

Reproduction: The biological process by which new individual organisms are produced.

Respiration: The process by which organisms exchange gases (oxygen and carbon dioxide) with their environment.

Species: The basic unit of classification, representing a group of organisms capable of interbreeding and producing fertile offspring.

Spicule: A needle-like structure found in the skeletons of certain invertebrates, such as sponges.

Stimulus: A signal or change in the environment that triggers a response in an organism.

Symbiosis: A close, long-term relationship between two different species, which can be mutualistic, commensal, or parasitic.

Taxonomy: The science of classifying organisms based on shared characteristics and evolutionary relationships.

Torpor: A temporary state of reduced metabolic activity, often used to conserve energy during harsh conditions (e.g., hibernation, daily torpor).

Toxin: A poisonous substance produced by some organisms (e.g., snake venom, plant toxins).

Trophic level: The position an organism occupies in a food chain (e.g., producer, primary consumer, secondary consumer).

Uptake: The absorption of nutrients or gases by an organism's cells or tissues.

Urinary system: The system responsible for removing waste products from the body through the production and excretion of urine.

Vertebrate: An animal with a backbone or spinal column (e.g., mammals, birds, reptiles).

Viviparous: Animals that give birth to live young rather than laying eggs (e.g., humans, most mammals).

Web: A structure built by some arthropods (e.g., spiders) to trap prey.

Wingspan: The distance from the tip of one wing to the tip of the other wing, used in the measurement of birds and bats.

Xenobiotic: A substance that is foreign to a living organism, often referring to toxic chemicals or pollutants.

Yeast: A type of fungus used in fermentation and as a model organism in scientific research.

Zoology: The branch of biology that studies animals, including their structure, behavior, classification, and distribution.

Zygote: The fertilized egg cell formed by the union of sperm and egg, which develops into an embryo.

INDEX

H

I

K

L

M

N

O

P

R

S

www.ingramcontent.com/pod-product-compliance
Lightning Source LLC
LaVergne TN
LVHW062226200926
848484LV00006B/1485
* 9 7 8 1 7 8 7 1 5 3 9 3 6 *